ORDERED RANDOM VARIABLES

ORDERED RANDOM VARIABLES

MOHAMMAD AHSANULLAH

VALERY B. NEVZOROV

Nova Science Publishers, Inc.
Huntington, New York

Senior Editors: Susan Boriotti and Donna Dennis
Office Manager: Annette Hellinger
Graphics: Wanda Serrano
Information Editor: Tatiana Shohov
Book Production: Cathy DeGregory, Jonathan Rose, Jennifer Vogt and Lynette Van Helden
Circulation: Ave Maria Gonzalez, Ron Hedges, Andre Tillman

Library of Congress Cataloging-in-Publication Data
Available upon request.

Includes bibliographical references and index.
ISBN 1-590333-024-2

227 Main Street, Suite 100
Huntington, New York 11743
Tele. 631-424-6682 Fax 631-425-5933
e-mail: Novascience@earthlink.net
Web Site: http://www.nexusworld.com/nova

Printed in the United States of America

To Masuda , Nisar,Tabassum and Omar
MA

To Ludmila and Igor
VN

ABOUT THE AUTHORS:

MOHAMMAD AHSANULLAH
Department of Management Sciences
Rider University
Lawrenceville, NJ, USA

VALERY B. NEVZOROV
Department of Mathematics and Mechanics
St. Petersburg State University
St. Petersburg, Russia

CONTENTS

PREFACE

Traditionally a statistician deals with a group of n independent random variables $X_1,X_2,\ldots,X_n$, having a common distribution function F (a sample of size n taken from a population distribution F). In this case the well-developed theory of sums of independent terms presents an important tool for treatment with observations $X_1,X_2,\ldots$ and different statistics based on X's. None the less, there are many situations when statisticians oblige to use dependent observations instead of the classical i.i.d. scheme. The most of these situations are connected with some orderings of the initial independent random variables. Both the authors of this book spent much time to investigate ordered random variables, which appear in different branches of Probability and Statistics. The results of these works are presented below. In the sequel we will discuss order statistics (chapter 1), extremes (chapter 2), ordered sums of random variables (chapter 3), records (chapter 4), induced random variables (chapter 5) and generalized order statistics (chapter 6). Indeed, each of these topics merits its own book. Moreover, some of such books, containing a lot of interesting material have already been published. Concerning these topics with respect to the corresponding numbers of authors and papers dealing with them, we can call order statistics as Goliath amongst other ordered random variables. It was David who happened to successfully try conclusions with this Goliath and the main result of this duel became a remarkable book "Order statistics" by Professor H.A. David, which was published in 1970 and its revised version was published in1981. We mention also "A first course in order statistics" by B.C. Arnold, N.Balakrishnan and H.N. Nagaraja (1992). Two editions (1978, 1987) of the book "The asymptotic theory of extreme order statistics" by J. Galambos and "Extremes and related properties of random sequences and processes" by M.R.Leadbetter, G.Lindgren and H.Rootzen (1983) will help a reader to study main methods and general results concerning the asymptotic behavior of extremes. The recent books "Record statistics" by M.Ahsanullah (1995), "Records" by B.C. Arnold, N.Balakrishnan and H.N.Nagaraja (1998), "Records. Mathematical Theory by V.B.Nevzorov (2000) give an elementary introduction to the theory of records. Lastly, a systematic exposition of the theory of generalized order statistics is given by U.Kamps (1995) in his book "A concept of generalized order statistics"

While we were writing our book we had the following intentions:

a) To collect in one place our results for ordered random variables scattered in many papers, published in different journals and volumes, some of which are not easily available or written in Russian.
b) To give an exposition of such topics as induced order statistics and order statistics based on sums of random variables, which now are ready , in our opinion, for a systematic presentation in the book form.
c) To locate all ordered random variables under a common roof and to compare results and methods for different types of orderings.

There are some general ideas, which are applied for all types of ordered random variables. For instance, in each chapter we try to express distributions of ordered random variables (which are dependent) via distributions of independent terms.

Since there are good presentations of order statistics, extremes, records and generalized order statistics, we restrict ourselves, discussing these topics, by giving our own results or upgrading the theory given in the books mentioned above. For the rest two topics we give more detailed exposition, which also in many ways based on results of authors.

While chapters 1 and 2 have chiefly applied polarity, containing useful formulae and tables for specific distributions, other chapters combine theoretical and practical directions. Each chapter includes an upgraded list of bibliography, which will help a reader to continue the studying of the concrete topics.

The book may be viewed as a research monograph with an extensive bibliography for probabilists and statisticians as well as for actuarial mathematicians, reliability engineers, meteorologists, hydrologists, economists, business and sport analysts. At the same time the simplicity of presentation allows it to be used as a textbook for graduate students.

We would like to express our gratitude to our family members for their constant encouragement, support and help.

The work of the first author was partially supported by Research Leave from Rider University. The work of the second author was partially supported by the Russian Fond of Basic Research (grants 99-01-00724 and 99-01-00732).

February 2001

Chapter 1

ORDER STATISTICS

We will consider in this chapter the (ordinary) order statistics. The basic properties of these statistics will be studied. These properties will be used to investigate inferences of several statistical distributions. Readers can find more detailed theory of ordered statistics in David (1981).

1.1. DEFINITIONS

Suppose $X_1, X_2, \ldots, X_n$ are n random variables with their corresponding cumulative distribution function (cdf) F_i, i =1, ..., n. We will consider first the case when F's are absolutely continuous. Let $X_{1,n}$ be the smallest of the $X_1, X_2, \ldots, X_n$, $X_{2,n}$ is the next of $X_1, X_2, \ldots, X_n$ greater than $X_{1,n}$ and $X_{n,n}$ is the largest of $X_1, X_2, \ldots, X_n$. Thus we obtain $X_{1,n} < X_{2,n} < \cdots < X_{n,n}$ as the order statistics corresponding to the random variables $X_1, X_2, \ldots, X_n$. We will call $X_{i,n}$ the ith order statistic of the random variables $X_1, X_2, \ldots, X_n$. Let $f_{1,2,\ldots,n}(x_1, x_2, \ldots, x_n)$ be the joint probability density function (pdf) of $X_1, X_2, \ldots, X_n$. For simplicity, we will assume that the random variables $X_1, X_2, \ldots, X_n$ are independent and identically distributed with pdf f(x) and cdf F(x).Let $F_{r,n}(x)$ be the cdf of $X_{r,n}$, then

$$F_{r,n}(x) = P(X_{r,n} \leq x) = P(\text{ at least r of the n X's are} \leq x)$$

$$= \sum_{i=r}^{n} c(i,n)\,(F(x))^{i}\,(1-F(x))^{n-i},\quad c(k,n) = \frac{n!}{k!(n-k)!}, 1 \leq k \leq n. \qquad (1.1.1)$$

Differentiating (1.1.1) with respect to x, we get the pdf $f_{r,n}$ of $X_{r,n}$ as

$$f_{r,n}(x) = r\,c(r,n)\,(F(x))^{r-1}\,(1-F(x))^{n-r}\,f(x). \qquad (1.1.2)$$

For r = 1, the pdf $f_{1,n}$ of $X_{1,n}$ is $f_{1,n}(x) = n\,(1-F(x))^{n-1}\,f(x)$, and for r = n, the pdf $f_{n,n}$ of $X_{n,n}$ is $f_{n,n}(x) = n(F(x))^{n-1}\,f(x)$

The joint cdf $F_{r,s,n}(x,y)$ of $X_{r,n}$ and $X_{s,n}$ for $r< s$ and for $-\infty<x<y<\infty$, is given by

$$F_{r,s,n}(x,y) = P(X_{r,n} \le x,\ X_{s,n} \le y)$$

$$= P(\text{ at least r of } X_1, X_2, \dots, X_n \le x \text{ and at least s of } X_1, X_2, \dots, X_n \le y)$$

$$=\sum_{j=s}^{n}\sum_{i=r}^{s-1} P(\text{exactly i of } X_1, X_2, \dots, X_n \le x \text{ and exactly j of } X_1, X_2, \dots, X_n \le y)$$

$$=\sum_{j=s}^{b}\sum_{i=r}^{j-1} c(i,j,n)(F(x))^{i}(F(y)-F(x))^{j-i}(1-F(y))^{n-j},$$

$$c(i,j,n)=\frac{n!}{i!(j-i)!(n-j)!}. \tag{1.1.3}$$

Differentiating $F_{r,s,n}(x,y)$ with respect to x and y, we get the joint pdf $f_{r,s,n}$ of $X_{r,n}$ and $X_{s,n}$ as

$$f_{r,s,n}(x,y)=r(s-r)\,c(r,s,n)(F(x))^{r-1}(F(y)-F(x))^{s-r-1}(1-F(y))^{n-s} f(x) f(y). \tag{1.1.4}$$

The joint pdf $f_{1,2,\dots,n}(x_1,x_2, \dots, x_n)$ of $X_{1,n}, X_{2,n},\dots,X_{n,n}$ is given as follows:

$$f_{1,2,\dots,n}(x_1,x_2,\dots, x_n) = n!\, f(x_1)f(x_2) \dots f(x_n), \text{ if } -\infty < x_1<x_2 < \dots< x_n < \infty,$$

and

$$f_{1,2,\dots,n}(x_1,x_2,\dots, x_n) = 0, \text{ otherwise.} \tag{1.1.5}$$

Using (1.1.5) we can obtain the joint pdf of any number of order statistics by integrating out the other variables. The joint pdf $f_{r_1,r_2,\dots,r_k}(x_1,x_2,\dots,x_k)$ of $X_{r_1,n}, X_{r_2,n},\dots,X_{r_k,n}$ for $1\le r_1 <r_2<\dots<r_k \le n$ and $-\infty < x_1<x_2 < \dots< x_n < \infty$ has the following expression:

$$f_{r_1,r_2,\dots,r_k}(x_1,x_2,\dots,x_k)=c(r_1,r_2,\dots,r_k,n)\prod_{i=1}^{k}(r_i-r_{i-1})!\{F(x_i)-F(x_{i-1})\}^{r_i-r_{i-1}-1} f(x_i)(\bar{F}(x_k))^{n-r_k}, \tag{1.1.6}$$

where $r_0=0$, $F(x_0) = 0$, $c(r_1,r_2,\dots,r_k,n)=\dfrac{n!}{r_1!(r_2-r_1)!\dots(r_k-r_{k-1})!(n-r_k)!}$ and $\bar{F}=1-F$.

Theorem 1.1.1.

For a random sample of size n from an absolutely continuous (with respect to Lebesgue measure) distribution with cdf as F(x) and pdf as f(x) , the conditional distribution of $X_{s,n}$ given $X_{r,n} = x$ (s > r) is the (s-r) th order statistics in a sample of size n-r from the distribution with pdf as $f_1(y) = f(y)/ (1-F(x))$.

Proof.

Using (1.1.2) and (1.1.3), we can write the conditional pdf of of $X_{s,n}$ given $X_{r,n} = x$ as

$$f_{s|r,n}(y|x) = \frac{(s-r)\,c(r,s,n)}{c(r,n)}\left[\frac{F(y)-F(x)}{1-F(x)}\right]^{s-r-1}\left[\frac{1-F(y)}{1-F(x)}\right]^{n-s}\frac{f(y)}{1-F(x)},\ \text{for } x\leq y.$$

Theorem 1.1.2.

The sequence of order statistics $\{ X_{j,n}, j =1,2,\ldots, n\}$ from an absolutely continuous (with respect to Lebesgue measure) distribution with cdf as F(x) and pdf as f(x) forms a Markov chain.

Proof.

From (1.1.6), we have the joint pdf of $X_{1,n}, X_{2,n}, \ldots, X_{r,n}$ as

$$f_{1,2,\ldots,r}(x_1, x_2,\ldots, x_r) = c(1,2,..,r,n)\, f(x_1)f(x_2)\ldots f(x_r)\, (\bar{F}(x_r))^{n-r},\ -\infty < x_1<x_2<\ldots,x_r< \infty.$$

The joint pdf of $X_{1,n}, X_{2,n}, \ldots, X_{r,n}, X_{s,n}$ (s>r) is

$$f_{1,2,\ldots,r,s}(x_1, x_2,\ldots, x_r, x_s) = c(1,2,..,r,s,n)\,)(s-r)\, f(x_1)f(x_2)\ldots f(x_r)\, f(x_s)$$
$$(F(x_s) - F(x_r))^{s-r-1}\, (\bar{F}(x_s))^{n-s},\ -\infty < x_1<x_2<\ldots,x_r< \infty.$$

Thus the conditional pdf of $X_{s,n}$ given $X_{1,n} = x_1, X_{2,n}= x_2,\ldots, X_{r,n}$ for

$$f_{s|1,2,\ldots,r,n}(x_s|x_1, x_2,\ldots, x_r) =$$
$$\frac{c(1,2,\ldots,r,s,n)(s-r)}{c(1,2,\ldots,r,n)}\left[\frac{F(x_s)-F(x_r)}{1-F(x_r)}\right]^{s-r-1}\left[\frac{1-F(x_s)}{1-F(x_r)}\right]^{n-s}\frac{f(x_s)}{1-F(x_r)},\ \text{for } x_r < x_s.$$
$$= 0.\ \text{otherwise.}$$

Since $c(1,2,\ldots,n) = r!\, c(r,n)$ and $c(1,2,\ldots,n) = r!\, c(r,s,n)$, we have for $x_r < x_s$

$$f_{s|1,2,\ldots,r,n}(x_s|x_1, x_2,\ldots, x_r) = \frac{c(r,s,n)(s-r)}{c(r,n)}\left[\frac{F(x_s)-F(x_r)}{1-F(x_r)}\right]^{s-r-1}\left[\frac{1-F(x_s)}{1-F(x_r)}\right]^{n-s}\frac{f(x_s)}{1-F(x_r)},$$
$$= f_{s|r,n}(x_s|x_r).$$

Thus the sequence of order statistics $\{ X_{j,n}, j=1,2,\ldots, n \}$ forms a Markov chain.

If the distribution of the random variable X is symmetric about 0, then $f(-x) = f(x)$ and $F(-x) = 1-F(x)$. In that case, it can easily be shown that $X_{r,n} \underline{\underline{d}} -X_{n-r+1,n}$ and jointly

$(X_{1,n}, X_{2,n},\ldots, X_{n,n}) \underline{\underline{d}} (-X_{n,n}, -X_{n-1,n},\ldots, -X_{1,n})$.

Theorem 1.1.3.

If $E(|X|^{k+\delta} < \infty$, k is any non negative integer and $\delta > 0$, then $E(|X_{r,n}|^k) < \infty$, for all r, $1 \le r \le n$.

Proof.

$$E(|X_{r,n}|^k) = r\,c(r,n)\int_{-\infty}^{\infty}|x|^k (F(x))^{r-1}(1-F(x))^{n-r} f(x)dx$$
$$= r\,c(r,n)\int_0^1 |F^{-1}(u)|^k u^{r-1}(1-u)^{n-r}du$$
$$\le r\,c(r,n)\{\int_0^1 |F^{-1}(u)|^{kp} du\}^{1/p}(\int_0^1 [u^{r-1}(1-u)^{n-r}]^q du)^{1/q},$$

by Holder's inequality, where $1/p+1/q = 1$, $p>1$. $q>1$.

$$= r\,c(r,n)\{\int_0^1 |F^{-1}(u)|^{k+\delta} du\}^{1/p}(\int_0^1 [(u))^{r-1}(1-u)^{n-r}]^q du)^{1/q}, \text{ where } k+\delta = kp$$
$$= r\,c(r,n)E|X|^{k+\delta}\ (\int_0^1 [u^{r-1}(1-u)^{n-r}]^q du)^{1/q} < \infty.$$

The following theorem (Sen (1959)) is an improvement of the Theorem 1.1.3.

Theorem 1.1.4.

Suppose that the random variable X has a continuous pdf f(x). If $E|X|^{\delta} < \infty$ for some $\delta > 0$, then $E(|X_{r,n}|^{\alpha} < \infty$, for all r and n satisfying the condition $\frac{\alpha}{\delta} \le r \le n+1-\frac{\alpha}{\delta}$.

We say that the k th moment of X exists if $E(|X|^k) < \infty$.

The mean of $X_{r,n}$ is defined as

$$\mu_{r,n}^{1} = E(X_{r,n}) = \int_{-\infty}^{\infty} r\,c(r,n)\,x(F(x))^{r-1}(1-F(x))^{n-r} f(x)\,dx. \quad (1.1.7)$$

The mth moment of $X_{r,n}$ is given as

$$\mu_{r,n}^{m} = E(X_{r,n}^{m}) = \int_{-\infty}^{\infty} rc(r,n)\, x^{m} (F(x))^{r-1} (1-F(x))^{n-r} f(x)\, dx. \tag{1.1.8}$$

The variance $\sigma_{r,r,n}$ of $X_{r,n}$ has the form

$$\sigma_{r.r.n} = \mu_{r,n}^{2} - (\mu_{r,n}^{1})^{2}. \tag{1.1.9}$$

The joint (m_1, m_2) th moment of $X_{r,n}$ and $X_{s,n}$ is

$$\mu_{r,s,n}^{m_1,m_2} = E(X_{r,n}^{m_1}\, X_{s,n}^{m_2}) = \int_{-\infty}^{\infty} \int_{-\infty}^{y} x^{m_1} y^{m_2} f_{r,s,n}(x,y)\, dx\, dy. \tag{1.1.10}$$

The covariance $\sigma_{r,s,n}$ of $X_{r,n}$ and $X_{s,n}$ is

$$\sigma_{r,s,n} = \mu_{r,s,n}^{2,2} - \mu_{r,n}^{1}\, \mu_{s,n}^{1} \quad . \tag{1.1.11}$$

The correlation coefficient $\rho_{r,s,n}$ between $X_{r,n}$ and $X_{s,n}$ is defined as

$$\rho_{r,s,n} = \frac{\sigma_{r,s,n}}{(\sigma_{r,r,n}\, \sigma_{s,s,n})^{1/2}}. \tag{1.1.12}$$

Srikantan (1962) proved the following relations for the moments.

Lemma 1.1.1.

For any function h(x), $(n-r)E(h(X_{r,n})) + r\, E(h(X_{r+1,n})) = nE(h(X_{r,n-1}))$, $1 \le r < n$, $n > 1$, provided the expectations exist.

Taking h(x) = x, we obtain the following relation between the moments:

$$(n-r)\, E(X_{r,n}^{m}) + rE(X_{r+1,n}^{m}) = nE(X_{r,n-1}^{m}),\ m > 0,\ 1 \le r \le n. \tag{1.1.13}$$

Example 1.1.1.

Suppose $X_1, X_2, \ldots, X_n$ are independent and identically distributed uniform random variables with $f(x) = 1$, $0 < x \le 1$,

$= 0$, otherwise.

The pdf of $X_{r,n}$ is $f_{r,n}(x) = r\, c(r,n) x^{r-1} (1-x)^{n-r}$, $0<x<1$,

$= 0$, otherwise.

The joint pdf of $X_{r,n}$ and $X_{s,n}$ is $f_{r,s,n}(x,y) = r(s-r)\, c(r,s,n)\, x^{r-1} (y-x)^{s-r-1}(1-x)^{n-s}$, $0<x<y<1$,

$= 0$, otherwise.

We have $\mu_{r,n}^{m} = r\,c(r,r,n)\int_0^1\int_0^y x^{m+r-1}(1-x)^{n-r}\,dx = \frac{(r+m-1)^{(m)}}{(n+m)^{(m)}}$ where $p^{(j)} = p(p-1)\ldots(p-j+1)$ and

$$\mu_{r,s,n}^{m_1,m_2} = r(s-r)c(r,s,n)\int_0^1\int_0^y x^{m_1}y^{m_2}x^{r-1}(y-x)^{s-r-1}(1-y)^{n-s}\,dx\,dy$$

$$= \frac{(m_1+r-1)^{(m_1)}(m_1+m_2+1)^{(m_2)}}{(n+m_1+m_2)^{(m_1+m_2)}}.$$

The variance of $X_{r,n}$ is

$$\sigma_{r,r,n} = \mu_{r,n}^{2} - (\mu_{r,n}^{1})^2 = \frac{r(n-r+1)}{(n+1)^2(n+2)},\ 1 \le r \le n.$$

The covariance of $X_{r,n}$ and $X_{s,n}$ is expressed as follows:

$$\mu_{r,s,n} = \frac{r(n-s+1)}{(n+1)^2(n+2)},\ 1 \le r \le s \le n,$$

and the corresponding correlation coefficient of $X_{r,n}$ and $X_{s,n}$ is given by

$$\rho_{r.,s,.n} = \left(\frac{r(n-s+1)}{s(n-r+1)}\right)^{1/2},1 \le r \le s \le n.$$

Szekely and Mori (1985) showed that for any distribution with finite variance

$$\rho_{r,s,n}^2 \le \frac{r(n-s+1)}{s(n-r+1)},$$

and the equality holds when F is uniform.

Example 1.1.2.

Suppose $X_1, X_2, \ldots, X_n$ are n independent and identically distributed exponential random variables with $f(x) = e^{-x}$, $0 < x < \infty$,

$= 0$, otherwise.

The joint pdf $f_{1,2,\ldots,n}(x_1, x_2, \ldots, x_n)$ of $X_{1,n}, X_{2,n}, \ldots, X_{n,n}$ is

$$f_{1,2,\ldots,n}(x_1,x_2,\ldots x_n) = n!\, e^{-(x_1+x_2+\ldots+x_n)},\quad 0 < x_1 < x_2 < \ldots < x_n < \infty,$$
$$= 0, \text{otherwise}.$$

Consider the standardized spacings,

$D_{j,n} = (n-j+1)(X_{j,n}-X_{j-1,n})$, $j = 1,2,\ldots, n$, with $X_{0,n} = 0$.

Then the joint pdf $f_{1,2,\ldots,n}(d_1, d_2,\ldots,d_n)$ of $D_{1,n}, D_{2,n}, \ldots, D_{n,n}$ is

$$f_{1,2,\ldots,n}(d_1,d_2,\ldots d_n) = e^{-(d_1+d_2+\ldots+d_n)}, 0 < d_1, d_2,\ldots, d_n < \infty,$$

$$= 0, \text{ otherwise}.$$

Thus $D_{1,n}, D_{2,n}, \ldots, D_{n,n}$ are independent and each has standard exponential distribution.

Equivalently we can express $X_{j,n}$ as a sum of independent random variables as follows:

$$X_{j,n} = \sum_{k=1}^{j} \frac{D_{k,n}}{n-k+1}.$$

Using this relation and noting that $E(D_{j,n}) = Var(D_{j,n}) = 1$ for all $j = 1,2,\ldots, n$, we get

$$E(X_{r,n}) = \sum_{j=1}^{r} \frac{1}{n-j+1},$$

$$Var(X_{r,n}) = \sum_{j=1}^{r} \frac{1}{(n-j+1)^2},$$

and

$$Cov(X_{r,n}, X_{s,n}) = \sum_{j=1}^{r} \frac{1}{(n-j+1)^2}, \quad 1 \le r < s \le n.$$

Example 1.1.3.

Assume that $X_1, X_2,\ldots, X_n$ are n independent and identically distributed logistic random variables with

$$f(x) = \frac{1}{(1+e^{-x})^2}, \ -\infty < x < \infty,$$

$= 0$, otherwise.

The moments of $X_{r,n}$ can be obtained by considering the following moment generating function $M_{r,n}(t)$ of $X_{r,n}$:

$$M_{r,n}(t) = Ee^{tX_{r,n}} = \int_{-\infty}^{\infty} r\,c(r,n)\,e^{tx} \frac{e^{-(n-r+1)x}}{(1+e^{-x})^{n+1}}\,dx$$

$$= \int_0^1 r\,c(r,n)\,u^{r+t-1}(1-u)^{n-r-t}\,du$$

$$= \frac{\Gamma(r+t)\Gamma(n-r+1-t)}{\Gamma(r)\Gamma(n-r+1)}, \quad 1 \le r \le n.$$

Using the moment generating function $M_{r,n}$ (t), it can easily be shown that

$E(X_{r,n}) = \Psi(r) - \Psi(n-r-1)$, where $\Psi(t) = \frac{d}{dt}\ln\Gamma(t)$

and

$Var(X_{r,n}) = \Psi'(r) - \Psi'(n-r+1)$.

Example 1.1.4.

Assume that $X_1, X_2, \ldots, X_n$ are n independent and identically distributed Pareto random variables with

$f(x) = \beta x^{-\beta-1}, \ 1 \le x < \infty,$

$= 0$, otherwise.

The kth moment $\mu_{r,n}^k$ of $X_{r,n}$ is

$$\mu_{r,n}^k = \int_1^{\infty} r\,c(r,n)\,x^k\,[1-x^{-\beta}]^{r-1}\,x^{-((n-r+1)\beta+1)}\,dx$$

$$= \int_0^1 r\,c(r,n)\,t^{n-r-\frac{k}{\beta}}(1-t)^{r-1}\,dt$$

$$=\frac{\Gamma(n+1)}{\Gamma(n-r+1)}\cdot\frac{\Gamma(n-r-\frac{k}{\beta}+1)}{\Gamma(n-\frac{k}{\beta}+1)}, \beta>k.$$

Thus,

$$E(X_{r,n})=\frac{\Gamma(n+1)}{\Gamma(n-r+1)}\frac{\Gamma(n-r-\frac{1}{\beta}+1)}{\Gamma(n-\frac{1}{\beta}+1)}$$

and

$$\mathrm{Var}(X_{r,n})=\frac{\Gamma(n+1)}{\Gamma(n-r+1)}\frac{\Gamma(n-r-\frac{2}{\beta}+1)}{\Gamma(n-\frac{2}{\beta}+1)}-\left(\frac{\Gamma(n+1)}{\Gamma(n-r+1)}\frac{\Gamma(n-r-\frac{1}{\beta}+1)}{\Gamma(n-\frac{1}{\beta}+1)}\right)^2.$$

From Theorem 1.1.1, we have the conditional pdf $f_{s|r,n}(y|x)$ of $X_{s,n}$ given $X_{r,n} = x$ $(1 \le r < s \le n)$ as

$$f_{s|r,n}(x,y)=(s-r)c(s-r,n-r)\left(\frac{\bar{F}(x)-\bar{F}(y)}{\bar{F}(x)}\right)^{s-r-1}\left(\frac{\bar{F}(y)}{\bar{F}(x)}\right)^{n-s}\frac{f(y)}{\bar{F}(x)},$$

$1 \le x < y < \infty$, $\bar{F}=1-F$. Substituting $F(x) = 1- x^{-\beta}$, we obtain on simplification

$$E(X_{s,n}|\,X_{r,n}=x)=x\frac{\Gamma(n-r+1)}{\Gamma(n-s+1)}\cdot\frac{\Gamma(n-r-\frac{2}{\beta}+1)}{\Gamma(n-r-\frac{1}{\beta})}.$$

Thus the expected value of $X_{r,n}$ and $X_{s,n}$ is

$$E(X_{r,n}X_{s,n})=\frac{\Gamma(n+1)}{\Gamma(n-s+1)}\cdot\frac{\Gamma(n-s-\frac{1}{\beta}+1)}{\Gamma(n-r-\frac{1}{\beta})}\cdot\frac{\Gamma(n-r-\frac{2}{\beta}+1)}{\Gamma(n-\frac{2}{\beta}+1)},$$

and the covariance of $X_{r,n}$ and $X_{s,n}$ is

$$\text{Cov}(X_{r,n}, X_{s,n}) = \frac{\Gamma(n+1)}{\Gamma(n-s+1)} \cdot \frac{\Gamma(n-s-\frac{1}{\beta}+1)}{\Gamma(n-r-\frac{1}{\beta})} \cdot \frac{\Gamma(n-r-\frac{2}{\beta}+1)}{\Gamma(n-\frac{2}{\beta}+1)} - E(X_{r,n})E(X_{s,n}).$$

Example 1.1.5.

Suppose $X_1, X_2, \ldots, X_n$ are n independent and identically distributed Type 1 extreme value distribution with

$$f(x) = e^{-x} e^{-e^{-x}}, -\infty < x < \infty.$$

Then,

$$E(X_{r,n}) = \int_0^{\infty} r\, c(r,n) \sum_{j=0}^{n-s} c(j, n-r)(-1)^j e^{-(r+j)t} \ln t \, dt$$

$$= r\, c(r,n) \sum_{j=0}^{n-r} (-1)^j c(j, n-r) \frac{(\gamma + \ln(r+j))}{r+j},$$

where $\gamma = 0.5772\ldots$, is the Euler's constant, and

$$E(X_{r,n}^2) = r\, c(r,n) \sum_{j=0}^{n-r} (-1)^j c(j, n-r) \int_{-\infty}^{\infty} x^2 e^{-x-(r+j)e^{-x}} dx$$

$$= r\, c(r,n) \sum_{j=0}^{n-r} (-1)^j c(j, n-r) \frac{(\gamma + \ln(r+j))^2 + \pi^2/6}{r+j}.$$

For $r = n$, we have $\text{Var}(X_{n,n}) = \pi^2/6$ for all n.

The expected value $E(X_{r,n} X_{s,n})$ of $X_{r,n}$ and $X_{s,n}$ is

$$E(X_{r,n} X_{s,n}) = \int_{-\infty}^{\infty} \int_{-\infty}^{y} r(s-r)c(r,s,n)\, xy e^{-re^{-x}} (e^{-e^{-y}} - e^{-e^{-x}})^{s-r-1} (1-e^{-e^{-y}})^{n-s} e^{-x} e^{-y} \, dx\, dy$$

$$= \int_{-\infty}^{\infty} \int_{-\infty}^{y} r(s-r)c(r,s,n) \sum_{i=0}^{s-r} \sum_{j=0}^{n-s} xy\, c(i, s-r) c(j, n-s)(-1)^{i+j} e^{-x-(r+i)e^{-x}} e^{-y-(s-r-i+j)e^{-y}} \, dx\, dy$$

$$= \int_{-\infty}^{\infty} \int_{-\infty}^{y} r(s-r)\, c(r,s,n) \sum_{i=0}^{s-r} \sum_{j=0}^{n-s} xy c(i, n-r)\, c(j, n-s)(-1)^{i+j}\, G(i, j, .r, .s, .n),$$

where

$$G(i,j,r,s,n) = \int_0^{\infty} \int_0^{y} e^{-x-(r+i)e^{-x}} e^{-y-(s-r-i+j)e^{-y}} \, dx\, dy.$$

White (1969) tabulated the means and variances of $X_{r,n}$ for all r and all n up to 20.

Leiblein (1953, 1962) and Leiblein and Zelen (1956) tabulated the covariances of the order statistics for n up to 6.

Example 1.1.6.

Suppose $X_1, X_2, \ldots, X_n$ are n independent and identically distributed as Power function distribution with

$$f(x) = \alpha\, x^{\alpha-1}, 0 < x < 1,$$

$$= 0, \text{ otherwise.}$$

Then

$$E(X_{r,n}^k) = \int_0^1 r\, c(r,n)\, \alpha\, x^{k-1} (x^\alpha)^r (1-x^\alpha)^{n-r} dx$$

$$= \frac{\Gamma(n+1)}{\Gamma(r)} \frac{\Gamma(r+\frac{k}{\alpha})}{\Gamma(n+\frac{k}{\alpha}+1)}.$$

In particular,

$$E(X_{r,n}) = \frac{\Gamma(n+1)}{\Gamma(r)} \frac{\Gamma(r+\frac{1}{\alpha})}{\Gamma(n+\frac{1}{\alpha}+1)}.$$

and

$$Var(X_{r,n}) = \frac{\Gamma(n+1)}{\Gamma(r)} \frac{\Gamma(r+\frac{2}{\alpha})}{\Gamma(n+\frac{2}{\alpha}+1)}. - \left(\frac{\Gamma(n+1)}{\Gamma(r)} \frac{\Gamma(r+\frac{1}{\alpha})}{\Gamma(n+\frac{1}{\alpha}+1)} \right)^2.$$

For r<s,

$$E(X_{r,n}X_{s,n}) = \int_0^1 \int_0^y \frac{\Gamma(n+1)}{\Gamma(r).} \frac{\alpha^2 x^{\alpha r} y^{\alpha-1} (y^\alpha - x^\alpha)^{s-r-1}}{\Gamma(s-r)\Gamma)\Gamma-s+1)} .(1-y^\alpha)^{n-s} dx\, dy$$

$$= \frac{\Gamma(n+1)}{\Gamma(r)} . \frac{\Gamma(r+\frac{1}{\alpha})}{\Gamma(s+\frac{1}{\alpha})} . \frac{\Gamma(s+\frac{2}{\alpha})}{\Gamma(n+1+\frac{2}{\alpha})}.$$

and

$$\mathrm{Cov}\,(X_{r,n}X_{s,n}) = \frac{\Gamma(n+1)}{\Gamma(r)}\frac{\Gamma(r+\frac{1}{\alpha})}{\Gamma(s+\frac{1}{\alpha})}\frac{\Gamma(s+\frac{2}{\alpha})}{\Gamma(n+1+\frac{2}{\alpha})} - \frac{\Gamma(n+1)\Gamma(r+\frac{1}{\alpha})\Gamma(n+1)\Gamma(s+\frac{1}{\alpha})}{\Gamma(r)\Gamma(r+\frac{1}{\alpha}+1)\Gamma(s)\Gamma(s+\frac{1}{\alpha}+1)},$$

Example 1.1.7.

Tukey (1960) introduced a family of distributions based on the uniformly distributed U=U([0,1]) random variable as follows:

$$X = \frac{U^{\lambda} - (1-U)^{\lambda}}{\lambda}, \ \lambda \neq 0$$

and

$$X = \ln U - \ln(1-U), \text{ for } \lambda = 0.$$

For $\lambda = 0$, X corresponds to logistic distribution (Example 1.1.3). We will consider the case $\lambda \neq 0$. Then

$$E(X_{r,n}) = \frac{1}{B(r,n-r+1)}\int_0^1 \frac{u^{\lambda}-(1-u)^{\lambda}}{\lambda} u^{r-1}(1-u)^{n-s}du,$$

$$= \frac{B(r+\lambda,n-r+1) - B(r,n-r+1+\lambda)}{\lambda B(r,n-r+1)}.$$

$$E(X_{r,n}^2) = \frac{1}{B(r,n-r+1)}\int_0^1 \left(\frac{u^{\lambda}-(1-u)^{\lambda}}{\lambda}\right)^2 u^{r-1}(1-u)^{n-r}du$$

$$= \frac{B(r+2\lambda,n-r+1) - 2B(r+\lambda,n-r+1+\lambda) + B(r,n-r+1+2\lambda\lambda}{\lambda^2 B(r,n-r+1)}$$

where

$$B(p,q) = \int_0^1 u^{p-1}(1-u)^{q-1}du.$$

The variance $\mathrm{Var}(X_{r,n})$ of $X_{r,n}$ is

$$\mathrm{Var}(X_{r,n}) = \frac{B(r+2\lambda,n-r+1) - 2B(r+\lambda,n-r+1+\lambda) + B(r,n-r+1+2\lambda)}{\lambda^2 B(r,n-r+1)} - (E(X_{r,n}))^2.$$

The expected value of $X_{r,n}\,X_{s,n}$ ($1 \leq r \leq s \leq n$) is

$$E(X_{r,n} X_{s,n}) = \frac{r(s-r)\,c(r,s,n)}{\lambda^2} \int_0^1 \int_0^v \{u^\lambda - (1-u)^\lambda\}\{v^\lambda - (1-v)^\lambda\} u^{r-1} (v-u)^{s-r} (1-v)^{n-s}\, du\, dv\,.$$

Writing $(\alpha)_{(i)} = \alpha(\alpha+1)...(\alpha+i-1)$ and using the relations

$\int_0^1 u^{p-1}(1-u)^{q-1}\, {}_2F_1(a,b;c;u)du = B(p,q)\; {}_3F_2(p,a,b;c,p+q;1)$ for p, q>0 ,b + q- p-a>0, where ${}_pF_q(\alpha_1,\alpha_2,...,\alpha_p;\beta_1,\beta_2,...,\beta_q;z) = \sum_{k=0}^{\infty} \frac{(\alpha_1)_{(k)}(\alpha_2)_{(k)}...(\alpha_p)_{(k)}}{(\beta_1)_{(k)}(\beta_1)_{(k)}...(\beta_1)_{(k)}} \frac{z^k}{k!}$,

and $B_s(p,q) = \frac{1}{B(p,q)} \int_0^s u^{p-1}(1-u)^{q-1}\, du$.

We get on simplification

$$E(X_{r,n} X_{s,n}) = \frac{r(s-r)c(r,s,n)}{\lambda^2} [\sum_{j=0}^{s-r-1} (-1)^j \binom{s-r-1}{j} \left\{ \frac{B(s+2\lambda, n-s+1)}{r+\lambda+j} - \frac{B(s+\lambda, n-s+\lambda+1)}{r+\lambda+j} \right\}$$

$-(r+j)^{-1}B(s+\lambda,n-s+1)\; {}_3F_2(s+\lambda,r+j,-\lambda;\, r+j+1,n+\lambda+1;1)$

$-(r+j)^{-1}B(s,n-s+\lambda+1)\; {}_3F_2(s,r+j,-\lambda;r+j+1,n+\lambda+1;1)]$

and

$$Cov(X_{r,n} X_{s,n}) = E(X_{r,n}X_{s,n}) - E(X_{r,n})\, E(X_{s,n}).$$

Raqab (1999) has tabulated the means, variances and covariances of these order statistics for $n \le 20$.

Example 1.1.8.

Suppose that $X_1, X_2, \ldots, X_n$ are i.i.d normal random variables with

$$f(x) = \frac{1}{\sqrt{2\pi}} e^{-\frac{u^2}{2}}, \; -\infty < x < \infty,$$

$= 0$, otherwise.

Then $F(x) = \int_{-\infty}^{x} \frac{1}{\sqrt{2\pi}} e^{-\frac{u^2}{2}} du$ and we have

$$E(X_{r,n}) = r\,c(r,r,n) \int_{-\infty}^{\infty} x(F(x))^{r-1}(1-F(x))^{n-r} f(x)\,dx, 1 \le r \le n,$$

$$Var(X_{r,n}) = \int_{-\infty}^{\infty} x^2 (F(x))^{r-1} (1-F(x))^{n-r} f(x)\,dx - \left[\int_{-\infty}^{\infty} x\ (F(x))^{r-1} (1-F(x))^{n-r} f(x)\,dx\right]^2,$$

$$\begin{aligned}\mathrm{Cov}(X_{r,n}X_{s,n}) &= r(s-r)c(r,s,n)\int_{-\infty}^{\infty}\int_{-\infty}^{y} xy(F(x))^{r-1}(F(y)-F(x))^{s-r-1}(1-F(y))^{n-s}\,dx\,dy\\ &\quad - rc(r,n)\int_{-\infty}^{\infty} x(F(x))^{r-1}(1-F(x))^{n-r} f(x)\,dx\\ &\quad - sc(s,n)\int_{-\infty}^{\infty} y(F(y))^{s-1}(1-F(y))^{n-s} f(y)\,dy,\quad 1\le r<s\le n.\end{aligned}$$

Teichroew(1956) gave tables of $E(X_{r,n})$, $Var(X_{r,n})$ and $Cov(X_{r,n}X_{s,n})$ for $n \le 20$.

Harter (1961) tabulated the values of $E(X_{r,n})$ for $n \le 100$. Tietjen, Kahaner and Beckman (1977) tabulated the values of $E(X_{r,n})$, $Var(X_{r,n})$ and $Cov(X_{r,n}X_{s,n})$ for n up to 50.The expected values of the order statistics are given in the following table for $n \le 10$.

n	r	$\mu_{r,n}$	N	r	$\mu_{r,n}$
2	1	-0.5642	7	6	0.7574
2	2	0.5642	7	7	1.3522
3	1	-0.8463	8	1	-1.4238
3	2	0.0000	8	2	-0.8522
3	3	0.8483	8	3	-0.4728
4	1	-1.0294	8	4	-0.1525
4	2	-0.2970	8	5	0.1525
4	3	0.2970	8	6	0.4728
4	4	1.0294	8	7	0.8522
5	1	-1.1630	8	8	1.4238
5	2	-0.4950	9	1	-1.4850
5	3	0.0000	9	2	-0.9323
5	4	0.4850	9	3	-0.5720
5	5	1.1630	9	4	-0.2745
6	1	-1.2672	9	5	0.0000
6	2	-0.6418	10	1	-1.5388
6	3	-0,2015	10	2	-1.0014
6	4	0.2015	10	3	-0,6561
6	5	0.6418	10	4	-0.3758
6	6	1.2672	10	5	-0.1227
7	1	-1.3522	10	6	0.1227
7	2	-0.7574	10	7	0.3758
7	3	-0.3527	10	8	0.6561
7	4	0.0000	10	9	1.0014
7	5	0.3527	10	10	1.5388

Table 1.1.1.. Expected values of the order statistics

Example 1.1.9.

Suppose that $X_1, X_2, \ldots, X_n$ are i.i.d Weibull random variables with cdf F(x) where

$$f(x) = c x^{c-1} e^{-x^c}, c > 0, 0 < x < \infty,$$

$$= 0, \text{ otherwise.}$$

The expected value, $E(X_{r,n})$ of $X_{r,n}$ ($1 \le r \le n$) is

$$E(X_{r,n}) = \int_{-\infty}^{\infty} r\, c(r,n)\, x (1 - e^{-x^c})^{r-1} e^{-(n-r+1)x^c} c x^{c-1} dx$$

$$= r\, c(r,n) \sum_{j=0}^{r-1} (-1)^j c(j,j,r-1) \int_0^{\infty} e^{-(n-r+1+j)x^c} c x^c dx$$

$$= r\, c(r,n) \Gamma(1 + \frac{1}{c}) \sum_{j=0}^{r-1} (-1)^j . \frac{c(j,j,r-1)}{(n-r+12+j)^{1+\frac{1}{c}}} .$$

Similarly we obtain

$$E(X_{r,n}^2) = r\, c(r,n) \Gamma(1 + \frac{2}{c}) \sum_{j=0}^{r-1} (-1)^j . \frac{c(j,j,r-1)}{(n-r+12+j)^{1+\frac{2}{c}}} .$$

Then,

$$Var(X_{r,n}) = rc(r,n) \Gamma(1 + \frac{2}{c}) \sum_{j=0}^{r-1} (-1)^j \frac{c(j,j,n-1)}{(n-r+1)^{1+\frac{2}{c}}} - (E(X_{rn}))^2 .$$

Leiblein (1955) and Weibull (1967) computed the means and the variances of the order statistics for n = 5(5) 20 and 1/c = 0.1(0.1) 1.0. Harter (1970) gave the means of order statistics for n up to 40 and c =0.5(0.5)4(1)8.

The expected value of $X_{r,n} X_{s,n}$ ($1 \le r < s \le n$) can be written as

$$E(X_{r,n} X_{s,n}) = r(s-r) c(r,s,n) c^2 \sum_{i=0}^{r-1} \sum_{j=0}^{s-r-1} (-1)^{s-r-1+i-j} i c(i,i,r) j c(j,j,s-r-1) H_c(i,j,r,n)$$

where

$$H_c(i,j,r,n) = \int_0^{\infty} \int_0^{y} e^{-(i+j+1)x^c} e^{-(n-r+1)y^c} x^c y^c \, dx\, dy .$$

Calculating the $H_c(i,j,r,n)$, the product moments and the covariance of $X_{r,n}$ and $X_{s,n}$ can be computed. Weibull (1967) computed the covariances of all order statistics for n = 5(5) 20 and 1/c = 0.1(0.1) 1.0.Adatia (1994) presented covariances of all order statistics for extended values of n.

Example 1.1.10.

Suppose that $X_1, X_2, \ldots, X_n$ are i.i.d Cauchy random variables with

$$f(x) = \frac{1}{\pi(1+x^2)} \quad -\infty < x < \infty.$$
$$= 0, \text{ otherwise.}$$

The mean and the variance of X do not exist. However $E|x|^\delta$ is finite for $0<\delta<1$ and by Theorem 1.1.3 the expected value of $E|X_{r,n}|$ is finite if $2 \le r \le n-1$. The expected value, $E(X_{r,n})$ of $X_{r,n}$ is

$$E(X_{r,n}) = r\,c(r,n) \int_{-\infty}^{\infty} \frac{x}{\pi}\left[\frac{1}{2}+\frac{arctg\,x}{\pi}\right]^{r-1}\left[\frac{1}{2}-\frac{arctg\,x}{\pi}\right]^{n-r}\frac{1}{1+x^2}dx$$

$$= -r\,c(r,n) \int_0^1 w^{r-1}(1-w)^{n-r}\cot(\pi w)\,dw$$

$$= -r\,c(r,n) \sum_{j=0}^{n-r}(-1)^j c(j,n-r) \int_0^1 w^{r+j-1}\cot(\pi w)\,dw$$

$$= r\,c(r,n) \sum_{j=0}^{n-r}(-1)^{j+1} c(j,n-r) \sum_{k=0}^{r+j-1}\frac{c(k,r+j)}{r+j}\int_0^1 B_k(w)\cot(\pi w)\,dw,$$

since

$$u^{m-1} = \frac{1}{m}\sum_{j=0}^{m-1} c(j,m)\,B_k(u)$$

and $B_k(u)$ is the kth Binomial polynomial given by

$$\frac{te^{ut}}{e^t-1} = \sum_{k=0}^{\infty} B_k(u)\frac{t^k}{k!}, \; |t| < 2\pi.$$

Similarly,

$$E(X_{r,n}) = r\,c(r,n) \int_{-\infty}^{\infty} \frac{x}{\pi}\left[\frac{1}{2}+\frac{arctg\,x}{\pi}\right]^{r-1}\left[\frac{1}{2}-\frac{arctg\,x}{\pi}\right]^{n-r}\frac{1}{1+x^2}dx$$

$$= -r\,c(r,n) \int_0^1 w^{n-r}(1-w)^{r-1}\cot(\pi w)\,dw$$

$$= -r\,c(r,n) \sum_{j=0}^{n-r}(-1)^j c(j,n-r) \int_0^1 (1-w)^{r+j-1}\cot(\pi w)\,dw$$

$$= r\,c(r,n) \sum_{j=0}^{n-r}(-1)^{j+1} c(j,n-r) \sum_{k=0}^{r+j-1}\frac{c(k,r+j)}{r+j}\int_0^1 B_k(1-w)\cot(\pi w)\,dw$$

$$= \mathrm{r\,c(r,n)} \sum_{j=0}^{n-r} (-1)^{j+1} c(j,n-r) \sum_{k=0}^{r+j-1} \frac{c(k,r+j)}{r+j} \int_0^1 B_k(w)\cot(\pi w)\,dw.$$

Thus adding the above two expressions of $E(X_{r,n})$ and using the relation

$$\int_0^1 B_{2m+1}(w)\cot(\pi w)dw = \frac{2(2m+1)(-1)^{m+1}}{(2\pi)^{2m+1}\zeta(2m+1)},$$

where

$$\zeta(s) = \sum_{k=1}^{\infty} k^{-s}, s > 1,$$

on simplification, we get

$$\mu_{r,n} = E(X_{r,n}) = \mathrm{r\,c(r,n)} \sum_{i=0}^{n-r} (-1)^i \mathrm{c(i,n-r)} \sum_{k=0}^{\left[\frac{r+i-2}{2}\right]} \frac{\mathrm{c(2k+1,r+i)}}{\mathrm{r+i}} \frac{2(2k+1)!(-1)^k}{(2\pi)^{2k+1}} \zeta(2k+1)$$

$$= \mathrm{r\,c(r,n)} \sum_{k=0}^{\left[\frac{n-2}{2}\right]} (-1)^k \frac{2(2k+1)!}{(2\pi)^{2k+1}} c_r(k)\zeta(2k+1),$$

where [m] is the largest integer contained in m , $c_r(0) = 0$ and

$$c_r(k) = \sum_{j=\max(0,2k+2-r)}^{n-r} (-1)^j c(j,n-r)c(2k+1,r+j)\frac{1}{r+j}.$$

The second moment $E(X_{r,n}^2)$, of $X_{r,n}$ exists if $3 \le r \le n-2$ and

$$E(X_{r,n}^2) = \frac{r\,c(r,n)}{\pi} \int_{-\infty}^{\infty} \frac{x^2}{1+x^2} \left(\frac{1}{2} + \frac{arctg\,\mathrm{x}}{\pi}\right)^{r-1} \left(\frac{1}{2} - \frac{arctg\,\mathrm{x}}{\pi}\right)^{n-r} dx$$

$$= \frac{r\,c(r,n)}{\pi} \int_{-\infty}^{\infty} \left(\frac{1}{2} + \frac{arctg\,\mathrm{x}}{\pi}\right)^{r-1} \left(\frac{1}{2} - \frac{arctg\,\mathrm{x}}{\pi}\right)^{n-r} dx - 1.$$

Thus,

$$\mathrm{Var}\,(X_{r,n}) = \frac{n}{\pi}[\mu_{r,n-1} - \mu_{r-1,n-1}] - 1 - (\mu_{r,n})^2$$

The expected value, $E(X_{r,n}X_{s,n})$ exists for $2 \le r < s \le n-1$, $n \ge 4$ and

$$\mu_{r,s,n} = E(X_{r,n}X_{s,n}) = r(s-r)c(r,s,n)I$$

where

$$I = \int_0^1 [\int_0^v \cot(\pi u)\cot(\pi v)\, u^{r-11}(v-u)^{s-r-1}(1-v)^{n-r} dv]du$$

Using the relation (Abramowitz and Stegun (1968), p.75)

$$\cot x = \frac{1}{x} + \sum_{j=1}^{\infty} (-1)^j \frac{2^{2j} B_{2j}}{(2j)!} x^{2j-1},\ |x| < \pi,$$

and completing the integral I, we get on simplification

$$\mu_{r,s,n} = r\,c(r,s,n) \sum_{j=0}^{\infty} \frac{(-1)^{j+1}}{\pi} \frac{(2\pi)^{2j}}{(2j)!(r+2j-1)} \frac{B_{2j}\mu_{s+2j-1,n+2j-1}}{c(r+2j-1,s+2j-1.n+2j-1)},$$

$B_k = B_k(0)$ are the Bernoulli numbers. Bernett (1966) computed the means, variances and covariances of the order statistics for $n \leq 20$.

Example 1.1.11.

Suppose that $X_1, X_2, \ldots, X_n$ are i.i.d gamma random variables with the pdf as $f_v(x)$ where

$$f_\alpha(x) = \frac{x^{\alpha-1} e^{-x}}{\Gamma(\alpha)},\ x>0.\ \alpha>0$$

We will assume α is a positive integer.

The corresponding cdf $F_\alpha(x)$ is given as

$$F_\alpha(x) = \frac{\Gamma_x(\alpha)}{\Gamma(\alpha)},\quad \Gamma_x(\alpha) = \int_0^x e^{-u} u^{\alpha-1}\, du$$

and it can be expressed as a partial sum of the Poisson probability distribution

$$F_\alpha(x) = \sum_{j=\alpha}^{\infty} \frac{e^{-x} x^j}{\Gamma(j+1)}.$$

The expected value $E(X_{r,n})$ of $X_{r,n}$ ($1 \leq r \leq n$) is

$$E(X_{r,n}) = \frac{r\,c(r,n)}{\Gamma(\alpha)} \int_0^\infty [1 - \sum_{j=0}^{\alpha-1} \frac{e^{-x} x^j}{j!}]^{r-1} [\sum_{j=0}^{\alpha-1} \frac{e^{-x} x^j}{j!}]^{n-r} e^{-x} x^\alpha dx .$$

The expected value $E(X_{r,n}^2)$ of $X_{r,n}$ is

$$E(X_{r,n}^2) = \frac{r\,c(r,n)}{\Gamma(\alpha)} \sum_{j=0}^{r-1} (-1)^j c(j, r-1) \sum_{k=0}^{(\alpha-1)(n-r+j)} b_k(r, n-r+j) \frac{\Gamma(\alpha+k+2)}{(n-r+j+1)^{\alpha+k+1}} .$$

The expected value $E(X_{r,n}X_{s,n})$ of $X_{r,n}X_{s,n}$ can be written as

$E(X_{r,n}X_{s,n})$

$$= \frac{r(s-n)c(r,s,n)}{(\Gamma(\alpha))^2} \int_0^\infty \int_u^\infty [1 - \sum_{j=0}^{\alpha-1} \frac{e^{-u} u^j}{j!}]^{r-1} [\sum_{j=0}^{\alpha-1} \frac{e^{-u} u^j}{j!} - \sum_{j=0}^{\nu-1} \frac{e^{-\nu} v^j}{j!}]^{s-r-1} [\sum_{j=0}^{\nu-1} \frac{e^{-\nu} v^j}{j!}]^{n-s} e^{-(u+v)} u^\alpha v^\alpha \, du dv$$

$$= \frac{r(s-r)\,c(r,s,n)}{(\Gamma\Gamma(\alpha)^2} \sum_{k=0}^{r-1} \sum_{l=0}^{s-r-1} (-1)^{k+1} c(k, r-1)\, c(l, s-r-1) I_1 ,$$

where

$$I_1 = \int_0^\infty e^{-c_1 u} \left[\sum_{j=0}^{\alpha-1} \frac{u^j}{j!} \right]^{c_1-1} u^\alpha \, I_2(u)\, du, \; c_1 = k+s-r-l$$

and

$$I_2(u) = \int_u^\infty e^{-c_2 u} \left[\sum_{j=0}^{\alpha-1} \frac{v^j}{j!} \right]^{c_2-1} v^\alpha \; dv, \; c_2 = n-s+l+1 .$$

Completing the integrals, we get on simplification

$$E(X_{r,n}X_{s,n}) = \frac{r(s-n)c(r,s,n)}{(\Gamma(\alpha))^2} \sum_{k=0}^{m-1} \sum_{l=0}^{s-r-1} \sum_{t=0}^{(\alpha-1)(c_2-1)} \sum_{m=0}^{t+\alpha} \sum_{p=0}^{(\alpha-1)(c_2-1)} c(k, r-1)\, c(l, r-s-1)$$

$$. b_t(\alpha-1, c_1-1) . b_p(\alpha-1, c_2-1) \frac{\Gamma(t+\alpha+1)\Gamma(p+\beta+m+1)}{m! c_2^{t+\alpha+m+1} (c_2+c_1)^{m+\alpha+p+1}} ,$$

where $b_p(a,b)$ is the coefficient of x^p in the expansion of $(\sum_{j=0}^{a-1} \frac{x^j}{j!})^b$.

Example 1.1.12.

Suppose that $X_1, X_2, \ldots, X_n$ are i.i.d generalized exponential distribution random variables with cdf F(x) where

$$f(x) = \theta e^{-x}(1-e^{-x})^{\theta-1}, \theta > 0, 0 < x < \infty,$$
$$= 0, \text{ otherwise.}$$

For $\theta = 1$, the above distribution corresponds to the exponential distribution. When θ is a positive integer, this distribution corresponds to the distribution of the maximum of random sample of size θ from the exponential distribution. The pdf of the rth order statistics, $X_{r,n}$ can be written as

$$f_{r,n}(x) = rc(r,n)\theta(1-e^{-x})^{(r-1)\theta}[1-(1-e^{-x})^{\theta}]^{n-r}(1-e^{-x})^{\theta-1}e^{-x}.$$

Expanding $[1-(1-e^{-x})^{\theta}]^{n-r}$ in powers of $(1-e^{-x})^{\theta}$, we can write

$$f_{r,n}(x) = rc(r,n)\theta\sum_{j=0}^{n-r}(-1)^j c(j,n-r)(1-e^{-x})^{(r+j)\theta-1}e^{-x}.$$

The first and second moment of $X_{r,n}$(r=1,2,...,n) are respectively

$$E(X_{r,n}) = \int_0^{\infty} rc(r,n)\theta x(1-e^{-x})^{(r-1)\theta}[1-(1-e^{-x})^{\theta}]^{n-r}(1-e^{-x})^{\theta-1}e^{-x}dx$$

$$= rc(r,n)\sum_{j=0}^{n-r}\frac{(-1)^j c(j,n-r)}{r+j}[\psi((r+j)\theta+1)+\gamma),$$

where

$$\psi(x) = \frac{d\ln\Gamma(x)}{dx}$$

is the digamma function and $\gamma\ (= -\psi(1))$ is the Euler's constant,

$$E(X^2_{r,n}) = \int_0^{\infty} rc(r,n)\theta x^2(1-e^{-x})^{(r-1)\theta}[1-(1-e^{-x})^{\theta}]^{n-r}(1-e^{-x})^{\theta-1}e^{-x}dx$$

$$= rc(r,n)\sum_{j=0}^{n-r}\frac{(-1)^j c(j,n-r)}{r+j}[\psi((r+j)\theta+1)+\gamma]^2$$

$$= rc(r,n)\sum_{j=0}^{n-r}\frac{(-1)^j c(j,n-r)}{r+j}\{[\psi((r+j)\theta+1)+\gamma]^2$$

$$+\psi'((r+j)\theta+1)\}+\pi^2/6.$$

For r = n, $E(X_{n,n}) = \psi(n\theta+1)+\gamma$,

$E(X^2_{r,n}) = (\psi(n\theta+1)+\gamma)^2 - \psi'(n\theta+1)+\pi^2/6$

and

$$\mathrm{Var}(X_{n,n}) = \frac{\pi^2}{6} - \psi'((n\theta+1)).$$

The joint pdf of $X_{r,n}$ and $X_{s,n}$, $1 \le r < s \le n$, can be written as

$$f_{r,s,n}(x,y) = r(s-r)c(r,s,n)\sum_{i=0}^{n-s}\sum_{j=0}^{s-r-1}(-1)^{i+j}c(j,s-r-1)\,c(i,n-s)(1-e^{-x})^{(r+j)\theta-1}$$

$$.(1-e^{-y})^{(s-r-j+r)\theta-1}e^{-x}e^{-y}, 0 < x < y < \infty.$$

Using the fact that

$$\int_0^\infty y(1-e^{-y})^{n\theta-1}e^{-y}dy = \frac{\psi(n\theta+1)+\gamma}{n\theta}$$

and

$$I_y = \int_0^y x(1-e^{-x})^{(r+j)\theta-1}e^{-x}dx$$

$$= \int_0^1 u^{(r+j)\theta-1}\{-\ln[1-u(1-e^{-y})]\}(1-e^{-y})^{(r+j)\theta}du = \sum_{k=1}^{\infty}\frac{(1-e^{-y})^{(r+j)\theta+k}}{k((r+j)\theta+k)},$$

we get

$$E(X_{r,n}X_{s,n}) = r(s-r)c(r,s,n)\int_0^\infty \sum_{i=0}^{n-s}\sum_{j=0}^{s-r-1}(-1)^{i+j}c(j,s-r-1)c(i,n-s)y(1-e^{-y})^{(s-r-j+r)\theta-1}I_y e^{-y}dy$$

$$= r(s-r)c(r,s,n)\sum_{i=0}^{n-s}\sum_{j=0}^{s-r-1}\sum_{k=1}^{\infty}(-1)^{i+j}c(j,s-r-1)c(i,n-s)\frac{\psi((s+i)\theta+\gamma}{k((r+j)\theta+k)((s+i)\theta+k)}.$$

Table 1.1.2: Means of Order Statistics

n	r	θ=0.5	θ=1.5	θ=2.0	θ=2.5	θ=3.0	θ=3.5	θ=4.0	θ=4.5	θ=5.0
1	1	0.613706	1.280372	1.500000	1.680372	1.833333	1.966087	2.083333	2.188309	2.283333
2	1	0.227411	0.727411	0.916667	1.077411	1.216667	1.339316	1.448810	1.547649	1.637698
2	2	1.000000	1.833333	2.083333	2.283333	2.450000	2.592857	2.717857	2.828968	2.928968
3	1	0.121489	0.529426	0.700000	0.848423	0.978968	1.095143	1.199639	1.294507	1.381324
3	2	0.439255	1.123382	1.350000	1.535387	1.692064	1.827662	1.947150	2.053934	2.150447
3	3	1.280372	2.188309	2.450000	2.657307	2.828968	2.975455	3.103211	3.216485	3.318229
4	1	0.076312	0.424724	0.582143	0.721747	0.845996	0.957460	1.058304	1.150259	1.234700
4	2	0.257022	0.843529	1.053571	1.228453	1.377886	1.508192	1.623644	1.727251	1.821197
4	3	0.621489	1.403235	1.646429	1.842321	2.006241	2.147132	2.270656	2.380617	2.479697
4	4	1.500000	2.450000	2.717857	2.928968	3.103211	3.251562	3.380729	3.495108	3.597740
5	1	0.052624	0.358923	0.506349	0.639163	0.758525	0.866314	0.964297	1.053964	1.136534
5	2	0.171065	0.687931	0.885317	1.052082	1.195879	1.322047	1.434334	1.535441	1.627365
5	3	0.385957	1.076927	1.305952	1.493009	1.650896	1.787409	1.907610	2.014964	2.111944
5	4	0.778511	1.620774	1.873413	2.075196	2.243137	2.386948	2.512686	2.624386	2.724866
5	5	1.680372	2.657307	2.928968	3.142411	3.318229	3.467716	3.597740	3.712789	3.815958
6	1	0.038580	0.313281	0.452742	0.580079	0.695471	0.800260	0.895901	0.983691	1.064723
6	2	0.122839	0.587133	0.774387	0.934586	1.073795	1.196581	1.306274	1.405329	1.495588
6	3	0.267516	0.889527	1.107179	1.287076	1.440048	1.572979	1.690455	1.795667	1.890918
6	4	0.504398	1.264327	1.504726	1.698942	1.861745	2.001839	2.124765	2.234262	2.332970
6	5	0.915567	1.798998	2.057756	2.263323	2.433834	2.579502	2.706647	2.819448	2.920813
6	6	1.833333	2.828968	3.103211	3.318229	3.495108	3.645359	3.775958	3.891457	3.994987
7	1	0.029542	0.279528	0.412421	0.535195	0.647260	0.749525	0.843189	0.929390	1.009120
7	2	0.092812	0.515798	0.694666	0.849381	0.984739	1.104674	1.212176	1.309493	1.398342
7	3	0.197907	0.765469	0.973687	1.147596	1.296436	1.426347	1.541519	1.644918	1.738703
7	4	0.360328	1.054937	1.285168	1.473049	1.631530	1.768488	1.889035	1.996666	2.093872
7	5	0.612451	1.421369	1.669395	1.868362	2.034407	2.176852	2.301562	2.412458	2.512293

Table 1.1.2 continued

n	r	θ=0.5	θ=1.5	θ=2.0	θ=2.5	θ=3.0	θ=3.5	θ=4.0	θ=4.5	θ=5.0
7	6	1.036814	1.950049	2.213101	2.421307	2.593604	2.740562	2.868681	2.982244	3.084221
7	7	1.966087	2.975455	3.251562	3.467716	3.645359	3.796158	3.927171	4.042992	4.146781
8	1	0.023369	0.253418	0.380760	0.499640	0.6088503	0.708942	0.800900	0.885728	0.964330
8	2	0.072749	0.462294	0.634051	0.784077	0.916124	1.033603	1.139211	1.235025	1.322652
8	3	0.153001	0.676312	0.876514	1.045296	1.190583	1.317888	1.431070	1.532898	1.625413
8	4	0.272749	0.914066	1.135644	1.318096	1.472858	1.607111	1.725602	1.831617	1.927520
8	5	0.447907	1.195809	1.434692	1.628001	1.790201	1.929865	2.052469	2.161716	2.260224
8	6	0.711177	1.556705	1.810216	2.012579	2.180930	2.325045	2.451018	2.562904	2.663535
8	7	1.145359	2.081164	2.347396	2.557550	2.731162	2.879068	3.007902	3.122024	3.224450
8	8	2.083333	3.103211	3.380729	3.597740	3.775958	3.927171	4.058495	4.174559	4.278543
9	1	0.018961	0.232536	0.355094	0.470592	0.577309	0.675497	0.765956	0.849577	0.927185
9	2	0.058636	0.420473	0.586084	0.732028	0.861181	0.976506	1.080449	1.174940	1.261490
9	3	0.122143	0.608670	0.801933	0.966247	1.108424	1.233444	1.344876	1.445321	1.536717
9	4	0.214719	0.811596	1.025675	1.203393	1.354899	1.486776	1.603459	1.708052	1.802806
9	5	0.345287	1.042153	1.273104	1.461475	1.620308	1.757529	1.878280	1.986074	2.083413
9	6	0.530003	1.318734	1.563962	1.761222	1.926116	2.067734	2.191819	2.302228	2.401672
9	7	0.801764	1.675691	1.933343	2.138257	2.308337	2.453701	2.580618	2.693242	2.794467
9	8	1.243530	2.197013	2.465696	2.677348	2.851970	3.000601	3.129983	3.244533	3.347302
9	9	2.188309	3.216485	3.495108	3.712789	3.891457	4.042992	4.174559	4.290813	4.394948
10	1	0.015700	0.215401	0.333773	0.446289	0.550798	0.647295	0.736421	0.818964	0.895686
10	2	0.048312	0.386758	0.546981	0.689319	0.815906	0.929315	1.031777	1.125087	1.210676
10	3	0.099935	0.555330	0.742499	0.902863	1.042282	1.165268	1.275140	1.374352	1.464748
10	4	0.173961	0.733127	0.940612	1.114142	1.262757	1.392521	1.507592	1.610915	1.704644
10	5	0.275855	0.929299	1.153269	1.337270	1.493112	1.628159	1.747259	1.853756	1.950050
10	6	0.414719	1.155007	1.392939	1.585681	1.747504	1.886899	2.009302	2.118392	2.216776
10	7	0.606859	1.427886	1.677977	1.878249	2.045191	2.188290	2.313498	2.424786	2.524937
10	8	0.885295	1.781893	2.042786	2.249689	2.421115	2.567448	2.695098	2.808295	2.909980
10	9	1.333088	2.300793	2.571424	2.784263	2.959083	3.108890	3.238705	3.353592	3.456633
10	1 0	2.283333	3.318229	3.597740	3.815958	3.994987	4.146781	4.278543	4.394948	4.499205

1.2 ORDER STATISTICS OF DISCRETE POPULATIONS

Let X be a discrete random variable which may take the values 0,1,2,...,with probability mass function (pmf) $p(k) = P_r(X = k)$ and the corresponding cumulative probability mass function (cmf) $P(k) = P_r(X \leq k)$. Suppose that $X_1, X_2, \ldots, X_n$ are n independent and identically distributed random variable with the same pmf as that of X. Let $X_{1,n} \leq X_{2,n} \leq \ldots \leq X_{n,n}$ be the corresponding order statistics. The probability mass function $p_{r,n}$ of $X_{r,n}$ can be written as

$$p_{r,n}(x) = P_r(X_{r,n} = x) = P_{r,n}(x) - P_{r,,n}(x-1)$$

$$\sum_{j=r}^{n} c(j,n)((P(x))^j - (1-P(x))^{n-j} - (P(x-1))^j (1-P(x-1))^{n-j}$$

Using the identity

$$\sum_{j=r}^{n} c(j,n)(P(x))^j (1-P(x))^{n-j} = \int_0^{P(x)} rc(r,n) u^{r-1} (1-u)^{n-r}\, du, 0 < u < 1,$$

we get

$$p_{r,n}(x) = \int_{P(x-1)}^{P(x)} r c(r,n) u^{r-1} (1-u)^{n-r}\, du \tag{1.2.1}$$

For r =1, we have the pmf of $X_{1,n}$ as

$$p_{1,n}(x) = n \int_{P(x-1)}^{P(x)} (1-u)^{n-1} du = (\overline{P}(x-1))^n - (\overline{P}(x))^n, \ \overline{P}(x) = 1 - P(x)$$

For r = n, we obtain the pmf of $X_{n,n}$ as

$$p_{n.n}(x) = \int_{P(x-1)}^{P(x)} nu^{n-1}\, du = (P(x))^n - P(x-1))^n$$

The joint joint pmf $p_{r,s,n}$ of $X_{r,n}$ and $X_{s,n}$ ($1 \leq r < s \leq n$) can be written as

$$p_{r,s,n}(x,y) = P_r(X_{r,n} = x, X_{s,n} = y)$$

$$=\sum_{i=0}^{r-1}\sum_{j=0}^{n-s}\sum_{u,w=0}^{u+w\le s-r-1} c(r-i-1,r+w,s-1-u,s+j,n)(P(x-1))^{r-1-i}\,(P(x))^{i+1+i})(P(y-1)-(P(y))^{s-u-r-w-1}$$

$$.(P(y))^{u+1+j}(1-P(y))^{n-s-j}.$$

On simplification, we obtain from the given expresiion,

$$p_{r,s,n}(x,y)=P_r(X_{r,n}=x,X_{s,n}=y)$$

$$=r(s-r)\,c(r,s,n)\int_{P(x-1)}^{P(x)}\int_{P(y-1)}^{P(y)}\alpha^{r-1}(\beta-\alpha)^{s-r-1}(1-\beta)^{n-s}\,d\alpha\,d\beta \tag{1.2.2}$$

If $r=1$ and $s=n$, then the joint pmf of $X_{1,n}$ and $X_{n,n}$ is

$$p_{1,n,n}(x,y)=P_r(X_{1,n}=x,X_{n,n}=y)$$

$$=n(n-1)\int_{P(x-1)}^{P(x)}\int_{P(y-1)}^{P(y)}(\beta-\alpha)^{n-2}\,d\alpha\,d\beta.$$

We define the range $W_{1,n}$ as $W_{1,n}=X_{n.n}-X_{1,n}$. It is easy to see that

$$P_r(W_{1,n}=0)=\sum_{u=0}^{\infty}(P_r(X=u))^n.$$

For $k>0$, we have

$$P_r(W_{1,n}=k)=\sum_{u=0}^{\infty}P_r(X_{1,n}=u,X_{n,n}=u+k)$$

$$=\sum_{u=0}^{\infty}[\{P(u+k),\text{-}P(u-1)\}^n-\{P(u+k)-P(u)\}^n$$

$$+\{P(u+k-1)-P(u)\}^n-\{P(u+k-1)-P(u-1)\}^n] \tag{1.2.3}$$

$$=\sum_{u=0}^{\infty}n(n-1)\int_{P(u-1)\,P(u)}^{P(u)}\int_{P(u+k-1)}^{P(u+k)}(\beta-\alpha)^{n-2}\,d\alpha\,d\beta$$

The joint pmf of $X_{r_1,n},X_{r_2,n},\ldots,X_{r_k,n}$ can be written as

$$p_{r_1,r_2,\ldots,r_k,n}(x_1,x_2,\ldots,x_k)$$

$$= c(r_1, r_2, \ldots, r_k, n) \int_D \left(\prod_{i=1}^{k} (r_i - r_{i-1})(u_i - u_{i-1})^{r_i - r_{i-1}} (1-u_k)^{n-r_k} du_1 du_2 \ldots du_k \right)$$

where $r_0 = 0, u_0 = 0$ and $\int_D = \int_{P(x_1-1)}^{P(x_1)} \int_{P(x_2-1)}^{P(x_2)} \ldots \int_{P(x_k-1)}^{P(x_k)}$.

Example 1.2.1.

Suppose $X_1, X_2, \ldots, X_n$ are n independent and identically distributed discrete uniform random variables with p(x)=1/k, x =1,2,…,k. We have P(x)= x/k, x=1,2,…, k, and

$$p_{r,n}(x) = \int_{P(x-1)}^{P(x)} r\, c(r,n)\, u^{r-1} (1-u)^{n-r} du$$

$$= \int_{(x-1)/k}^{x/k} r\, c(r,n)\, u^{r-1} (1-u)^{n-r} du .$$

For r = 1 and r = n, we have respectively

$$p_{1,n}(x) = \left(\frac{k+1-x}{k} \right)^n - \left(\frac{k-x}{k} \right)^n \quad , x = 1,2,\ldots,k$$

and

$$p_{n,n}(x) = \left(\frac{x}{k} \right)^n - \left(\frac{x-1}{k} \right)^n \quad , x = 1,2,\ldots,k.$$

The first two moments of $X_{1.n}$ are

$$E(X_{1,n}) = \sum_{y=1}^{k} y \left(\left(\frac{k+1-y}{k} \right)^n - \left(\frac{k-y}{k} \right)^n \right)$$

$$= \sum_{i=1}^{k} \left(\frac{k+1-i}{k} \right)^n$$

and

$$E(X_{1,n}^2) = \sum_{y=1}^{k} y^2 \left(\left(\frac{k+1-y}{k} \right)^n - \left(\frac{k-y}{k} \right)^n \right)$$

$$= \sum_{i=1}^{k}(2i-1)\left(\frac{k+1-i}{k}\right)^{n} .$$

For the range $W_{1,n}$, we have

$$P(W_{1,n} = 0) = (1/k)^{n-1} \text{ and}$$

$$P(W_{1,n} = w) = \frac{k-w}{k^n}\{(w+1)^n - 2w^n + (w-1)^n\}, w = 1,2,..k-1$$

For k=4 and n=5, $P(W_{15}= 0) = 4/1024$, $P(W_{15}= 1) = 90/1024$, $P(W_{15}= 2) = 360/1024$ and $P(W_{15} = 3) = 570/1024$.

Example 1.2.2.

Suppose $X_1, X_2, ..., X_n$ are n independent and identically distributed discrete geometric random variables, G(p), with $p(x) = p\, q^{k-1}$, $k = 1,2,\ldots$, $0<p<1$, $p + q = 1$. We have $P(x) = 1-q^x$ and $p_{r,n}(x) = \int_{1-q^{x-1}}^{1-q^x} r\, c(r,n)\, u^{r-1}(1-u)^{n-r}\, du$

For r = 1, we obtain

$$p_{1,n}(x) = n\int_{1-q^{x-1}}^{1-q^x}(1-u)^{n-1}\, du, \quad x = 1,2,....$$

$$= q^{n(x-1)} - q^{nx} = (1-q^n)\, q^{n(x-1)}, \; x=1,2,\ldots$$

Thus $X_{1,n}$ is $G(1-q^n)$ and the first two moments of $X_{1,n}$ are

$$E(X_{1,n}) = (1-q^n)\sum_{x=1}^{\infty} x\, q^{n(x-1)} = \frac{1}{1-q^n},$$

$$E(X_{1,n}^2) = (1-q^n)\sum_{x=1}^{\infty} x^2\, q^{n(x-1)} = \frac{1+q^n}{(1-q^n)^2}.$$

For the range $W_{1,n}$, we obtain

$$E(W_{1,n} = 0) = \sum_{x=1}^{\infty} p^n q^{n(x-1)} = \frac{p^n}{1-q^n}$$

and for k>0,

$$P_r(W_{1,n} = k)$$

$$= \sum\{(q^{u-1}(1-q^{k+1}))^n - (q^u(1-q^k))^n - (q^u(1-q^{k-1}))^n - (q^{u-1}(1-q^k))^n\}$$

$$=\frac{(1-q^{k+1})^n + q^n(1-q^{k-1})^n - q^n(1-q^k)^n - (1-q^k)^n}{1-q^n}$$

Example 1.2.3.

Suppose X_1, X_2, ...,X_n are n independent and identically distributed binomial random variables, B(n,p) with $p(X_i = x) = c(n,p)\, p^x (1-p)^{n-x}$, x =1,2,...,n, 0<p<1. We have

$$P(x) = P(X \le x\} = \sum_{k=0}^{x} c(k,n) p^k (1-p)^{n-k}, x = 0,1,2...,n$$

and

$$P_{r,n}(x) = \sum_{k=r}^{n} c(k,n)(P(x))^k (1-P(x))^{n-k}, x = 0,1.2,....,n$$

For r =1, we obtain

$$P_{1,n}(x) = \sum_{k=1}^{n} c(k,n)(P(x))^k (1-P(x))^{n-k}, x = 0,1.2,....,n$$
$$= 1-(1-P(x))^n, x = 0,1,2,...,n.$$

For r = n, we get

$$P_{n,n}(x) = \sum_{k=n}^{n} c(k,n)(P(x))^k (1-P(x))^{n-k}, x = 0,1.2,....,n$$
$$= (P(x))^n, x = 0,1,2,...,n.$$

We can write the first two moments of $X_{i,n}$ as

$$\mu_{i,n} = \sum_{x=0}^{n-1} \{1 - P_{i,n}(x)\}, \ 1 \le i \le n,$$

$$\mu_{i,n}^{(2)} = 2\sum_{x=1}^{n-1} x\{1 - P_{i,n}(x)\} + \mu_{i,n}, \ 1 \le i \le n.$$

Example 1.2.4.

Suppose X_1, X_2, ...,X_n are n independent and identically distributed Poisson random variables with $p(X_i = x) = e^{-\lambda}\lambda^x / x!$, x = 0,1,2,..., 0<λ< ∞. We have

$$P(x) = P(X \le x\} = \sum_{k=0}^{x} e^{-\lambda}\lambda^k / k!, x = 0,1,2,...$$

and

$$P_{i,n}(x)=\sum_{k=i}^{n} c(k,n)\,(P(x))^{k}\,(1-P(x))^{n-k},\ x=0,1,2,\ldots,n$$

We can write the first two moments of $X_{\cdot i,n}$ as

$$\mu_{i,n}=\sum_{x=0}^{\infty}\{1-P_{i,n}(x)\},\ 1\le i\le n,$$

and

$$\mu_{i,n}^{(2)}=2\sum_{x=1}^{\infty}x\{1-P_{i,n}(x)\},+\mu_{i,n},\ 1\le i\le n.$$

1.3. Estimation of Parameters

We will consider here the best (in the sense of minimum variance) linear unbiased estimates (BLUEs) of location and scale parameters. Suppose X has an absolutely continuos distribution function of the form $F((x-\mu)/\sigma)$, $-\infty<\mu<\infty$, $\sigma>0$. Further assume

$$E(X_{r,n})=\mu+\alpha_r\sigma\ ,r=1,2,\ldots,n$$

$$Var(X_{r,n})=v_{rr}\sigma^2,r=1,2,\ldots,n$$

$$Cov(X_{r,n}X_{s,n})=v_{rs}\sigma^2,1\le r<s\le n.$$

Let $X' = (X_{1,n},X_{2,n},\ldots,X_{n,n})$. Then we can write

$$E(X)=\mu\,1+\sigma\,\alpha \tag{1.3.1}$$

where $1'=(1,1,\ldots,1)'$, $\alpha'=(\alpha_1,\alpha_2,\ \ldots,\ \alpha_n)$

and $Var(X)=\sigma^2\Sigma$, where Σ is a matrix with elements v_{rs}, $1\le r\le s\le n$.

Then the BLUEs of the location and scale parameters μ and σ are

$$\hat{\mu}=\frac{1}{\Delta}\{\alpha'\Sigma^{-1}\alpha 1'\Sigma^{-1}-\alpha'\Sigma^{-1}1\alpha'\Sigma^{-1}\}X \tag{1.3.2}$$

and

$$\hat{\sigma}=\frac{1}{\Delta}\{1'\Sigma^{-1}1\alpha'\Sigma^{-1}-1'\Sigma^{-1}\alpha 1'\Sigma^{-1}\}X \tag{1.3.3}$$

where

$$\Delta = (\alpha'\Sigma^{-1}\alpha)(1'\Sigma 1) - (\alpha'\Sigma^{-1}1)^2. \tag{1.3.4}$$

The variance and the covariance of these estimators are given as

$$\text{Var}(\hat{\mu}) = \frac{\sigma^2(\alpha'\Sigma^{-1}\alpha)}{\Delta}, \tag{1.3.5}$$

$$\text{Var}(\hat{\sigma}) = \frac{\sigma^2(1'\Sigma^{-1}1)}{\Delta} \tag{1.3.6}$$

and

$$\text{Cov}(\hat{\mu}, \hat{\sigma}) = -\frac{\sigma^2(\alpha'\Sigma^{=1}1)}{\Delta}. \tag{1.3.7}$$

For symmetric distribution $\alpha_j = -\alpha_{n-j+1}$, $1'\Sigma^{-1}\alpha = 0$ and $\Delta = (\alpha'\Sigma^{-1}\alpha)(\underline{1}'\Sigma^{-1}\underline{1})$. The best linear unbiased estimates of μ and σ for the symmetric case are

$$\hat{\mu}^* = \frac{1'\Sigma^{-1}X}{1'\Sigma^{-1}1} \tag{1.3.8}$$

and

$$\hat{\sigma}^* = \frac{\alpha'\Sigma^{-1}X}{\alpha'\Sigma^{-1}\alpha}. \tag{1.3.9}$$

The corresponding covariance of the estimates is zero and the their variances are

$$Var(\hat{\mu}^*) = \frac{\sigma^2}{1'\Sigma^{-1}1} \tag{1.3.10}$$

$$Var(\hat{\sigma}^*) = \frac{\sigma^2}{\alpha'\Sigma^{-1}\alpha}. \tag{1.3.11}$$

We can use the above formulas to obtain the BLUEs of the location and scale parameters for any distribution numerically provided the variances of the order statistics exist. For some distributions the BLUEs of the location and scale parameters can be expressed in simplified form.

The following Lemma (see Garybill (1983, p.198)) will be useful to find the inverse of the covariance matrix.

Lemma 1.3.1.

Let $\Sigma = (\sigma_{r,s})$ be a mxm matrix with $\sigma_{rs} = \sigma_{sr} = c_r\, d_s$, $1 \le r \le s \le n$. Then Σ^{-1}, the inverse is given by $\Sigma^{-1} = (\sigma^{r,s})$, where

$$\sigma^{1,1} = \frac{c_2}{c_1(c_2\, d_1 - c_1\, d_2)}, \ \sigma^{n,n} = \frac{d_{i-1}}{d_i(c_i d_{i-1} - c_{i-1} d_i)}, \sigma^{i+1,i} = \sigma^{,ii+1} = -\frac{1}{c_{i+1} d_i - c_i\, d_{i+1}},$$

$$\sigma^{i,i} = \frac{c_{i+1} d_{i-1} - c_{i-1}\, d_{i+1}}{(c_i d_{i-1} - c_{i-1}\, d_i)(c_{i+1} d_i - c_i\, d_{i+1})}, i = 2,\ldots,n-1 \ \text{ and } \sigma^{i,j} = 0, |i-j| > 1.$$

The Lemma can be verified easily by showing $\Sigma\Sigma^{-1} = I$.

Example 1.3.1.

Suppose $X_1, X_2, \ldots, X_n$ are n independent and identically distributed uniform random variables with f(x) as

$f(x) = 1/\sigma$, $\mu - \sigma/2 \le x \le \mu + \sigma/2$, $-\infty < \mu < \infty$, $\sigma > 0$.

= 0, otherwise.

From example 1.1.1, we have

$$E(X_{r,n}) = \mu + \sigma\left(\frac{r}{n+1} - \frac{1}{2}\right), \ \mathrm{Var}(X_{r,n}) = \frac{r(n-r+1)}{(n+1)^2(n+2)}\,\sigma^2, \ r = 1,2,\ldots,n$$

and

$$Cov(X_{r,n}\, X_{s,n}) = \frac{r(n-s+1)}{(n+1)^2(n+2)}\sigma^2.$$

We can write $\mathrm{Cov}(X_{r,n}\, X_{s,n}) = \sigma^2 c_r\, d_s, c_r = \frac{r}{(n+1)^2}, d_s = \frac{n-s+1}{n+2}, 1 \le r \le s \le n.$

Using Lemma 1.3.1, we obtain

$$\sigma^{j.j} = 2(n+1)(n+2), j = 1,2,\ldots,n,$$
$$\sigma^{i,j} = -(n+1)(n+2), j = i+1, i = 1.2,\ldots,n-1,$$
$$\sigma^{i,j} = 0, |i-j| > 1.$$

It can easily be verified that $\underline{1}'\Sigma^{-1} = ((n+1)(n+2)\ 0\ 0 \ldots\ldots.(n+1)(n+2))$, $\underline{1}'\Sigma^{-1}\underline{1} = 2(n+1)(n+2)$, $\underline{1}'\Sigma^{-1}\alpha = 0$, $\alpha'\Sigma^{-1} = (-\frac{(n+1)(n+2)}{2}\ 0\ 0 \ldots.. \frac{(n+1)(n+2)}{2})$ and $\alpha'\Sigma^{-1}\alpha = \frac{(n-1)(n+2)}{2}$.

The BLUEs of the location and scale parameters of μ and σ are

$$\hat{\mu} = \frac{\underline{1}'\Sigma^{-1}X}{\underline{1}'\Sigma^{-1}\underline{1}} = \frac{X_{1,n} + X_{n,.n}}{2}$$

and

$$\hat{\sigma} = \frac{\alpha'\Sigma^{-1}X}{\alpha'\Sigma^{-1}\alpha} = \frac{(n+1)(X_{n.n} - X_{1,n})}{n-1}.$$

The corresponding covariance of the estimators is zero and their variances are

$$Var(\hat{\mu}) = \frac{\sigma^2}{\underline{1}'\Sigma^{-1}\underline{1}} = \frac{\sigma^2}{2(n+1)(n+2)}$$

and

$$Var(\hat{\sigma}) = \frac{\sigma^2}{\alpha'\Sigma^{-1}\alpha} = \frac{\sigma^2}{(n-1)(n+2)}.$$

Suppose the smallest r_1 and the largest r_2 observations are missing. Then considering these $X_{r_1+1,n}, X_{r_2+2,n}, \ldots, X_{n-r_2,n}$ order statistics, it can be shown that the inverse of the corresponding covariance matrix is

$$(n+1)(n+2)\begin{pmatrix} \frac{r_1+2}{r_1+1} & -1 & 0 & 0 & 0 & \ldots. & 0 \\ -1 & 2 & -1 & 0 & 0 & \ldots. & 0 \\ 0 & -1 & 2 & -1 & 0 & \ldots. & 0 \\ 0 & 0 & -1 & 2 & -1 & \ldots. & 0 \\ . & . & . & . & . & . & 0 \\ 0 & 0 & 0 & 0 & 0 & \ldots & -1 \\ 0 & 0 & 0 & 0 & 0 & \ldots & \frac{r_2+2}{r_2+1} \end{pmatrix}.$$

The BLUEs of μ and σ are respectively

$$\hat{\mu}^* = \frac{(n-2r_2-1)X_{r_1+1,n} + (n-2r_1-1)X_{n-r_2,n}}{2(n-r_i-r_2-1)}$$

and

$$\hat{\sigma}^* = \frac{n+1}{n-r_1-r_2-1}(X_{n-r_2,n} - X_{r_1+1,n}).$$

The variances and the covariance of the estimators are

$$Var(\hat{\mu}^*) = \frac{(r_1+1)(n-2r_2-1)+(r_2+1)(n-2r_1-1)}{4(n+2)(n+1)(n-r_1-r_2-1)}\sigma^2,$$

$$Var(\hat{\sigma}^*) = \frac{r_1+r_2+2}{(n+2)(n-r_1-r_2-1)}\sigma^2$$

and

$$Cov(\hat{\mu}^*, \hat{\sigma}^*)$$
$$= \frac{1}{2(n+1)(n+2)}\left[(n-2r_1-1)(r_2+1)(n-r_2)-(n-2r_2-1)(r_1+1)(n-r_1)-2(r_2-r_1)(r_1+1)(r_2+1)\right]$$

Note that If $r_1 = r_2 = r$, then $\hat{\mu}^* = \frac{X_{r,n} + X_{n-r,n}}{2}$ and $Cov(\hat{\mu}^*, \hat{\sigma}^*) = 0$.

Example 1.3.2.

Suppose $X_1, X_2, \ldots, X_n$ are n independent and identically distributed negative exponential random variables with

$f(x) = (1/\sigma) \exp(-(x-\mu)/\sigma)$, $-\infty < \mu < x < \infty$, $0 < \sigma < \infty$,

$= 0$, otherwise.

From example 1.1.2, we have

$$E(X_{r,n}) = \mu + \sigma\sum_{j=1}^{r}\frac{1}{n-j+1}, \quad Var(X_{r,n}) = \sigma^2\sum_{j=1}^{r}\frac{1}{(n-j+1)^2}, \ r = 1,2,\ldots,n,$$

and

$$\text{Cov}(X_{r,n}\ X_{s,n}) = \sigma^2 \sum_{j=1}^{r} \frac{1}{(n-j+1)^2}, 1 \le r \le s. \le n.$$

We can write $\text{Cov}(X_{r,n}\ X_{s,n}) = \sigma^2 c_r\ d_s, c_r = \sum_{j=1}^{r} \frac{1}{(n-j+1)^2}, d_s = 1, 1 \le r \le s \le n.$

Using Lemma 1.3.1, we obtain

$\sigma^{j,j} = (n-j)^2 + (n-j-1)^2, j = 1,2,\ldots,n,$

$$\sigma^{j+1,j} = \sigma^{j,j+1} = (n-j)^2, j = 1,\ \ i,j = 1,2,..\ ,n\text{ -1},$$

and $\sigma^{i,j} = 0$ for $|i-j| > 1,\ \ i,j = 1,2,..\ ,n.$

The BLUEs of the location and scale parameters of μ and σ are respectively

$$\hat{\mu} = \frac{nX_{1,n} - \overline{X}_n}{n-1}$$

and

$$\hat{\sigma} = \frac{n(\overline{X}_n - X_{1,n})}{n-1}.$$

The corresponding variances and the covariance of the estimators are

$$Var(\hat{\mu}) = \frac{\sigma^2}{n(n-1)},$$

$$Var(\hat{\sigma}) = \frac{\sigma^2}{n-1}$$

and

$$Cov(\hat{\mu}, \hat{\sigma}) = -\frac{\sigma^2}{n(n-1)}.$$

Suppose the smallest r_1 and the largest r_2 observations are missing, Then considering the order statistics $X_{r_1+1,}, X_{r_2+2}, \ldots, X_{n-r_2}$, it can be shown that the corresponding BLUEs of μ and σ are

$$\hat{\mu}^* = X_{r_1+1,n} - \alpha_{r_1+1}\ \hat{\sigma}^*,\ \alpha_{r_1+1} = \frac{1}{\sigma} E(X_{r_1+1,n} - \mu) = \sum_{i=1}^{r_1+1} \frac{1}{n-i+1}$$

and

$$\hat{\sigma}^* = \frac{1}{n-r_2-r_1-1}\left\{\sum_{i=r_1+1}^{n-r_2} X_{i,n} - (n-r_1)X_{r_1+1,n} + r_2 X_{n-r_2,n}\right\}.$$

The variances and the covariance of the estimators are

$$Var(\hat{\mu}^*) = \sigma^2\left[\frac{\alpha_{r_1}^2}{n-r_2-r_1-1} + \sum_{i=1}^{r_1+1}\frac{1}{(n-i+1)^2}\right],$$

$$Var(\hat{\sigma}^*) = \frac{\sigma^2}{n-r_1-r_2-1}$$

and

$$Cov(\hat{\mu}^*, \hat{\sigma}^*) = -\frac{\alpha_r \sigma^2}{n-r_2-r_1-1}.$$

Sarhan and Greenberg (1957) have prepared tables of the coefficients of the BLUEs and the variances and covariances of $\hat{\mu}^*$ and $\hat{\sigma}^*$ for n up to 10.

Example 1.3.3.

Suppose $X_1, X_2, \ldots, X_n$ are n independent random variables having power function distribution with

$$f(x) = \frac{\alpha}{\sigma}\left(\frac{x-\mu}{\sigma}\right)^{\alpha-1}, \quad -\infty < \mu < x < \infty, 0 < \sigma < \infty \quad 0 < \alpha < \infty$$

= 0, otherwise.

From example 1.1.6, we have

$$E(X_{r,n}) = \mu + \alpha_r \sigma, \quad \mathrm{Cov}(X_{r,n}\, X_{s,n}) = c_r\, d_s\, \sigma^2, 1 \le r \le s \le n,$$

where

$$c_r = \alpha_r = \frac{\Gamma(n+1)}{\Gamma(r)} \cdot \frac{\Gamma(r+\frac{1}{\alpha})}{\Gamma(n+1+\frac{1}{\alpha})}, \; r = 1,2,\ldots,n$$

$$d_r = \frac{1}{\alpha_r \beta_r} - \alpha_r$$

$$\beta_r = \frac{\Gamma(n+1+2\alpha^{-1})}{\Gamma(n+1)} \frac{\Gamma(r)}{\Gamma(r+2\alpha^{-1})}, \quad r = 1,2, \ldots, n.\backslash$$

Using Lemma 1.3.1, it can be shown (see Kabir and Ahsanullah (1974)) that

$$\sigma^{ii} = \{(\alpha(i-1)+1) + i\alpha(i\alpha+2)\}\beta \text{ , r =1,2,...,n.}$$

$$\sigma^{i\,i+1,} = \sigma^{i+1\,i} = -i(i\alpha+1)\beta \text{ , i = 1,2,...,n-1.}$$

$$\sigma^{i,j} = 0 \text{ for} |i-j| > 1.$$

The BLUEs of the location and scale parameters of μ and σ are

$$\hat{\mu} = M_1 \left[\sum_{i=1}^{n-1} P_i X_{i,n} - c_n^{-1} \sum_{i=1}^{n-1} c_i P_i X_{n.n} \right]$$

and

$$\hat{\sigma} = M_2 \left[\sum_{i=1}^{n-1} - P_i X_{i,n} + \sum_{i=1}^{n-1} P_i X_{n.n} \right],$$

where

$$M_1 = c_n M_2, M_2 = (c_n \sum_{i=i}^{n-1} P_i X_{i,n} - \sum_{i=0}^{n-1} c_i P_i)^{-1}, P_1 = (\beta+1)\beta_1, P_j = (\alpha-1)\beta_j, j = 2,\ldots,n-1.$$

The corresponding variances and the covariance of the estimators are

$$\text{Var}(\hat{\mu}) = M_1 \sigma^2,$$

$$\text{Var}(\hat{\sigma}) = \frac{T_n}{\Delta_n} \sigma^2$$

and

$$\text{Cov}(\hat{\mu}, \hat{\sigma}) = -\frac{1}{\Delta_n d_n} \sigma^2, \; T_n = \sum_{i=1}^{n} P_i, \Delta_n = \frac{c_n}{d_n} T_n - \frac{1}{d_n^2}.$$

Example 1.3.4.

Suppose $X_1, X_2, \ldots, X_n$ are n independent and identically distributed Pareto random variables with

$$f(x)=\frac{\alpha}{\sigma}\left(\frac{x-\mu}{\sigma}\right)^{-1-\alpha}, \quad -\infty<\mu<x<\infty, 0<\sigma<\infty \quad 0<\alpha<\infty$$

= 0, otherwise.

Using example 1.1.4, we get

$$E(X_{i,n})=\mu+(c_i-1)\sigma, \quad \alpha>\frac{1}{n-i+1} \text{ and}$$

$$Cov(X_{i,n}X_{j,n})=c_j\left(\frac{1}{c_i d_i}-c_i\right)\sigma^2, \; 1\le i\le j\le n, \text{ if}$$

$$\alpha>\max\left(\frac{2}{n-i+1}, \frac{1}{n-j+1}\right), i\le j, i, j=1,2,\ldots,n, \text{ where}$$

$$c_i=\frac{\Gamma(n+1)\Gamma(n+1-i-\alpha^{-1})}{\Gamma(n+1-\alpha^{-1})\Gamma(n+1-i)} \text{ and } d_i=\frac{\Gamma(n+1-2\alpha^{-1})\Gamma(n+1-i)}{\Gamma(n+1-i-2\alpha^{-1})\Gamma(n+1)}, \; i=1,2,\ldots,n.$$

It can be shown (see Kulldorff and Vannman (1973)) that the BLUEs of the location and scale parameter of μ and σ are

$$\hat{\mu}=X_{1,n}-(c_1-1)\hat{\sigma}$$

and

$$\hat{\sigma}=M_2\left[\sum_{i=1}^{n-1}-P_i X_{i,n}+\sum_{i=1}^{n-1}P_i X_{n.n}\right],$$

where

$$M_2=(c_n\sum_{i=i}^{n-1}P_i X_{i,n}-\sum_{i=0}^{n-1}c_i P_i)^{-1}, P_1=D-(\alpha+1)d_1, P_j=-(\alpha+1)d_j, j=2,\ldots,n-1, P_n=(\alpha-1)d_n$$

and

$$D=(\alpha+1)\sum_{i=1}^{n-1}d_i-(\alpha-1)d_n$$

The corresponding variances and the covariance of the estimators are

$$\mathrm{Var}(\hat{\mu}) = E\sigma^2,$$
$$\mathrm{Var}(\hat{\sigma}) = ((n\alpha - 1)^2 E - 1)\sigma^2$$

and

$$\mathrm{Cov}(\hat{\mu}, \hat{\sigma}) = \frac{(n\alpha - 1)(n\alpha - 2) - E}{(n\alpha - 2)E}\sigma^2$$

where $E = n\alpha(\alpha - 2) - \dfrac{(n\alpha - 2)^2}{n\alpha - 2 - D}$

It is difficult for many distributions to express the BLUEs of the location and scale parameters based on order statistics in closed form. However we can obtain numerically the coefficients of the BLUEs. We will present these estimators for the generalized exponential distribution.

Example 1.3.5.

Suppose that $X_1, X_2, \ldots, X_n$ are i.i.d generalized exponential distribution with pdf f(x) as

$$f(x) = \theta e^{-x}(1 - e^{-x})^{\theta - 1}, \theta > 0, 0 < x < \infty$$
= 0, otherwise.

Then from Example 1.1.12, we have

$$E(X_{r,n}) = rc(r,n)\sum_{j=0}^{n-r}\frac{(-1)^j c(j, n-r)}{r+j}[\psi((r+j)\theta + 1) + \gamma]$$

where $\psi(x) = \dfrac{d \ln \Gamma(x)}{dx}$ is the digamma function and γ (= $-\psi(1)$) is the Euler's constant.

The means of the order statistics for $n \leq 10$ are given in example 1.1.12. Note that

$$E(X^2_{r,n}) = rc(r,n)\sum_{j=0}^{n-r}\frac{(-1)^j c(j, n-r)}{r+j}[\psi((r+j)\theta + 1) + \gamma]^2$$

$$Var(X_{r,n}) = rc(r,n)\sum_{j=0}^{n-r}\frac{(-1)^j c(j,n-r)}{r+j}\{[\psi((r+j)\theta+1)+\gamma]^2$$

$$+\psi'((r+j)\theta+1)\}+\pi^2/6,$$

$$E(X_{r,n}X_{s,n})=r(s-r)c(r,s,n)\sum_{i=0}^{n-s}\sum_{j=0}^{s-r-1}\sum_{k=1}^{\infty}(-1)^{i+j}c(j,s-r-1)c(i,n-s)\frac{\psi((s+i)\theta+\gamma}{k((r+j)\theta+k)((s+i)\theta+k)}.$$

and

$$Cov(X_{r,n}X_{s,n})=E(X_{r,n}X_{s,n})-E(X_{r,n})E(X_{s,n}).$$

The following table gives the variances and covariances of the order statistics for $n \leq 5$ and for $\theta = 0.5(.5)\ 5.0$.

Table 1.3.1 Variances and Covariances of the Order Statistics

n	r	s	θ=0.5	θ=1.5	θ=2.0	θ=2.5	θ=3.0	θ=3.5	θ=4.0	θ=4.5	θ=5.0
1	1	1	0.710132	1.154576	1.250000	1.314576	1.361111	1.396209	1.423611	1.445592	1.463611
2	1	1	0.121817	0.336510	0.395833	0.438418	0.470278	0.494938	0.514559	0.530527	0.543766
2	1	2	0.149223	0.305765	0.340276	0.363560	0.380274	0.392837	0.402614	0.410437	0.416835
2	2	2	1.000000	1.361111	1.423611	1.463611	1.491389	1.511797	1.527422	1.539768	1.549768
3	1	1	0.039469	0.169118	0.212222	0.244567	0.269450	0.289081	0.304917	0.317940	0.328826
3	1	2	0.049425	0.151782	0.178889	0.197931	0.211950	0.222671	0.231121	0.237946	0.243570
3	1	3	0.057561	0.142465	0.162222	0.175658	0.185343	0.192640	0.198329	0.202885	0.206615
3	2	2	0.219196	0.436104	0.481389	0.511506	0.532930	0.548931	0.561329	0.571214	0.579279
3	2	3	0.251592	0.412208	0.441384	0.460185	0.473289	0.482939	0.490337	0.496188	0.500930
3	3	3	1.154576	1.445592	1.491389	1.520096	1.539768	1.554087	1.564977	1.573536	1.580440
4	1	1	0.016961	0.105254	0.139162	0.165530	0.186270	0.202881	0.216428	0.227660	0.237110
4	1	2	0.021554	0.093679	0.115761	0.131789	0.1438315	0.153168	0.160600	0.166649	0.171663
4	1	3	0.025386	0.087499	0.104196	0.115914	0.124528	0.131105	0.136282	0.140459	0.143898
4	1	4	0.028619	0.083492	0.096986	0.106250	0.112964	0.118041	0.122007	0.125189	0.127797
4	2	2	0.082504	0.229161	0.264718	0.289116	0.306810	0.320201	0.330676	0.339088	0.345990
4	2	3	0.096274	0.215067	0.240092	0.256705	0.268502	0.277300	0.284109	0.289533	0.293954
4	2	4	0.107929	0.205781	0.224446	0.236565	0.245049	0.251314	0.256128	0.259941	0.263036
4	3	3	0.289469	0.486413	0.522320	0.545479	0.561634	0.573538	0.582670	0.589897	0.595757
4	3	4	0.321035	0.467521	0.491699	0.506970	0.517481	0.525154	0.530998	0.535596	0.539307
4	4	4	1.250000	1.491389	1.527422	1.549768	1.564977	1.575996	1.584347	1.590893	1.596163
5	1	1	0.008586	0.073401	0.101386	0.123806	0.141772	0.156342	0.168331	0.178340	0.186804
5	1	2	0.011030	0.064924	0.083510	0.097377	0.107976	0.116288	0.122960	0.128423	0.132973
5	1	3	0.013103	0.060414	0.074742	0.085066	0.092782	0.098739	0.103465	0.107301	0.110475
5	1	4	0.014879	0.057496	0.069300	0.077616	0.083742	0.088424	0.092112	0.095088	0.097539
5	1	5	0.016413	0.055406	0.065501	0.072498	0.077596	0.081464	0.084493	0.086927	0.088925
5	2	2	0.039236	0.146069	0.175376	0.196022	0.211241	0.222886	0.232068	0.239486	0.245600
5	2	3	0.046303	0.136421	0.157890	0.172505	0.183045	0.190989	0.197184	0.202147	0.206210
5	2	4	0.052367	0.130103	0.146880	0.158046	0.165984	0.171908	0.176494	0.180148	0.183127
5	2	5	0.057615	0.125539	0.139120	0.148005	0.154252	0.158879	0.162441	0.165267	0.167563
5	3	3	0.119697	0.263008	0.292571	0.312106	0.325939	0.336236	0.344194	0.350526	0.355684
5	3	4	0.134454	0.251574	0.273423	0.287554	0.297426	0.304706	0.310295	0.314720	0.318309
5	3	5	0.147264	0.243221	0.259755	0.270271	0.277540	0.282862	0.286926	0.290130	0.292720

5	4	4	0.341011	0.517041	0.546681	0.565484	0.578465	0.587960	0.595207	0.600919	0.605536
5	4	5	0.370346	0.501527	0.521975	0.534745	0.543471	0.549807	0.554615	0.558385	0.561419
5	5	5	1.314576	1.520096	1.549768	1.568049	1.580440	1.589393	1.596163	1.601463	1.605723

Numerically we can obtain the coefficients BLUEs of the location and scale parameters. Their numerical values are given in the following tables:

1.4. Order Statistics from Extended Sample

Suppose $X_1,\ldots,X_n$ are n independent and identically distributed random variables from a distribution F. Let $X_{1,n}\le\ldots\le X_{n,n}$ be the corresponding order statistics. We consider m additional independent observations from F with $X_{1,n+m}\le\ldots\le X_{n+m,n+m}$ as corresponding order statistics in the extended sample. Sometimes one needs to consider sets of some order statistics $X_{i(1),n},\ldots,X_{i(k),n}$ together with additional r.v.'s $X_{n+1},\ldots,X_{n+m}$. For example, Wilks (1959) and Gumbel (1961) investigated such r.v. N(n,k) as a number of additional observations $X_{n+1},\ldots,X_{n+N(n,k)}$ which one must have to exceed $X_{n-k+1,n}$..The two-dimensional distributions of order statistics $X_{k,n}$ and $X_{l,m+n}$ are more complicated than these of order statistics $X_{k,n}$ and $X_{l,n}$ because for continuous distributions $X_{k,n}$ and $X_{l,m+n}$ can coincide with nonzero probability unlike $X_{k,n}$ and $X_{l,n}$.

Suppose that d.f. F is continuous. Compare order statistics $X_{k,n}$ and $X_{l,m+n}$ based respectively on sets $X_1,\ldots,X_n$ and $X_1,\ldots,X_n,X_{n+1},\ldots,X_{n+m}$, $1\le k\le n$, $1\le l\le n+m$.Evidently, $P\{X_{k,n}>X_{l,m+n}\}=1$ if $l<k$, and $P\{X_{l,m+n}>X_{k,n}\}=1$ if $l>m+k$. Consider the intermediate case $k\le l\le m+k$ and calculate probabilities $p_{k,l}=P\{X_{k,n}=X_{l,m+n}\}$. Recalling the formula for distributions of $X_{k,n}$ we obtain that $p_{k,l}(x)=P\{(l-k)$ of r.v.'s $X_{n+1},\ldots,X_{n+m}$ are less than $X_{k,n}$ $\}$

$$=\int_{-\infty}^{\infty}\binom{m}{l-k}(F(x))^{l-k}(1-F(x))^{m-l+k}\,dP\{X_{k,n}\le x\}$$

$$=\frac{m!n!}{(l-k)!(m-l+k)!(k-1)!(n-k)!}\int_{-\infty}^{\infty}(F(x))^{l-1}(1-F(x))^{n+m-l}\,dF(x)$$

$$=\frac{m!n!}{(l-k)!(m-l+k)!(k-1)!(n-k)!}\int_0^1 u^{l-1}(1-u)^{n+m-l}\,du$$

$$=\frac{m!n!(l-1)!(n+m-l)!}{(l-k)!(m-l+k)!(k-1)!(n-k)!(n+m)!}=\binom{m}{l-k}\frac{k^{(l-k)}(n+1-k)^{(m+k-l)}}{(n+1)^{(m)}},\quad(1.4.1)$$

where $x^{(s)}=x(x+1)\ldots(x+s-1)$.

Table 1.3.2: Coefficients for the BLUE of μ

n	r	θ=0.5	θ=1.5	θ=2.0	θ=2.5	θ=3.0	θ=3.5	θ=4.0	θ=4.5	θ=5.0
2	1	1.294350	1.657740	1.785720	1.893430	1.986490	2.068430	2.141650	2.207860	2.268280
2	2	-0.294349	-0.657740	-0.785715	-0.893433	-0.986487	-1.068430	-1.141650	-1.207860	-1.268280
3	1	1.191480	1.445680	1.539300	1.619780	1.690470	1.753550	1.81054	1.862530	1.910340
3	2	-0.119381	-0.197104	-0.221611	-0.243057	-0.262474	-0.280343	-0.296945	-0.312467	-0.327048
3	3	-0.072099	-0.248574	-0.317687	-0.376725	-0.427999	-0.473212	-0.513597	-0.550064	-0.583293
4	1	1.154400	1.327160	1.394030	1.453390	1.506850	1.555450	1.600010	1.641150	1.679340
4	2	-0.088505	-0.068646	-0.056189	-0.046862	-0.040095	-0.035237	-0.031769	-0.029354	-0.027708
4	3	-0.038135	-0.121889	-0.154811	-0.183411	-0.208653	-0.231227	-0.251654	-0.270297	-0.287453
4	4	-0.027764	-0.136626	-0.183026	-0.223120	-0.258098	-0.288989	-0.316587	-0.341498	-0.364179
5	1	1.138080	1.249880	1.295990	1.339110	1.379410	1.417040	1.452240	1.485210	1.516190
5	2	-0.079319	-0.014206	0.015139	0.038218	0.056513	0.071285	0.083421	0.093593	0.102232
5	3	-0.029035	-0.064226	-0.076007	-0.086529	-0.096178	-0.105129	-0.113474	-0.121308	-0.128698
5	4	-0.016457	-0.082902	-0.112067	-0.137627	-0.160191	-0.180317	-0.198455	-0.214945	-0.230052
5	5	-0.013273	-0.088551	-0.123052	-0.153171	-0.179549	-0.202879	-0.22373	-0.24249	-0.29675
6	1	1.130050	1.194710	1.224180	1.254380	1.284330	1.313380	1.341270	1.367920	1.393340
6	2	-0.075883	0.014046	0.052287	0.082385	0.106388	0.126013	0.142381	0.156282	0.168261
6	3	-0.025841	-0.033094	-0.032244	-0.032017	-0.032405	-0.033313	-0.034555	-0.036041	-0.037676
6	4	-0.012668	-0.052103	-0.068566	-0.083106	-0.096107	-0.107833	-0.118516	-0.128330	-0.137404
6	5	-0.008425	-0.060516	-0.085354	-0.107343	-0.126801	-0.144155	-0.159779	-0.173956	-0.186925
6	6	-0.007232	-0.063042	-0.090298	-0.114304	-0.135403	-0.154091	-0.170801	-0.185880	-0.199601
7	1	1.125820	1.152890	1.168610	1.188270	1.209790	1.231930	1.253990	1.275580	1.29655
7	2	-0.074418	0.030611	0.073977	0.107845	0.134874	0.156969	0.175430	0.191220	0.204901
7	3	-0.024587	-0.014275	-0.005359	0.001705	0.0071185	0.011237	0.014436	0.016881	0.018800
7	4	-0.011147	-0.033003	-0.040692	-0.047595	-0.053973	-0.059885	-0.065422	-0.070631	-0.075556
7	5	-0.006550	-0.042057	-0.058704	-0.073517	-0.086703	-0.098554	-0.109285	-0.119066	-0.128061
7	6	-0.004808	-0.046470	-0.067723	-0.086703	-0.103551	-0.118583	-0.132115	-0.144392	-0.155613
7	7	-0.004310	-0.047695	-0.070107	-0.090006	-0.107555	-0.123115	-0.137032	-0.149593	-0.161017
8	1	1.123510	1.119820	1.123910	1.134730	1.149270	1.165690	1.182940	1.200400	1.217740
8	2	-0.073715	0.041134	0.087574	0.123547	0.152133	0.175483	0.195000	0.211687	0.226183

Table 1.3.2 continued

n	r	θ=0.5	θ=1.5	θ=2.0	θ=2.5	θ=3.0	θ=3.5	θ=4.0	θ=4.5	θ=5.0
8	3	-0.024195	-0.001955	0.012430	0.024014	0.033251	0.040643	0.046702	0.051688	0.055897
8	4	-0.010406	-0.020293	-0.021756	-0.023251	-0.024950	-0.026769	-0.028690	-0.030652	-0.032652
8	5	-0.005724	-0.029528	-0.039960	-0.049273	-0.057659	-0.065287	-0.072279	-0.078716	-0.084691
8	6	-0.003759	-0.034487	-0.050132	-0.064167	-0.076683	-0.087909	-0.098047	-0.107280	-0.115739
8	7	-0.002976	-0.037038	-0.055454	-0.072027	-0.086783	-0.099960	-0.111815	-0.122568	-0.132397
8	8	-0.002739	-0.037655	-0.056616	-0.073578	-0.088578	-0.101894	-0.113809	-0.124561	-0.134340
9	1	1.122060	1.092840	1.086950	1.090190	1.098790	1.110400	1.123600	1.137600	1.151920
9	2	-0.073167	0.048247	0.096527	0.133651	0.163019	0.186900	0.206904	0.223980	0.238785
9	3	-0.024093	0.006552	0.024785	0.039526	0.051334	0.060967	0.068896	0.075573	0.081310
9	4	-0.010101	-0.011366	-0.008263	-0.005832	-0.004149	-0.003026	-0.002329	-0.001975	-0.001890
9	5	-0.005317	-0.020620	-0.026311	-0.031441	-0.036155	-0.040568	-0.044711	-0.048598	-0.052241
9	6	-0.003238	-0.025804	-0.036906	-0.046890	-0.055840	-0.063929	-0.071291	-0.078024	-0.084256
9	7	-0.002364	-0.028805	-0.043184	-0.056191	-0.067830	-0.078255	-0.087647	-0.096203	-0.104029
9	8	-0.001942	-0.030366	-0.046533	-0.061182	-0.074254	-0.085941	-0.096467	-0.106003	-0.114713
9	9	-0.001833	-0.030678	-0.047067	-0.061828	-0.074915	-0.086546	-0.096954	-0.106347	-0.114890
10	1	1.121560	1.070340	1.055660	1.052310	1.055800	1.063290	1.073060	1.084120	1.000900
10	2	-0.073410	0.053206	0.102643	0.140353	0.170008	0.194055	0.214143	0.231251	0.394366
10	3	-0.023855	0.012683	0.033752	0.050688	0.064305	0.075444	0.084667	0.092549	-0.045241
10	4	-0.010018	-0.004826	0.001701	0.007069	0.011261	0.014551	0.017153	0.019177	0.037559
10	5	-0.005142	-0.014053	-0.016058	-0.017954	-0.019842	-0.021809	-0.023743	-0.025688	-0.002776
10	6	-0.002936	-0.019303	-0.026776	-0.033474	-0.039580	-0.045108	-0.050215	-0.054911	-0.024938
10	7	-0.002043	-0.022523	-0.033491	-0.043432	-0.052341	-0.060374	-0.067669	-0.074335	-0.032683
10	8	-0.001557	-0.024452	-0.037608	-0.049600	-0.060334	-0.069967	-0.078640	-0.086523	-0.452438
10	9	-0.001329	-0.025463	-0.039811	-0.052903	-0.064618	-0.075098	-0.084533	-0.093088	0.105391
10	10	-0.001274	-0.025605	-0.040010	-0.053062	-0.064662	-0.074980	-0.084219	-0.092554	-0.087336

Table 1.3.3: Coefficients for the BLUE of σ

n	r	θ=0.5	θ=1.5	θ=2.0	θ=2.5	θ=3.0	θ=3.5	θ=4.0	θ=4.5	θ=5.0
2	1	-1.294350	-0.904223	-0.857143	-0.829241	-0.810811	-0.79774	-0.787993	-0.780446	-0.774431
2	2	1.294350	0.904223	0.857143	0.829241	0.810811	0.797740	0.787993	0.780446	0.774431
3	1	-1.362930	-0.877263	-0.815518	-0.778385	-0.753616	-0.73593	-0.722675	-0.712374	-0.704139
3	2	0.688937	0.427519	0.388323	0.363669	0.346722	0.334361	0.324951	0.317549	0.311575
3	3	0.673993	0.449744	0.427194	0.414716	0.406894	0.401569	0.397724	0.394825	0.392565
4	1	-1.414700	-0.854967	-0.781048	-0.736357	-0.706486	-0.685143	-0.669143	-0.656718	-0.646787
4	2	0.502610	0.261925	0.222025	0.196614	0.179072	0.166264	0.156506	0.148844	0.142661
4	3	0.443196	0.296886	0.278676	0.267756	0.260475	0.255271	0.251372	0.248336	0.245911
4	4	0.468890	0.296156	0.280346	0.271986	0.266939	0.263608	0.261265	0.259537	0.258214
5	1	-1.453540	-0.836922	-0.753168	-0.702473	-0.668629	-0.644488	-0.626430	-0.612420	-0.601248
5	2	0.410329	0.178792	0.138742	0.113333	0.095909	0.083264	0.073701	0.066217	0.060213
5	3	0.341924	0.214686	0.195482	0.183377	0.175047	0.168967	0.164331	0.160683	0.157743
5	4	0.337179	0.223991	0.212048	0.205298	0.200987	0.198005	0.195825	0.194163	0.192852
5	5	0.364105	0.219453	0.206896	0.200465	0.196687	0.194251	0.192573	0.191357	0.190440
6	1	-1.483330	-0.822047	-0.730184	-0.674618	-0.637609	-0.611278	-0.591624	-0.576419	-0.564319
6	2	0.354567	0.129157	0.089417	0.064428	0.047464	0.035255	0.026083	0.018961	0.013286
6	3	0.283101	0.164342	0.144054	0.131020	0.121954	0.115321	0.110252	0.106262	0.103036
6	4	0.270310	0.176255	0.164204	0.156881	0.151961	0.148414	0.145738	0.143647	0.141966
6	5	0.275898	0.178602	0.169380	0.164445	0.161426	0.159407	0.157974	0.156902	0.156076
6	6	0.299450	0.173690	0.163129	0.157844	0.154804	0.152881	0.151578	0.150648	0.149955
7	1	-1.506650	-0.809512	-0.710810	-0.651208	-0.611610	-0.583526	-0.562624	-0.546482	-0.533658
7	2	0.316836	0.096305	0.057097	0.032741	0.016367	0.004700	-0.003982	-0.010699	-0.016020
7	3	0.244178	0.130560	0.109506	0.095874	0.086390	0.079466	0.074175	0.070035	0.066690
7	4	0.228046	0.143388	0.130682	0.122642	0.117116	0.113064	0.109981	0.107554	0.105600
7	5	0.227406	0.147936	0.139106	0.134002	0.130670	0.128328	0.126585	0.125232	0.124154
7	6	0.235037	0.147929	0.140206	0.136245	0.133912	0.132402	0.131361	0.130604	0.130032
7	7	0.255148	0.143394	0.134214	0.129703	0.127155	0.125566	0.124504	0.123756	0.123205
8	1	-1.525420	-0.798758	-0.694175	-0.631159	-0.589409	-0.55988	-0.537963	-0.521082	-0.507694
8	2	0.289532	0.073048	0.034474	0.010823	-0.004909	-0.016017	-0.024217	-0.030508	-0.035473
8	3	0.216443	0.106388	0.084837	0.070876	0.061198	0.054152	0.048794	0.044616	0.041254

Table 1.3.3 continued

n	r	$\theta=0.5$	$\theta=1.5$	$\theta=2.0$	$\theta=2.5$	$\theta=3.0$	$\theta=3.5$	$\theta=4.0$	$\theta=4.5$	$\theta=5.0$
8	4	0.198351	0.119552	0.106194	0.097563	0.091556	0.087138	0.083762	0.081094	0.078945
8	5	0.195126	0.125111	0.116025	0.110468	0.106711	0.103998	0.101951	0.100344	0.099053
8	6	0.197772	0.126819	0.119714	0.115796	0.113333	0.111643	0.110409	0.109469	0.108722
8	7	0.205490	0.125932	0.119174	0.115820	0.113908	0.112705	0.111894	0.111321	0.110900
8	8	0.222700	0.121908	0.113756	0.109812	0.107614	0.106262	0.105370	0.104747	0.104293
9	1	-1.540450	-0.789379	-0.679689	-0.613709	-0.570134	-0.539406	-0.516647	-0.499158	-0.485325
9	2	0.268433	0.055734	0.017872	-0.005073	-0.020172	-0.030707	-0.038455	-0.044351	-0.048960
9	3	0.195498	0.088299	0.066443	0.052307	0.042580	0.035507	0.030194	0.026049	0.022722
9	4	0.176339	0.101541	0.087648	0.078568	0.072228	0.067555	0.063975	0.061160	0.058896
9	5	0.171640	0.107610	0.098116	0.092107	0.087950	0.084919	0.082606	0.080786	0.079304
9	6	0.172042	0.110191	0.103091	0.098927	0.096178	0.094233	0.092785	0.091652	0.090756
9	7	0.175742	0.110663	0.104595	0.101393	0.099457	0.098159	0.097224	0.096529	0.095983
9	8	0.182930	0.109437	0.103370	0.100435	0.098802	0.097803	0.097150	0.096696	0.096368
9	9	0.197822	0.105904	0.098555	0.095045	0.093112	0.091938	0.091169	0.090639	0.090256
10	1	-1.553420	-0.781133	-0.666895	-0.598327	-0.553179	-0.521429	-0.497977	-0.479997	-0.394417
10	2	0.252561	0.042410	0.005252	-0.017017	-0.031495	-0.041507	-0.048801	-0.054299	-0.195346
10	3	0.178831	0.074291	0.052217	0.038058	0.028361	0.021357	0.016118	0.012027	0.128760
10	4	0.159196	0.087465	0.073177	0.063770	0.057192	0.052359	0.048659	0.045767	0.023250
10	5	0.153749	0.093844	0.083934	0.077527	0.073041	0.069765	0.067244	0.065265	0.036996
10	6	0.152788	0.096887	0.089587	0.085093	0.082064	0.079846	0.078180	0.076857	0.041255
10	7	0.154662	0.098092	0.092194	0.088879	0.086754	0.085286	0.084206	0.083382	0.036904
10	8	0.158474	0.097959	0.092596	0.089871	0.088266	0.087222	0.086488	0.085945	0.395129
10	9	0.165074	0.096648	0.091100	0.088476	0.087048	0.086195	0.085646	0.085276	-0.098374
10	10	0.178087	0.093537	0.086837	0.083670	0.081946	0.080908	0.080237	0.079776	0.065565

Table 1.3.4: Variances and Covariances of the BLUEs of μ and σ in terms of σ^2

n	θ=0.5	θ=1.5	θ=2.0	θ=2.5	θ=3.0	θ=3.5	θ=4.0	θ=4.5	θ=5.0
2	0.177010	0.846824	1.186230	1.510020	1.816730	2.107000	2.382110	2.643430	2.892280
	1.379430	0.888009	0.836738	0.807915	0.789633	0.777070	0.767940	0.761017	0.755596
	-0.278224	-0.673748	-0.814630	-0.932532	-1.033700	-1.122190	-1.200810	-1.271510	-1.335740
3	0.045538	0.311218	0.458452	0.601668	0.738834	0.869559	0.994045	1.112710	1.226000
	0.736919	0.434505	0.406012	0.390611	0.381137	0.374784	0.370256	0.366879	0.364270
	-0.070475	-0.249436	-0.318512	-0.376941	-0.427280	-0.471375	-0.510546	-0.545756	-0.577714
4	0.018020	0.168910	0.259014	0.347987	0.433885	0.516140	0.594719	0.669782	0.741566
	0.513205	0.285269	0.265053	0.254427	0.248035	0.243827	0.240874	0.238699	0.237036
	-0.027655	-0.135909	-0.181063	-0.219674	-0.253087	-0.282407	-0.308471	-0.331896	-0.353154
5	0.008757	0.108751	0.172412	0.236059	0.297900	0.357335	0.414241	0.468689	0.520810
	0.396950	0.211468	0.195651	0.187514	0.182708	0.179591	0.177430	0.175857	0.174664
	-0.013370	-0.087716	-0.120984	-0.149720	-0.174691	-0.196643	-0.216169	-0.233724	-0.249648
6	0.004824	0.077128	0.125806	0.174997	0.223042	0.269357	0.313782	0.356334	0.397104
	0.324899	0.167599	0.154546	0.147949	0.144110	0.141652	0.139967	0.138751	0.137836
	-0.007338	-0.062312	-0.088509	-0.111347	-0.131267	-0.148809	-0.164423	-0.178461	-0.191196
7	0.002894	0.058215	0.097345	0.137259	0.176423	0.214266	0.250619	0.285474	0.318890
	0.275549	0.138590	0.127447	0.121897	0.118711	0.116693	0.115323	0.114342	0.113610
	-0.004391	-0.047089	-0.068615	-0.087538	-0.104102	-0.118709	-0.131719	-0.143419	-0.154032
8	0.001849	0.045891	0.078448	0.111937	0.144926	0.176871	0.207595	0.237076	0.265354
	0.239494	0.118015	0.108276	0.103488	0.100770	0.099066	0.097919	0.097104	0.096500
	-0.002800	-0.037154	-0.055374	-0.071514	-0.085687	-0.098202	-0.109355	-0.119386	-0.128485
9	0.001241	0.037352	0.065130	0.093921	0.122381	0.149993	0.176576	0.202101	0.226593
	0.211935	0.102680	0.094020	0.089810	0.087445	0.085976	0.084994	0.084302	0.083791
	-0.001876	-0.030262	-0.046025	-0.060086	-0.072470	-0.083419	-0.093181	-0.101962	-0.109927
10	0.000865	0.031157	0.055316	0.080530	0.105533	0.129831	0.153245	0.175739	0.179048
	0.190148	0.090819	0.083015	0.079260	0.077169	0.075882	0.075028	0.074429	0.060916
	-0.001306	-0.025257	-0.039125	-0.051577	-0.062572	-0.072304	-0.080985	-0.088795	-0.080771

Note: In Table 1.3.4, for each n, the first row gives the variance of $\hat{\mu}$, the second row gives the variance of $\hat{\sigma}$ and the third row gives the covariance of $\hat{\mu}$ and $\hat{\sigma}$.

Now under additional restriction that the initial X's have a density function f we calculate conditional probabilities $p_1(x)=P\{X_{l,n+m}=X_{k,n}\,|X_{k,n}=x\}$ and $p_2(x)= P\{X_{l,n+m}=X_{k,n}\,|X_{l,n+m}=x\}$ for all x such that f(x)>0. Since distribution density functions of $X_{k,n}$ and $X_{l,n+m}$ are $f_{k,n}(x)=(n!/(k-1)!(n-k)!)(F(x))^{k-1}(1-F(x))^{n-k}f(x)$ and $f_{l,m+n}(x)= ((n+m)!/(l-1)!(n+m-l)!)\,(F(x))^{l-1}\,(1-F(x))^{m+n-l}\,f(x)$ respectively, we have

$$p_2(x) = p_1(x)f_{k,n}(x)/\, f_{l,m+n}(x)$$

$$=(n!(l-1)!(n+m-l)!/(k-l)!(n-k)!(n+m)!)\ (F(x))^{k-l}\ (1-F(x))^{l-m-k}\ p_1(x). \qquad (1.4.2)$$

The following equality holds :

$$p_1(x) = P\{(l-k) \text{ of r.v.'s } X_{n+1},\dots,X_{n+m} \text{ are less than } x\} =(F(x))^{l-k}(1-F(x))^{n+m+k-l}\,. \qquad (1.4.3)$$

It implies from (1.4.1) - (1.4.3) that

$$P\{X_{l,n+m} = X_{k,n}\,|X_{l,n+m} = x\} = \frac{m!n!(l-1)!(n+m-l)!}{(l-k)!(m-l+k)!(k-l)!(n-k)!(n+m)!}$$

$$= P\{X_{l,n+m} = X_{k,n}\}\,.$$

It means that the event $\{X_{k,n}-X_{l,n+m}= 0\}$ and the order statistic $X_{l,n+m}$ are independent and the conditional distribution of the spacing $X_{k,n}- X_{l,n+m}$ given that $X_{l,n+m}= x$ has an atom $p_{k,l}$ at the point x. Note that the value $p_{k,l}$ is positive if $k \le l \le m+k$, and $p_{k,l}= 0$ otherwise.

It appears that conditional distribution $H_{k,l}(y,x) = P\{X_{k,n}- X_{l,n+m} <y|\, X_{l,n+m} = x\}$ of the spacing $X_{k,n}-X_{l,n+m}$ anywhere has density function $h_{k,l}(y,x)$ if $1\le l <k\le n$ and $m+k< l \le n+m$. If $k\le l \le m+k$ then $h_{k,l}(y,x)\, h_{k,l}(y,x)=\partial H_{k,l}(y,x)/\partial y$ exists if $y\neq 0$ and $H_{k,l}(y,x)$ has an atom $p_{k,l}$ at the point y=0.

It is easy to see that $h_{k,l}(y,x)=g_{k,l}(x+y,x)/f_{l,n+m}(x)$, $y\neq 0$, where $g_{k,l}(u,v)= \partial^2 P\{X_{k,n}<u,\ X_{l,n+m} < v\}/\partial u\partial v$. We consider three options.

1\. Let $1\le l<k$. In this case $P\{X_{l,n+m} <X_{k,n}\}=1$ and $g_{k,l}(u,v)= 0$ for $v\ge u$. Let us now consider v<u. To find $g_{k,l}(u,v)$ for v<u we can apply the well known approach which was used to derive single and joint densities of order statistics taken from the same sample.

```
-----------|---|   ------|---|   -------->
           v   v+Δ1      u   u+Δ2
```

We must calculate $P\{v \le X_{l,n+m} \le v+\Delta_1, u \le X_{k,n} \le u+\Delta_2\}/\Delta_1\Delta_2$ and let Δ_1 and Δ_2 tend to zero. It is enough to propose that only one of X's can lie in each of intervals $[v, v+\Delta_1]$ and $[u,u+\Delta_2]$. Denote A={l of $X_1,\ldots,X_{n+m}$ are less than v, one of them belongs to $[v, v+\Delta_1]$ and $k-l$ are greater than $v+\Delta_1$, $k-l$ of $X_1,\ldots,X_n$ are less than u, one of them belongs to $[u,u+\Delta_2]$ and n-k are greater than $u+\Delta_2$}, B={one of values $X_1,\ldots,X_n$ belongs to $[v, v+\Delta_1]$}, C={one of values $X_{n+1},\ldots,X_{n+m}$ belongs to $[v,v+\Delta_1]$}, D_r={r of values $X_1,\ldots,X_n$ belong to $(-\infty,v)$, one belongs to $[v, v+\Delta_1]$, k-r-2 belong to $(v+\Delta_1,u)$, one belongs to $[u,u+\Delta_2]$, n-k belong to $(u+\Delta_2, \infty)$} and E_s={s of values $X_{n+1},\ldots,X_{n+m}$ belong to $(-\infty,v)$, m-s belong to $(v+\Delta_1, \infty)\backslash[u,u+\Delta_2]$}. Then $AB = \bigcup_{r=\max(1,l-m)}^{l} D_{r-1} E_{l-r}$, where D_{r-1} and E_{l-r} are independent. It means that

$$P\{AB\} = \sum_{r=\max(1,l-m)}^{l} P\{D_{r-1}\}P\{E_{l-r}\},$$

Where

$$P\{D_{r-1}\}=(n!/(r-1)!(k-r-1)!(n-k)!)(F(v))^{r-1}(F(v+\Delta_1)-F(v))(F(u)-F(v+\Delta_1))^{k-r-1}(F(u+\Delta_2)-F(u))(1-F(u+\Delta_2))^{n-k}$$

and

$$P\{E_{l-r}\}=(m!/(l-r)!(m-l+r)!)(F(v))\ (F(v))^{l-r}\ (1-F(v+\Delta_1)-F(u+\Delta_2)+F(u))^{m-l+r}.$$

Analogously

$$P\{AC\} = \sum_{r=\max(1,l-m+1)}^{l} P\{F_{r-1}\}P\{G_{l-r}\},$$

where F_r={ r of values $X_1,\ldots,X_n$ belong to $(-\infty,v)$, k-r-1 belong to $(v+\Delta_1,u)$, one belongs to $[u,u+\Delta_2]$, n-k belong to $(u+\Delta_2, \infty)$} , G_s={s of values $X_{n+1},\ldots,X_{n+m}$ belong to $(-\infty,v)$, one belongs to $[v,v+\Delta_1]$ and m-s-1 belong to $(v+\Delta_1, \infty)\backslash[u,u+\Delta_2]$} and hence

$$P\{F_{r-1}\}=(n!/(r-1)!(k-r)!(n-k)!)(F(v))^{r-1}(F(u)-F(v+\Delta_1))^{k-r}\ (F(u+\Delta_2)-F(u))(1-F(u+\Delta_2))^{n-k}$$

and

$$P\{G_{l,r}\}= (m!/(l-r)!(m-l+r-1)!)(F(v))^{l-r}\ (F(v+\Delta_1)-F(v))(1-F(v+\Delta_1)-F(u+\Delta_2)+F(u))^{m-l+r-1}$$

Combining these arguments we have the following expressions:

$$g_{k.l}(u,v) = \sum_{r=\max(1,l-m)}^{l} \frac{m!n!}{(r-1)!(k-r-1)!(n-k)!(l-r)!(m-l+r)!} (F(v))^{l-1}\ (F(u)-$$

$F(v))^{k-r-1}(1-F(v))^{m-l+r}(1-F(u))^{n-k}f(v)f(u)+$

$$\sum_{r=\max(1,l-m+1)}^{l} \frac{m!n!}{(r-1)!(k-r)!(n-k)!(l-r)!(m-l+r-1)!}(F(v))^{l-1}(F(u)-F(v))^{k-r}$$

$$.(1-F(v))^{m-l+r-1}(1-F(u))^{n-k}f(v)f(u),\ u>v, \tag{1.4.4}$$

and

$$h_{k,l}(y,x)=\sum_{r=\max(1,l-m)}^{l} \frac{m!n!(l-1)!(n+m-l)!}{(r-1)!(k-r-1)!(n-k)!(l-r)!(m-l+r)!(n+m)!}$$

$$(F(x+y)-F(x))^{k-r-1}(1-F(x))^{r-n}(1-F(x+y))^{n-k}f(x+y)+$$

$$\sum_{r=\max(1,l-m+1)}^{l} \frac{m!n!(l-1)!(n+m-l)!}{(r-1)!(k-r)!(n-k)!(l-r)!(m-l+r-1)!(n+m)!}(F(x+y)-F(x))^{k-r}$$

$$.(1-F(x))^{r-n-1}(1-F(x+y))^{n-k}f(x+y),\ y>0. \tag{1.4.5}$$

Example 1.4.1.

Suppose $F(x)=1-e^{-x}$, $0<x<\infty$, then (1.4.5) can be rewritten as

$$h_{k,l}(y,x)=\frac{m!n!(l-1)!(n+m-l)!}{(n-k)!(n+m)!}\exp(-(n-k+1)y)\Big(\sum_{r=\max(1,l-m)}^{l}$$

$$\frac{1}{(r-1)!(k-r-1)!(l-r)!(m-l+r)!}(1-\exp(-y))^{k-r-1}+$$

$$\sum_{r=\max(1,l-m+1)}^{l} \frac{1}{(r-1)!(k-r)!(l-r)!(m-l+r-1)!}(1-\exp(-y))^{k-r}\Big),\ x>0,y>0. \tag{1.4.6}$$

The RHS of (1.4.6) does not depend on x. It means that r.v.'s $X_{k,n}-X_{l,n+m}$ and $X_{l,n+m}$ are independent in the exponential case if $1\le l<k$.

Let $l>m+k$. In this situation $P\{X_{l,n+m}>X_{k,n}\}=1$ and $g_{k.l}(u,v)=0$ for $v\le u$. Consider the case v>u. The similar arguments show that then

$$g_{k.l}(u,v)=\sum_{r=l-m-k}^{\min(l-k,n-k)} \frac{m!n!}{(r-1)!(k-1)!(n-k-r)!(l-k-r)!(m-l+k+r)!}$$

$$.(F(u))^{k-1}(F(v))^{l-k-r}(F(v)-F(u))^{r-1}(1-F(v))^{n+m-l}f(u)f(v)$$

$$+\sum_{r=l-m-k}^{\min(l-k,n-k+1)}\frac{m!n!}{(r-1)!(k-1)!(n-k-r+1)!(l-k-r)!(m-l+k+r-1)!}$$

$$(F(u))^{k-1}(F(v))^{l-k-r}$$

$$(F(v)-F(u))^{r-1}(1-F(v))^{n+m-l}f(u)f(v),\ u<v. \qquad (1.4.7)$$

In this case

$$h_{k,l}(y,x)=\frac{m!n!(l-1)!(n+m-l)!}{(k-1)!(n+m)!}(F(x+y))^{k-1}(F(x))^{1-1}f(x+y)$$

$$(\sum_{r=l-m-k}^{\min(l-k,n-k)}\frac{1}{(r-1)!(n-k-r)!(l-k-r)!(m-l+k+r)!}(F(x))^{l-k-r}(F(x)-$$

$$F(x+y))^{r-1}+\sum_{r=l-m-k+1}^{\min(l-k,n-k+1)}\frac{1}{(r-1)!(n-k-r+1)!(l-k-r)!(m-l+k+r-1)!}$$

$$.(F(x))^{l-k-r}(F(x)-F(x+y))^{r-1},\ y<0. \qquad (1.4.8)$$

If $F(x) = \min(e^x,1)$, then the RHS of (1.4.8) is transformed as follows:

$$h_{k,l}(y,x)=\frac{m!n!(l-1)!(n+m-l)!}{(k-1)!(n+m)!}e^{yk}$$

$$.(\sum_{r=l-m-k}^{\min(l-k,n-k)}\frac{1}{(r-1)!(n-k-r)!(l-k-r)!(m-l+k+r)!}(1-e^y)^{r-1}+$$

$$\sum_{r=l-m-k+1}^{\min(l-k,n-k+1)}\frac{1}{(r-1)!(n-k-r+1)!(l-k-r)!(m-l+k+r-1)!}(1-e^y)^{r-1}),\ x<0. \qquad (1.4.9)$$

The RHS of (1.4.9) does not depend on x . It means that r.v.'s $X_{k,n}$--$X_{l,n+m}$ and $X_{l,n+m}$ are independent in this case .

Let $k\leq l\leq k+m$. The analogous approach gives us the following equalities:

$$g_{k.l}(u,v)=\sum_{r=1}^{\min(l-k,n-k)}\frac{m!n!}{(r-1)!(k-1)!(n-k-r)!(l-k-r)!(m-l+k+r)!}$$

$$(F(u))^{k-1}(F(v))^{l-k-r}(F(v)-F(u))^{r-1}(1-F(v))^{n+m-l}f(u)f(v)+\sum_{r=1}^{\min(l-k,n-k+1)}$$

$$\frac{m!n!}{(r-1)!(k-1)!(n-k-r+1)!(l-k-r)!(m-l+k+r-1)!}(F(u))^{k-1}(F(v))^{l-k-r}$$

$$.(F(v)-F(u))^{r-1}(1-F(v))^{n+m-l}f(u)f(v),\ \text{if } u<v \qquad (1.4.10)$$

and

$$g_{k,l}(u,v)=\sum_{r=\max(1,l-m)}^{k-1}\frac{m!n!}{(r-1)!(k-r-1)!(n-k)!(l-r)!(m-l+r)!}(F(v))^{l-1}$$

$$.(F(u)-F(v))^{k-r-1}(1-F(v))^{m-l+r}(1-F(u))^{n-k}f(v)f(u)+$$

$$\sum_{r=\max(1,l-m+1)}^{k}\frac{m!n!}{(r-1)!(k-r)!(n-k)!(l-r)!(m-l+r-1)!}(F(v))^{l-1}(F(u)-F(v))^{k-r}$$

$$.(1-F(v))^{m-l+r+1}(1-F(u))^{n-k}f(v)f(u),\ \text{if } u>v. \qquad (1.4.11).$$

It implies from (1.4.10) and (1.4.11) that

$$h_{k,l}(y,x)=\sum_{r=1}^{\min(l-k,n-k)}\frac{m!n!(l-1)!(n+m-l)!}{(r-1)!(k-1)!(n-k-r)!(l-k-r)!(m-l+k+r)!(n+m)!}$$

$$(F(x+y))^{k-1}$$

$$.(F(x)-F(x+y))^{r-1}(F(x))^{l-k+r}f(x+y)+$$

$$\sum_{r=1}^{\min(l-k,n-k+1)}\frac{m!n!(l-1)!(n+m-l)!}{(r-1)!(k-1)!(n-k-r+1)!(l-k-r)!(m-l+k+r-1)!(n+m)!}$$

$$(F(x+y))^{k-1}$$

$$.(F(x)-F(x+y))^{r-1}(F(x))^{l-k-r}f(x+y)\ \text{if } y<0, \qquad (1.4.12)$$

and

$$h_{k,l}(y,x)=\sum_{r=\max(1,l-m)}^{k-1}\frac{m!n!(l-1)!(n+m-l)!}{(r-1)!(k-r-1)!(n-k)!(l-r)!(m-l+r)!(n+m)!}$$

$$.(F(x+y)-F(x))^{k-r-1}(1-F(x))^{r-n}(1-F(x+y))^{n-k}f(x+y)$$

$$+\sum_{r=\max(1,l-m+1)}^{k} \frac{m!n!(l-1)!(n+m-l)!}{(r-1)!(k-r)!(n-k)!(l-r)!(m-l+r-1)!(n+m)!} (F(x+y)-F(x))^{k-r}$$

$$.(1-F(x))^{r-n-1} (1-F(x+y))^{n-k} f(x+y), \text{ if } y>0. \tag{1.4.13}$$

We suppose in (1.14.10)-(1.4.13) that $\sum_{r=i}^{j} = 0$, if $j < i$.

Recalling that the conditional d.f. $H_{k,l}(y,x) = P\{X_{k,n} - X_{l,n+m} \le y | X_{l,n+m} = x\}$ has an atom $p_{k,l} = (n!m!(l-1)!(n+m-l)!/(l-k)!(m-l+k)!(k-1)!(n-k)!(n+m)!)$ one can write that

$$H_{k,l}(y,x) = \int_{-\infty}^{y} h_{k,l}(t,x)dt + p_{k,l} 1_{\{y \ge x\}}. \tag{1.4.14}$$

Easy to see that the RHS of (1.4.12) does not depend on x if $F(x)=\min\{e^x,1\}$ and the RHS of (1.4.13) does not depend on x, if $F(x) = \max(0,1-e^{-x}\}$. Therefore viewing all three situations we can formulate the following results.

Theorem 1.4.1.

If $F(x)=\max(0,1-e^{-x}\}$ then for any $1\le k \le n$, $1\le l \le n+m$, $n \ge 1$, $m \ge 1$, random variables $\max(0, X_{k,n} - X_{l,n+m}$ and $X_{l,n+m}$ are independent.

Theorem 1.4.2.

If $F(x) = \min(e^x, 1)$ then for any $1 \le k \le n$, $1 \le l \le n+m$, $n \ge 1$, $m \ge 1$, random variables $\min(0, X_{k,n} - X_{l,n+m}$ and $X_{l,n+m}$ are independent.

Corollary 1.4.1.
If $m = 0$ we obtain from theorem 1.4.1 the well-known property of the exponential distribution: independence of $\max(0, X_{k,n} - X_{l,n})$ and $X_{l,n}$.

Suppose that $X_1, X_2, \ldots$ are independent r.v.'s with a common continuous distribution function F. First of all let us recall the following well-known result (compare , for example, with theorems 2.4.1 and 2.4.2 in Arnold, Balakrishnan and Nagaraja (1992)).

Theorem 1.4.3.
Let $X_1, X_2, \ldots, X_n$ be a random sample from a population with a continuous distribution function F. Then conditional distributions of vectors of order statistics $(X_{1,n}, \ldots, X_{l-1,n})$ and $(X_{l+1,n}, \ldots, X_{n,n})$ given that $X_{l,n}=x$ are independent and coincide correspondingly with the distribution of vector $(Y_{1,l-1}, \ldots, Y_{l-1,l-1})$ from a sample of size l-1 from a population with d.f.
$G(u)=P\{Y \le x\}=F(u)/F(x)$, $u<x$,
and with distribution of vector $(V_{1,n-l}, \ldots, V_{n-l,n-l})$ *from a sample of size* $n-l$ *from a population with d.f.*

$H(u)=(F(u)-F(x))/(1-F(x)),\ u>x$.

Due to the statement of Theorem 1.4.3 the conditional distribution of order statistics $X_{1,n+m},\dots,X_{n+m,n+m}$ under condition that $X_{l,n+m}= x$ coincides with the unconditional distribution of order statistics based on independent r.v.'s

$Y_1,\dots,Y_{l-1},\ Z,\ V_1,\dots,V_{n-+m-l}$,

where Y's have d.f. G, V's have d.f. H and $P\{Z=x\}=1$. Note that $P\{Y<x\}=P\{V>x\}=1$. Symmetry arguments implies that under condition that $X_{l,n+m}=x$ the initial sample $X_1,\dots,X_n$ with equal probabilities can be presented by any group $W_1,\dots,W_n$ of n r.v.'s taken from

$Y_1,\dots,Y_{l-1},\ Z,\ V_1,\dots,V_{n+m-l}$.

Hence an order statistic $X_{k,n}$ can coincide with order statistics $Y_{k,r}$ or $V_{s,s+n-k}$ under different values of r ($l-1\geq r\geq k$) and s ($\max\{0,k-l\}\leq s\leq \min\{k,\ m-l+k\}$) ,where Z is denoted as $V_{0,n-k}$. Note that

$A_r=\{X_{k,n} = Y_{k,r}\}$

takes place if a set $W_1,\dots,W_n$ includes exactly r of l-1 random variables $Y_1,\dots,Y_{l-1}$. The probability q_r of this event is given as follows:

$q_r =P\{A_r\}=(l-1)!(n+m-l+1)!n!m!/r!(l-1-r)!(n-r)!(m+r-l+1)!(n+m)!$

Analogously the event

$B_s=\{ X_{k,n} = V_{s,s+n-k}\},s>0,$

occurs if a set $W_1,\dots,W_n$ includes exactly (n-k+s) of r.v.'s $V_1,\dots,V_{n+m-l}$ and its probability p_s is given by the following equality:

$p_s =P\{ B_s\}=l!(n+m-l)!n!m!/(k-s)!(l -k+s)!(n-k+s)!(m+k-l-s)!(n+m)!$.

One more situation $B_0=\{X_{k,n}=V_{0,n-k} \}=\{X_{k,n} =Z\}=\{X_{k,n}=x\}$ means that a set $W_1,\dots,W_n$ includes Z, (k-1) of r.v.'s $Y_1,\dots,Y_{l-1}$ and (n-k) of V's. Hence,

$p_0=P\{B_0\}= (l-1)!(n+m-l)!n!m!/(k-1)!(l -k)!(n-k)!(m+k-l)!(n+m)!$.

Therefore one immediately gets that conditional distribution of $X_{k,n}$ given that $X_{l,n+m} = x$ presents a mixture of distributions of order statistics $Y_{k,r}$ and $V_{s,s+n-k}$ taken with probabilities q_r ($\min\{l-1,n\}\geq r\geq k$) and p_s ($\max\{0,k-l\}\leq s\leq\min\{k,m-l+k\}$):

$$P\{X_{k,n}<y|X_{l,n+m}=x\}= \sum_r q_s\, P\{Y_{k,r}<y\}+p_0 1\{y>x\}+\sum_s p_s\, P\{V_{s,s+n-k} <x\}, \quad (1.4.15)$$

where sums are taken over all possible values of r ($l-1\geq r\geq k$) and s ($\max\{1,k-l\}\leq s\leq m-l+k$) it is easy to see that the first sum in (1.4.15) is omitted if $k>l-1$. If $k>l$ then

$$P\{X_{k,n}<y|X_{l,n+m}=x\}=\sum_s p_s\, P\{V_{s,s+n-k}<x\}.$$

Analogous arguments show that conditional distribution of vector $(X_{l,n},\ldots,X_{n,n})$ given that $X_{l,n+m}=x$ can be represented as a mixture of random vectors

$$V_0=(V_{0,n-l},\ldots,V_{n-l,n-l}),\ldots,V_t=(V_{t,n+t-l},\ldots,V_{n+t-l,n+t-l}),$$

where $t=\min(m,l)$, taken with weights $p_0,\ldots,p_t$. We can rewrite this fact in the following short form :

$$(X_{l,n},\ldots,X_{n,n}\,|X_{l,n+m}=x)\ \underline{\underline{d}}\ \sum_{s=0}^{t}\ {}_{\text{mixt.}}\ p_s V_s\,. \tag{1.4.16}$$

The exponential distribution.

Consider now a sequence of independent r.v.'s $X_1,X_2,\ldots$ with the same standard exponential distribution . In this case r.v.'s Y's and V's determined in theorem 3 have distribution functions $G(u)=P\{X\le u|X\le x\}=(1-\exp(-u))/(1-\exp(-x))$, $0<u<x$, and

$H(u)=P\{X\le u|X>x\}=(\exp(-x)-\exp(-u))/\exp(-x)=1-\exp(x-u)$, $u>x$, correspondingly.

Note that $H(u)$ is a d.f. of a random variable $x+\xi$, where ξ has the standard exponential distribution. It means that

$$(X_{l,n},\ldots,X_{n,n}\,|X_{l,n+m}=x)\stackrel{d}{=}\sum_{s=0}^{l}\ {}_{\text{mixt.}}\ p_s(x+\xi_{s,n+s-l},\ldots,x+\xi_{n+s-l,n+s-l})$$

$$=x+\sum_{s=0}^{l}\ {}_{\text{mixt.}}\ p_s(\xi_{s,n+s-l},\ldots,\xi_{n+s-l,n+s-l}). \tag{1.4.17}$$

Equality (1.4.11) implies that

$$(X_{l,n}-X_{l,n+m}\ldots,X_{n,n}-X_{l,n+m}\,|X_{l,n+m}=x)\ \underline{\underline{d}}\ \sum_{s=0}^{l}\ {}_{\text{mixt.}}\ p_s(\xi_{s,n+s-l},\ldots,\xi_{n+s-l,n+s-l}). \tag{1.4.18}$$

The RHS of (1.4.18) does not depend on x and r.v.'s Y's, which determine order statistics $X_{1,n+m},\ldots,X_{l-1,n+m}$. Therefore in the exponential case distribution of vector of spacings $(X_{l,n}-X_{l,n+m},\ldots,X_{n,n}-X_{l,n+m})$ does not depend on order statistics $X_{1,n+m},\ldots,X_{l,n+m}$. It is not difficult to see that all these spacings are nonnegative. The same method applies to prove that random variables

$T_k=\max\{0, X_{k,n} - X_{l,n+m}\}$, $k=1,...,l-1$,

also do not depend on $X_{l,n+m}$. In fact it means that a more general result than Theorem 1.4.1 is true.

Theorem 1.4.4.
Let $F(x)=1-\exp\{-(x-a)/b\}$, $x>a$. *Then a vector of spacings* $(T_1,...,T_n)$ *does not depend on order statistics* $X_{1,n+m},X_{2,n+m},...,X_{l,n+m}$.

The following corollaries of theorem 4 are evident.

Corollary 1.4.2.

If $m=0$ *then the statement of theorem 4 coincides with the well-known property of the exponential spacings.*

Corollary 1.4.3.

Differences $(T_2-T_1,...,T_n-T_{n-1})$ *do not depend on order statistics* $X_{1,n+m},X_{2,n+m},...,X_{l,n+m}$.

Corollary 1.4.4.

Spacings $X_{n,n} -X_{n-1,n}$,$X_{n-1,n} -X_{n-2,n},...,X_{l+2,n}-X_{l+1,n}$,$X_{l+1,n}-X_{l,n}$ *do not depend on order statistics* $X_{1,n+m},X_{2,n+m},...,X_{l,n+m}$, $l=1,2,...,n-1$.

One more important result is a partial case of Corollary 1.4.4.

Corollary 1.4.5.

Spacings $X_{n,n} -X_{n-1,n}$, $X_{n-1,n} -X_{n-2,n},...$, $X_{2,n}-X_{1,n}$ *do not depend on order statistic* $X_{1,n+m}$.

Of course, all results given above for the exponential distribution can be reformulated with evident modifications for distribution functions

$F(x)= \exp((x-a)/b\}$, $x<a$.

Joint moments of extended exponential order statistics

The independence property of the exponential spacings helps us to find joint moments of order statistics $X_{k,n}$ and $X_{l,n+m}$, if $k\geq l$. David and Rogers (1980) gave some expressions for covariances of overlapping order statistics in terms of the product moments of usual order statistics from the same parent distributions. They also got exact values for correlations between moving medians for the normal distribution. Note also that some moment characteristics for moving minima for the uniform distribution were given in Inagaki (1980).

Let us consider the standard exponential distribution with $F(x)=1-\exp(-x)$, $x>0$. We know that spacings $(X_{l,n}-X_{l,n+m},\dots,X_{n,n}-X_{l,n+m})$ do not depend on the order statistics $X_{1,n+m},\dots,X_{l,n+m}$. Hence if $r\geq l$ then r.v.'s
$T_r=(X_{r,n}-X_{l,n+m})$ and $X_{l,n+m}$ are independent. From the equality

$$0=\mathrm{cov}(T_r,X_{l,n+m})=\mathrm{cov}(X_{r,n},X_{l,n+m})-\mathrm{Var}(X_{l,n+m})$$

one gets that

$$\mathrm{cov}(X_{r,n},X_{l,n+m})=\mathrm{Var}(X_{l,n+m}).$$

From classical formulae for expectations and variances of exponential order statistics (see, for example, David (1981))

$$EX_{l,n+m}=\frac{1}{n+m}+\frac{1}{n+m-1}+\dots+\frac{1}{n+m-l+1},\quad \mathrm{Var}\,X_{l,n+m}=\sum_{k=n+m-l+1}^{n+m}\frac{1}{k^2},$$

One obtains that

$$\mathrm{Cov}(X_{r,n},X_{l,n+m})=\sum_{k=n+m-l+1}^{n+m}\frac{1}{k^2}$$

and

$$EX_{r,n}X_{l,n+m}=\mathrm{Var}\,X_{l,n+m}+EX_{l,n+m}EX_{r,n}$$

$$=\sum_{k=n+m-l+1}^{n+m}\frac{1}{k^2}+\sum_{k=n+m-l+1}^{n+m}\frac{1}{k}\sum_{k=n-r+1}^{n}\frac{1}{k}.$$

More complicated joint moments of exponential order statistics $X_{r,n}$ and $X_{l,n+m}$ also can be obtained by this method.

Dependence of extended extremal order statistics

Amongst the results formulated above we got the independence of maximal order statistic $X_{1,n+m}$ and spacing $X_{1,n}-X_{1,n+m}$ in the exponential case. In regression form the latter fact can be rewritten as

$$E(X_{1,n}-X_{1,n+m}\,|X_{1,n+m}=x)=c,$$

where c is a positive constant. What types of positive functions

$$\varphi(x)=E(X_{1,n}-X_{1,n+m}\,|X_{1,n+m}=x) \tag{1.4.19}$$

can we obtain for other parent distributions? We suppose that (1.4.19) is true for all x from some interval (γ,η) where $-\infty \le \gamma < \eta \le \infty$.

Applying (1.4.5) for the given situation one can see that conditional distribution of $X_{1,n}$ given that $X_{1,n+m}=x$ is a mixture of two distributions:

$$P\{X_{1,n}<y|X_{1+m,n+m}=x\}= p_0 1\{y>x\}+p_1\, P\{V_{1,n}<y\},$$

where $p_0=n/(n+m)$ and $p_1=m/(n+m)$. Then

$$\varphi(x)=E(X_{1,n}-X_{1,n+m}\,|X_{1,n+m}=x)=nx/(n+m)-(m/(n+m))\int_x^\infty vd(1-H(v))^n-x=$$

$$\frac{nm}{n+m}\frac{1}{(1-F(x))^n}\int_x^\infty v(1-F(v))^{n-1}dF(v)-\frac{mx}{n+m}. \tag{1.4.20}$$

Equality (1.4.20) allows to obtain different characterizations of distribution functions F based on a form of regression $E(X_{1,n}-X_{1,n+m}\,|X_{1,n+m}=x)$.

Denote $\overline{F}(v)=1-F(v)$. Then

$$\varphi(x)=-\frac{mx}{n+m}-\frac{nm}{n+m}\frac{1}{{}^n(x)}\int_0^\infty v(\overline{F}(v))^{n-1}d\overline{F}(v).$$

$$=\frac{m}{n+m}\int_x^\infty\left(\frac{\overline{F}(u)}{\overline{F}(x)}\right)^n du \tag{1.4.21}$$

Suppose that $\varphi(x)$ is differentiable then $\overline{F}(x)$ is also differentiable. It is easy to show that (1.4.21) implies the following equality:

$$n(n+m)\,\overline{F}'(x)\,\varphi(x)=-\overline{F}(x)((n+m)\varphi'(x)+m),$$

from which we have that

$$\overline{F}(x)=\exp\{c-\int_\gamma^x\{(\varphi'(u)+\frac{m}{n+m})/n\varphi(u)\}du$$

and

$$F(x)=1-\exp\{c-\int_{\gamma}^{x}\{(\varphi'(u)+\frac{m}{n+m})/n\varphi(u)\}du, \qquad (1.4.22)$$

where c is a constant.

Substituting any positive φ, one can find the corresponding function. Note that F can be df if $\varphi'(x)+\frac{m}{m+n}>0$ for $-\infty \leq \gamma \leq x \leq \eta \leq \infty$.

One of the simplest situations is considered in Theorem 1.4.5.

Theorem 1.4.5.

The necessary and sufficient condition for $\varphi(x) = \alpha x + \beta$ is

(i) $F(x) = 1- e^{-\frac{m(x-\mu)}{n(m+n)\beta}}$, if $\alpha = 0$, $\beta>0$ where $x > \mu$ and μ is an arbitrary real number,

(ii) $F(x)=1-(1-x)^{-\frac{1}{n}(\frac{m}{(m+n)\alpha}+1)}$, if $-\frac{m+n}{m}<\alpha<0, \beta=-\alpha, 0<x<1$,

(iii) $F(x)=1-(x-1)^{-\frac{1}{n}(\frac{m}{(m+n)\alpha}+1)}$, if $\alpha>0$, $\beta=-\alpha$, $1<x<\infty$.

Proof.

To prove the statements of the theorem, we need only to use the relations (1.4.21) and (1.4.22).

1.5. MISCELLANEOUS

Best Linear Unbiased Predictor of the sth order statistic.

Let $X_{1,n}< X_{2,n}<\ldots< X_{n,n}$ denote the order statistics of a random sample of size n from an absolutely continuous distribution function F. We will assume that F belongs to a location-scale family with location parameter μ and scale parameter σ. Suppose we observe $X' = (X_{1,n}, X_{2,n},\ldots, X_{r,n})$ and we are interested to predict $X_{s,n}$ where $1\leq r < s \leq n$. Note that $X_{s,n}$ can be considered as the life length of a (n-s+1) out of n system of independent components with independent life lengths. The best (in the sense of minimum variance) linear unbiased predictor (BLUP) $\hat{X}_{s,n}$ of $X_{s,n}$ is given by

$$\hat{X}_{s,n} = \hat{\mu}^* + \alpha_{s.n}\hat{\sigma}^* + w'\Sigma^{-1}(X - \hat{\mu}^* 1 - \alpha\,\hat{\sigma}^*) \qquad (1.5.1)$$

where $\hat{\mu}^*$ and $\hat{\sigma}^*$ are the BLUEs of μ and σ based on the r order statistics $X_{1,n}, X_{2,n}, \ldots, X_{r,n}$, with $w' = (w_{1s}, w_{2s}, \ldots, w_{rs})$, $w_{js} = Cov(X_{j,n}, X_{s,n})/\sigma^2$, $1' = (1,1,\ldots,1)$,

$\alpha' = (\alpha_{1,n}, \alpha_{2,n}, \ldots, \alpha_{r,n})$ and $\alpha_{j,n} = E(X_{j,n} - \mu)/\sigma$, j=1,2,…,r.

If $w_{js} = c_j\, d_s$ then using Lemma 1.3.1, it can easily be shown that

$w'\Sigma^{-1} = (0,0,\ldots,d_s/d_r)$. Thus the BLUP of $X_{s,n}$ is

$$\hat{X}_{s,n} = \hat{\mu}^* + \alpha_{s.n}\hat{\sigma}^* + \frac{c_s}{c_r}(X_{s,n} - \hat{\mu}^* - \alpha_{r,n}\hat{\sigma}^*). \qquad (1.5.2)$$

Example 1.5.1.
Consider the two parameter uniform distribution, U(μ- σ/2, μ+σ/2)).. Using the example 1.3.1, we obtain from (1.5.1)

$$\hat{X}_{s,n} = \hat{\mu}^* + \alpha_{s.n}\hat{\sigma}^* + \frac{n-s+1}{n-r+1}(X_{r,n} - \hat{\mu}^* - \alpha_{r,n}\hat{\sigma}^*), \text{ where}$$

$$\alpha_{r,n} = \frac{r}{n+1} - \frac{1}{2},\ c_r = \frac{r}{(n+1)^2},\ d_s = \frac{n-s+1}{n+2},$$

$$\hat{\mu}^* = \frac{(n-1)X_{r,n} - (n-2r+1)X_{1,n}}{2(r-1)} \text{ and } \hat{\sigma}^* = \frac{(n+1)(X_{r,n} - X_{1,n})}{r-1}.$$

On simplification, we obtain

$$\hat{X}_{s,n} = \frac{n-s+1}{n-r+1}X_{r,n} + \frac{s-r}{n-r+1}(\hat{\mu}^* - \frac{1}{2}\hat{\sigma}^*).$$

Example 1.5.2.

Consider the two parameter exponential distribution, E(μ, σ)).

From example 1.3.2, we have $d_s/d_r = 1$ and using (1.5.1), we get

$$\hat{X}_{s,n} = X_{r,n} + (\alpha_{s,n} - \alpha_{r,n})\hat{\sigma}^*, \text{ where}$$

$$\alpha_{r,n} = \sum_{i=1}^{r} \frac{1}{n-i+1}, \alpha_{s,n} = \sum_{I=1}^{S} \frac{1}{n-i+1} \text{ and } \hat{\sigma}^* = \frac{1}{r-1}\left\{\sum_{i=1}^{r} X_{i,n} - nX_{1,n} + (n-r)X_{r,n}\right\}$$

Thus on simplification, we obtain

$$\hat{X}_{s,n} = X_{r,n} + \left(\sum_{i=r+1}^{s} \frac{1}{n-i+1}\right)\hat{\sigma}^*.$$

For Pareto and Power Function distributions the covariance w_{ij} of the random variables $X_{I,n}$ and $X_{j,n}$ can be expressed as $w_{js} = c_j\ d_s$ and the BLUP of $X_{s,n}$ can similarly be obtained.

Raqab and Nagaraja (1995) and Raqab (19970) considered maximum likelihood predictors of order statistics.

Characterizations

Order statistics are widely used to characterize various distributions. There are many characterizations of the exponential distributions based on order statistics and by the spacings of order statistics. Galambos and Kotz (1978) and Azlarov and Volodin (1996) have given many interesting characterizations of the exponential distribution. We have seen that for the exponential distribution n $X_{1,n}$ and X have identical distribution. The equality in distribution of n $X_{1,n}$ and X will lead to the following functional equation

$$\left(\overline{F}(x/n)\right)^n = \overline{F}(x) \tag{1.5.3}$$

The solution of the equation for absolutely continuous F(x) with two n relatively prime to each other and all x will give $\overline{F}(x) = e^{-\sigma x}$, $x \geq 0$. Since F(x) is a distribution function, we must have $\sigma > 0$. A detailed discussions and proof of similar results see Desu (1971) and Ahsanullah (1977). It is known (see Galambos and Kotz (1978)), Theorem 3.2.2, p.39) that for a fixed n, $n \geq 2$, if $nX_{1,n} \stackrel{d}{=} X$ and $\lim F(x)/x = \sigma$, $0<\sigma<\infty$, as $x \to 0^+$, then $F(x) = 1 - e^{-\sigma x}$, $x \geq 0$. Suppose n is a an geometric random variable with $P(n=k) = p(1-p)^{k-1}$, k=1,2,..., and 0<p<1. It is known (see Feller (1966), p. 54) that a geometric sum of independent and identically distributed random variables having the same distribution, is again exponentially distributed.Kakosyan et al ((1964), pp 77-78) have shown that for an integer valued random variable n, the equality of the distributions of

$nX_{1,n}$ and X_1 characterizes the exponential distribution. Kakosyan et al ((1984), p.81) raised the question : given $P(N=k)= p(1-p)^{k-1}$, $k=1,2,\ldots$, $0<p<1$, whether the equality of the distributions $p\sum_{i=1}^{N} X_i$ and $NX_{1,N}$ characterizes the exponential distribution. As a partial answer to the question we will give the following theorem (for details, see Ahsanullah (1988)).

We will call a distribution function F " new better than used" (NBU), if

$\overline{F}(x+y) \le \overline{F}(x)\,\overline{F}(y), x, y \ge 0$

and F is " new worse than used " (NWU), if

$\overline{F}(x+y) \ge \overline{F}(x)\,\overline{F}(y), x, y \ge 0.$

We say that $X \in C_1$, if its distribution function is either NBU or NWU.

Theorem 1.5.1.

Let $X_1, X_2, \ldots$ be independent and identically distributed non- negative random variables with pdf f(x), x>0 and M is an interger valued random variable independent of the X's with $P(M=k) = p(1-p)^{k-1}$, $k=1,2,\ldots$ and $0<p<1$. Then the following two properties are equivalent

(a) X's are exponentially ditributed with $F(x)=1-e^{-\sigma x}, \sigma > 0$.

(b) $p\sum_{j=1}^{M} X_j \overset{d}{=} MX_{1,M}$, $F(x)<1$ for all $x>0$, $X_j \in C_1$, $E(X_j) < \infty$,

$\lim_{x \to 0^+} F(x)/x = \sigma,\ \sigma > 0.$

Proof.

It is easy to verify that (a) $\Rightarrow$ (b). We will proof here (b) $\Rightarrow$ (a).

Let $\phi(s)$, $\phi_1(s)$ and $\phi_2(s)$ be the Laplace transform of X_1, $p\sum_{j=1}^{M} X_j$ and $MX_{1,M}$ respectively. We have for s>0,

$$\phi_1(s) = E(e^{-sp\sum_{j=1}^{M} X_j}) = \sum_{k=1}^{\infty} p(1-p)^{k-1}(\phi(ps))^k$$

$$= \frac{p\phi(ps)}{1-(1-p)\phi(ps)} \tag{1.5.4}$$

and

$$\phi_2(s) = \sum_{k=1}^{\infty} \int_0^{\infty} pq^{k-1} e^{-skx} k(\bar{F}(x))^{k-1} f(x)\,dx$$

$$= 1 - s\sum_{k=1}^{\infty} \int_0^{\infty} pq^{k-1} e^{-sx} (\bar{F}(\tfrac{x}{k}))^k \,dx. \qquad (1.5.5)$$

Equating (1.5.4) and (1.5.5) and letting $s \to 0^+$, we get on simplification

$$\sum_{k=1}^{\infty} \int_0^{\infty} pq^{k-1} (\bar{F}(\frac{X}{k}))^k \,dx = \sum_{k=1}^{\infty} \int_0^{\infty} pq^{k-1} (\bar{F}(x))\,dx$$

i.e.

$$\sum_{k=1}^{\infty} pq^{k-1} \int_0^{\infty} \{(\bar{F}(\tfrac{x}{k}))^k - \bar{F}(x)\}dx = 0. \qquad (1.5.6)$$

Since $X \in C_1$, we must have

$$\bar{F}(u) = (\bar{F}(\tfrac{u}{k}))^k \qquad (1.5.7)$$

for almost all u and all k, k=1,2,...

Since $\lim_{x \to 0^+} F(x)/x = \sigma,\ \sigma > 0.$, it follows from (1.5.7) that

$$\bar{F}(u) = e^{-x\sigma}, x > 0, \sigma > 0. \qquad (1.5.8)$$

Puri and Rubin (1970) proved that if F is absolutely continuous then the identical distribution of $X_{2,2}$ - $X_{1,2}$ and X characterizes the exponential distribution. Rossberg (1972) proved that if F is non-lattice, then the identical distribution of $X_{k+1,n}$ - X_k, k>1 and $X_{1,n-k}$ characterizes the exponential distribution. Ahsanullah (1974) proved that if F is absolutely continuous with strictly increasing distribution function for all x>0, then the identical distributions of the statistics $X_{s_i,n} - X_{r,n}$ and $X_{s_i-r,n-r}$ for any fixed r and two distinct numbers s_1 and s_2, $1 \le r < s_1 < s_2 \le n$, characterize the exponential distribution. We shall define the standardized spacings as $D_{r,n}$ = (n-r) ($X_{r+1,n}$ -$X_{r,n}$), $1 \le r <$n, with $D_{0,n}$ = $nX_{1,n}$ and $D_{n,n}$ = 0. It is known (Ahsanllah (1978)) that if $X \in C_1$, then the identical distribution of $D_{I,n}$ and $D_{k,n}$ characterizes the exponential distribution. Rossberg (1972) proved that if F is continuous, and if $X_{s,n}$ and $\sum_{i=k}^{m} c_i X_{i,n}$ with $c_k \ne 0, c_m \ne 0$ and $\sum_{i=k}^{m} c_i = 0$, s >m, are independent, then F is exponential. .Ferguson (1967) gave a complete solution to the problem of determining all dfs for which

$$E(X_{k+m,n} \mid X_{k,n}) = a X_{k,n} + b, \ m=1,\ 1<m \le n\text{-}k. \qquad (1.5.9)$$

Nagaraja (1988) showed that if $E(X_{m+1,n}| X_{m,n})$ and $E(X_{m,n}| X_{m+1,n})$ are both linear then the parent distribution is exponential. Wesolowski and Ahsanullah (1997) proved that for k=2, the condition (1.5.9) characterizes the exponential distribution for $a = 1$ and characterizes the Power Function and Pareto distribution for $a > 1$ and for $a<1$ respectively.

There are several characterizations of the uniform distribution based on the spacings of the order statistics. For the uniform distribution, U(0,1), the spacings $U_{j,n} = X_{j+1,n} - X_{j,n}$ with j =1,..., n, and $U_{0,n} = X_{1,n}$ are identically distributed as U(0,1). Saleh (1976) gave a characterization of the uniform distribution using the expected values of $U_{j,n}$. We will say that F is "sub-additive" if $F(x+y) \geq F(x) + F(y)$, $x,y \geq 0$ and F is super additive if $F(x+y) \leq F(x) + F(y)$. We will say that F belongs to class C_2 if F is either sub additive or super additive. Huang, Arnold and Ghosh (1979) showed that if F belongs to C_2, then the identical distribution of $U_{j,n}$ and $U_{0,n}$ characterizes the uniform (U(0,1)) distribution. The following (Ahsanullah (1989)) is a characterization result of the uniform distribution,(U(0,1)).

Theorem 1.5.2.

Let X be bounded symmetric random variable having an absolutely continuous (with respect to Lebesgue measure) distribution function F with $\inf\{x| F(x) >0\} = 0$. We assume without a loss of generality that F(1) = 1. Then the following two properties are equivalent:

(i) X is distributed as U(0,1).

(ii) $X_{n,n} - X_{1,n}$ and $X_{n-1,n}$ are identically distributed and the cdf of X, F belongs to C_2 for some n >1.

Proof.

First we proof that (i) implies (ii).
The pdf f_1 of $X_{n,n} - X_{1,n}$ is given by

$$f_1(v) = \int_0^{1-v} n(n-1)(F(u+v)-F(u))^{n-2} f(u) f(u+v)\, du, \quad 0<v<1. \qquad (1.5.10)$$

The above reduces to the pdf of $X_{n-1,n}$ if X is distributed as U(0,1).

Next we proof that (ii) implies (i). Writing $F_1(v) = \int_0^v f_1(u)\, du$, we get on simplification

$$F_1(v_0) = \int_0^{1-v_0} n(F(u+v_0)-F(u))^{n-1} f(u)\, du + (1-F(1-v_0))^n, \quad 0<v<1. \qquad (1.5.\ 11)$$

The distribution function of $X_{n-1,n}$ is

$$F_2(v_0) = n(F(v_0))^{n-1} - (n-1)(F(v_0))^n, \quad 0<v<1. \qquad (1.5.12)$$

Since X is symmetric and F(1) =1, we must have $F(x) = \overline{F}(1-x)$ and hence on equating

(1.5.11) and (1.5.12), we obtain for all v, $0 \le v \le 1$,

$$\int_0^{1-v} n(F(u+v)-F(u))^{n-1} f(u)\,du = n(F(1-v))^{n-1}(1-F(v)),\quad 0\le v\le 1$$

i.e.

$$\int_0^{1-v} q(u,v)\,f(u)\,du = 0,\quad \text{for all } v,\quad 0\le v\le 1, \tag{1.5.13}$$

where $q(u,v) = (F(u+v)) - F(u))^{n-1} - (F(v))^{n-1}$.

If F is sub additive, then $q(u,v) \ge 0$, and (1.5.13) to hold we must have $q(u,v) = 0$ for all v, $0 \le v \le 1$ and for all u, $0 \le u \le 1-v$. Now $q(u,v) = 0$ implies that
$F(u+v) = F(u) + F(v)$. (1.5.14)

The only continuous solution of (1.5.14) with $0 \le v \le 1$, $0 \le u \le 1-v$ and the boundary condition $F(0) = 0$ and $F(1) = 1$ is

$$F(x) = x,\ 0 \le x \le 1. \tag{1.5.15}$$

If F is sup additive, then we will get equation (1.5.14) and hence (1.5.15).

It is known that for the uniform distribution (U(0,1)) X and $X_{1,n} / X_{2,n}$ are identically distributed. However this not a characteristic property of the uniform distribution. The identical distribution of X and $X_{1,n} / X_{2,n}$ characterizes a family of distribution (Power function distribution) which includes uniform distribution. Ahsanullah and Bairamov (1999) has recently proved the following characterization of the uniform distribution.

Theorem 1.5.3.
Let X be a positive random variable having a non decreasing absolutely continuous probability distribution F. Then X has the probability density function

$$f(x) = 1/a,\ 0<x<a$$
$$= 0,\ \text{otherwise,}$$

if and only if

$\sum_{i=1}^{n} \frac{X_i}{X_{n,n}} \stackrel{d}{=} 1+\sum_{i=1}^{n-1} U_i$, for two consecutive values of n = m and m+1 (m >1), where U_j, j =1,2,…, m, are independently and identically distributed as U(0,1).

Representations for order statistics via independent random variables

Indeed, order statistics $X_{1,n} \le X_{2,n} \le \ldots \le X_{n,n}$ are dependent random variables, but there are some useful representations, which enable us to express distributions of order statistics via distributions of sums of independent random variables. Such kind of results are collected below. In fact, the first of these results (representation 1.5.1) was given in example 1.1.2.

Representation 1.5.1.

Let $Z_{1,n} \le \ldots \le Z_{n,n}$, n=1,2,...,-

be exponential order statistics , based on a sequence of independent random variables $Z_1, Z_2, \ldots$, having the common standard exponential distribution function H(x)=max(0,1-exp(-x)). Then for any n=1,2,... the following relation holds:

$$(Z_{1,n}, Z_{2,n}, \ldots, Z_{n,n}) \stackrel{d}{=} (\nu_1/n, \nu_1/n+\nu_2/(n-1), \ldots, \nu_1/n+\nu_2/(n-1)+\ldots+\nu_n), \qquad (1.5.\ 16)$$

where $\nu_1, \nu_2, \ldots$ are independent random variables, having the same exponential d.f. H(x).

The next representation has been obtained for the uniform order statistics.

Representation 1.5.2.

Let $U_{1,n} \le \ldots \le U_{n,n}$, n=1,2,..., be order statistics, corresponding to the uniform U(0,1) distribution. Then for any n=1,2,...

$$(U_{1,n}, \ldots, U_{n,n}) \stackrel{d}{=} (S_1/S_{n+1}, \ldots, S_n/S_{n+1}), \qquad (1.5.17)$$

where

$$S_m = \nu_1 + \nu_2 + \ldots + \nu_m,\ m=1,2,\ldots,$$

and $\nu_1, \nu_2, \ldots$ are independent random variables introduced in the previous representation.

The results given above show that exponential and uniform order statistics are expressed in terms of sums of independent random variables having the standard exponential distribution. Evidently, for any random variable X with a continuous d.f. F, the probability integral transformation U=F(X) produces a uniformly U([0,1]) distributed random variable U . Since this transformation does not change the order of X's, the values $U_{1,n}, \ldots, U_{n,n}$ of order statistics in a sequence $U_i = F(X_i)$ coincide with values $F(X_{1,n}), \ldots, F(X_{n,n})$. This property gives us the following representations involving a more wide class of order statistics.

Representation 1.5.3.

Let $X_{1,n} \le \ldots \le$ be the order statistics corresponding to a continuous distribution function F, and let $U_{1,n} \le \ldots \le U_{n,n}$ be the uniform order statistics defined in representation 1.5.2. Then

$$(F(X_{1,n}),\ldots,F(X_{n,n})) \stackrel{d}{=} (U_{1,n},\ldots,U_{n,n})\,,\ n=1,2,\ldots. \qquad (1.5.18)$$

Corollary 1.5.1.

Let $G(s)=\inf\{x: F(x)\geq s\}$ be the inverse of the function F. Then it follows from representation 1.5.3 that

$$(X_{1,n},\ldots,X_{n,n}) \stackrel{d}{=} (G(U_{1,n}),\ldots,G(U_{n,n})). \qquad (1.5.19)$$

It is interesting to note that (1.5.19) holds for any distribution function F, while (1.5.18) is valid for only continuous distribution functions.

The following statement enables us to compare distributions of order statistics from different populations.

Corollary 1.5.2.

Let order statistics $X_{1,n}\leq \ldots \leq X_{n,n}$ and $Y_{1,n} \leq\ldots\leq Y_{n,n}$ correspond to continuous distribution functions F and H respectively. Then

$$(X_{1,n},\ldots,X_{n,n}) \stackrel{d}{=} (G(H(Y_{1,n})),\ldots,G(H(Y_{n,n}))), \qquad (1.5.20)$$

where G is the inverse function of F .

Combining (1.5.16) and (1.5.20) one obtains that a distribution of any order statistics $X_{k,n}$, corresponding to arbitrary distribution function F, can be expressed via independent random variables $\nu_1.\nu_2,\ldots$, having the standard exponential distribution, as follows:

$$_{k,n} \stackrel{d}{=} G(1-\exp(-(\nu_1/n+\nu_2/(n-1)+\ldots+\nu_k/(n-k+1)))), \qquad (1.5.21)$$

where $G(s) = \inf\{x: F(x)\geq s\}$.

There is one more useful representation for the uniform order statistics $U_{1,n}\leq\ldots\leq U_{n,n}$.

Representation 1.5.4.

Let $S_m= \nu_1 +\nu_2 +\ldots+\nu_m$, m=1,2,... denote sums of independent random variables $\nu_1,\nu_2,\ldots$, defined in representation 1.5.1. Then

$$(U_{1,n},\ldots,U_{n,n}) \stackrel{d}{=} (S_1,\ldots, S_n \mid S_{n+1}=1),\ n=1,2,\ldots, \qquad (1.5.22)$$

that is the distribution of the uniform order statistic $U_{k,n}$ coincides with the conditional distribution of sums S_k given that S_{n+1} takes the fixed value 1.

The next representation of the distribution of the uniform order statistics $U_{1,n} \le \ldots \le U_{n,n}$ via distributions of products of independent uniformly U([0,1]) distributed random variables $W_1, W_2, \ldots$ also is convenient in a number of situations.

Representation 1.5.5.
The following relation holds for any n=1,2,... :

$$\{U_{k,n}\}_{k=1}^{n} \stackrel{d}{=} \{W_k^{1/k} W_{k+1}^{1/(k+1)} \ldots W_n^{1/n}\}_{k=1}^{n}. \qquad (1.5.23)$$

It turns out sometimes that one needs to deal with ordered observations taken from different populations. Let us assume that the independent random variables $X_1, X_2, \ldots$ have distribution functions $F_1, F_2, \ldots$ respectively and let $X_{1,n} \le \ldots \le X_{n,n}$, n = 1,2,..., be the corresponding order statistics. Firstly we consider the case of independent random variables having exponential distributions with arbitrary scale parameters. Nevzorov (1984) has obtained the following result.

Representation 1.5.6.

Let $X_{k,n}$, $1 \le k \le n$, be order statistics based on a sequence of independent random variables $X_1, \ldots, X_n$ with distribution functions

$_k(x) = \max\{0, 1-\exp(-\lambda_k x)\}$, $\lambda_k > 0$, k=1,2,...,n,

respectively. Then

$$\{Z_{k,n}\}_{k=1}^{n} \stackrel{d}{=} \sum_{k=1}^{n} {}_{\text{mixt.}} p_l \{\nu_1/(\lambda_{\tau(1)}+\ldots+\lambda_{\tau(n)})+\nu_2/(\lambda_{\tau(2)}+\ldots+\lambda_{\tau(n)})+ +\nu_k/(\lambda_{\tau(k)}+\ldots+\lambda_{\tau(n)})\}_{k=1}^{n}, \qquad (1.5.24)$$

where $\nu_1, \nu_2, \ldots$ are given in representation 1.5.1,

Let $\sum_{l=1}^{n!} {}_{\text{mixt.}} p_l \{W_l\}_{k=1}^{n}$ denotes a vector , distribution of which is a mixture with weights

$p_l = \lambda_1 \ldots \lambda_n / (\lambda_{\tau(1)}+\ldots+\lambda_{\tau(n)})(\lambda_{\tau(2)}+\ldots+\lambda_{\tau(n)}) \ldots \lambda_{\tau(n)}$

of distributions of n-dimensional vectors $W_l = (W_{l,1}, \ldots, W_{l,n})$ with components

$W_{l,k} = \nu_1/(\lambda_{\tau(1)}+\ldots+\lambda_{\tau(n)})+ +\nu_k/(\lambda_{\tau(k)}+\ldots+\lambda_{\tau(n)})$, $1 \le k \le n$, and this mixture includes n! vectors, corresponding to all n! permutations (r(1),...,r(n)) of integers 1,2,...,n.

A more simple form of (1.5.24) was suggested by Tikhov (1991). He has used antiranks D(1),...,D(n), defined by equalities

$$\{D(r)=m\}=\{X_{r,n}=X_m\},\ 1\le r\le n,\ 1\le m\le n.$$

The usage of antiranks helped Tikhov to rewrite (1.5.24) as follows:

$$\{Z_{k,n}\}_{k=1}^{n} \overset{d}{=}$$

$$\{\nu_1/(\lambda_{D(1)}+...+\lambda_{D(n)})+\nu_2/(\lambda_{D(2)}+...+\lambda_{D(n)})+...+\nu_k/(\lambda_{D(k)}+...+\lambda_{D(n)})\}_{k=1}^{N}, \qquad (1.5.25)$$

the vectors $(\nu_1,..., \nu_n)$ and (D(1),...,D(n)) being independent.

The next representation is true for arbitrary distribution functions $F_1,F_2,\ldots$. Introduce the indicators $I_k(x)$ such that $I_k(x)=1$, if $X_k<x$, and $I_k(x)=0$, if $X_k\ge x$, k=1,...,n. Note that for any x, the random indicators $I_k(x)$, k=1,2,..., are independent.

Representation 1.5.7.

For any x and k = 1,2,...,n the following relation holds:

$$P\{X_{k,n}<x\}=P\{I_1(x)+ +I_n(x) \ge k\}. \qquad (1.5.26)$$

The following examples are due to Nevzorov (2000)

Example 1.5.1.

Exponential Distribution. Let $Z_{1,n} \le \ldots \le Z_{n,n}$ be order statistics corresponding to the i.i.d. random variables $Z_1, Z_2,..., Z_n$ with pdf $f(x) = e^{-x}$, x>0.Then, we know from Representation 1.5.1.that:

$$(Z_{1,n},Z_{2,n},..,Z_{n,n}) \overset{d}{=} (\nu_1/n,\nu_1/n+\nu_2/(n-1),...,\nu_1/n+\nu_2/(n-1)+...+\nu_n).$$

where $\nu_1, \nu_2,...$ are independent random variables, having the exponential d.f.

$$H(x) = \max\{ 0, 1-\exp(-x)\}.$$

Since $\nu_1,\nu_2,...$ are i.i.d. exponential with $E(\nu_k)=1$ and $Var((\nu_k),=1$, k=1,2,, we have

$$E(Z_{k,n})=E(\frac{\nu_1}{n}+\frac{\nu_2}{n-1}+...+\frac{\nu_k}{n-k})$$

$$=\frac{1}{n}+\frac{1}{n-1}+...+\frac{1}{n-k}$$

$$\text{Var}(Z_{k,n}) = Var(\frac{\nu_1}{n}+\frac{\nu_2}{n-1}+...+\frac{\nu_k}{n-k})$$

$$=\frac{1}{n^2}+\frac{1}{(n-1)^2}+...+\frac{1}{(n-k)^2} \text{ and}$$

$$\text{Cov}(Z_{r,n} Z_{s,n}) = E(Z_{r,n} - E(Z_{r,n})(Z_{s,n}-E(Z_{s,n}))$$

$$=\frac{1}{n^2}+\frac{1}{(n-1)^2}+...+\frac{1}{(n-k)^2}, r \le s.$$

Example 1.5.2.
The Standard Power Function Distribution. Let $X_{1,n} \le ... \le X_{n,n}$ be the order statistics corresponding to the i.i.d. random variables X_1, $X_{2,...,}$ x_n with pdf f(x) = $\alpha x^{\alpha-1}$, 0<x<1.Then, we know from Representation 1.5.1, we have

$$X_{k,n} \stackrel{d}{=} (U_{k,n})^{1/\alpha} \text{ and } E(X^m_{k.n})=E(U_{k,n})^{m/\alpha} = \frac{\Gamma(n+1)\Gamma(k+\frac{m}{\alpha})}{\Gamma(n+\frac{m}{\alpha}+1)\Gamma)\Gamma)}, k= 1,2,..., n.$$

Thus $E(X_{k.n}) = \frac{\Gamma(n+1)\Gamma(k+\frac{1}{\alpha})}{\Gamma(n+\frac{1}{\alpha}+1)\Gamma(k)}$ and $E(X^2_{k.n}) = \frac{\Gamma(n+1)\Gamma(k+\frac{2}{\alpha})}{\Gamma(n+\frac{2}{\alpha}+1)\Gamma(k)}$.

Hence

$$\text{Var}(X_{k.n}) = \frac{\Gamma(n+1)\Gamma(k+\frac{2}{\alpha})}{\Gamma(n+\frac{2}{\alpha}+1)\Gamma(k)} - (\frac{\Gamma(n+1)\Gamma(k+\frac{1}{\alpha})}{\Gamma(n+\frac{1}{\alpha}+1)\Gamma(k)})^2$$

In a similar way , using the equality in distribution

$(X_{r,n})^\gamma (X_{s,n})^\delta \stackrel{d}{=} (U_{r,n})^{\gamma/\alpha} (U_{s,n})^{\delta/\varepsilon}$, $1 \le r < s \le n$, we obtain

$$E\{X_{r,n}X_{s,n}\} = E\{(U_{r,n})^{1/\alpha} (U_{s,n})^{1/\alpha}\} = \frac{\Gamma(n+1)\Gamma(r+\frac{1}{\alpha})\Gamma(s+\frac{2}{\alpha})}{\Gamma(r)\Gamma(s+\frac{1}{\alpha})\Gamma(n+1+\frac{2}{\alpha})}$$

Thus

$$\text{Cov}\,(X_{r,n}X_{s,n}) = \frac{\Gamma(n+1)\Gamma(r+\frac{1}{\alpha})\Gamma(s+\frac{2}{\alpha})}{\Gamma(r)\Gamma(s+\frac{1}{\alpha})\Gamma(n+1+\frac{2}{\alpha})} - \frac{\Gamma(n+1)\Gamma(r+\frac{1}{\alpha})}{\Gamma(n+\frac{1}{\alpha}+1)\Gamma(r)}\,\frac{\Gamma(n+1)\Gamma(s+\frac{1}{\alpha})}{\Gamma(n+\frac{1}{\alpha}+1)\Gamma(s)},\ r \leq s.$$

REFERENCES

1. Abramowitz, M. and Stegun, I (1972) . Handbook of Mathematical Functions.Dover, New York, NY.

2. Aczel, J. (1966) . Lectures on Functional Equations and Their Applications. Academic Press, New York, NY.

3. Ahsanullah, M. and Kabir, A.B. M.L. (1974). A characterization of the power function distribution. Canad. J. Statist. 2, 95-98.

4. Ahsanullah, M. (1975). A characterization of the exponential distribution.In G.P. Patil, S. Kotz and J. Ord, Eds., Statistical Distributions in Scientific Work. Vol.3, 71-88..Dordrecht-Holland: D. Reidal Publishing Company.

5. Ahsanullah, M.(1976). On a characterization of the exponential distribution by order statistics. J.Appl. Prob. 13,818-882.

6. Ahsanullah, M. (1977). A characteristic property of the exponential distribution. Ann. of Statist.5, 580-582.

7. Ahsanullah, M. (1978 a). A characterization of the exponential distribution by spacing. J. Appl Prob. 15, 650-653.

8. Ahsanullah, M. (1978 b). On a characterizations of the exponential distributions by spacings.Ann. Inst. Statist. Math. 30,A, 163-166.

9. Ahsanullah, M. (1981). On characterizations of the exponential distribution by spacings. Statische Hefte, 22, 316-320.

10. Ahsanullah, M. (1987). Two characterizations of the exponential distribution. Comm. Statist. Theory Meth. 16(2), 375-381.

11. Ahsanullah, M. (1988). On a conjecture of Kakosisyan, Klebanov and Melamed. Statistiche Hefte, 29, 151-157.

12. Ahsanullah,M. and Bairamov, I.(1995). Distributional relations between order statistics and the sample itself and a characterization of exponential distribution. J. Appl. Stat. Sc.10(1) 1-16.

13. Ahsanullah,M. and Bairamov, I.(1999).A characterization of uniform distribution. Istatistik, 2(2) 145-151.

14. Ahsanullah M. and Nevzorov V.B. (1995). Independence of order statistics from extended sample. – *In : Appl. Statistical Science I, 23-32.* (eds: M.Ahsanullah, D.S.Bhoj).

15. Ahsanullah M. and Nevzorov V.B. (2000). Generalized spacings of order statistics from extended sample. J. Stat. Plan. Inf. 85,75-83.

16. Ahsanullah M. and Nevzorov V.B. (1997). Spacings of order statistics from extended samples. *In : Appl. Statistical Science IV, 251-257.* (eds: M.Ahsanullah, F.Yildirim).

17. Ahsanullah,M. and Rahim, M.A. (1973). Simplified estimates of the parameters of the Double exponential distribution. Based on optimum order statistics from a middle censored sample. Naval Research Logistic Quarterly. 20, 745-75.

18. Ahsanullah, M. and Rahman, M. (1972). A characterization of the exponential distribution. J.Appl. prob. 9, 457-461.

19. Arnold, B. C., Balakrishnan, N., and Nagaraja, H. N. (1992). A first course in Order Statistics. Wiley, New York.

20. Azlarov, T. A. and Volodin, N. A. (1986). Characterization problems associated with the exponential distribution. Springer-Verlag, New York.

21. Bairamov, I.G.and Gebizliogiu, O.L (1997). On the ordered random vectors in a norm sense. J.Appl. Stat. Sc. 6,1,77-86.

22. Balakrishnan, N. Order Statistics from discrete distributions. Comm.Statist.-Theor. Meth.15(3), 657-675.

23. Balakrishnan,N. and Cohen, A.C. (1991). Order Statistics and Inference: Estimation Methods.Academic Press.

24. Bennet,V.D. (1966). Order statistics estimators of the location of the Cauchy distribution. J Amer. Statist. Assoc., 61,1205-1218, correction, 63, 383-385.

25. Bose, R.C. and Gupta, S.S. (1959). Moments of order statistics from a normal distribution. Biometrika, 46, 433-440.

26. David, H.A. (1981). Order Statistics. Second Edition. John Wiley and Sons, Inc., New York, NY.

27. David H.A., Rogers M.P. (1983). Order statistics in overlapping samples, moving order Statistics.and U-statistics.- Biometrika *70*, 245-249.

28. Deheuvels, P. (1984). The characterization of distributions by order statistics and record values - an unified approach. J. Appl. Prob. 21, 326-334.

29. Desu, M.M. (1971). A characterization of the exponential distribution by order statistics. Ann. Math. Statist. 42,837-838.

30. Feller, W. (1966). An Introduction to Probability Theory and its Applications. Vol. II, Wiley, New York.

31. Gajek, L. and Gather, U. (1991). Moment inequalities for order statistics with applications to characterizations of distributions. Metrika, 38, 357-367.

32. Galambos,J. (1975).Characterizations of probability distributions by properties of order statistics 1. In G.P. Patil, S. Kotz and J. Ord, Eds., Statistical Distributions in Scientific Work. Vol.3, 71-88. Dordrecht-Holland: D. Reidal Publishing Company.

33. Galambos, J. and Kotz,S. (1978). Characterizations o Probability Distributions Lectur Notes in Mathematics 675.Berlin and New York. Springer-Verlag.

34. Galambos, J. (1984). Introductory Probability Theory. Marcel Dekker, Inc. New York, NY.

35. Gather, U. (1988). On a characterizations of the exponential distribution by properties of order statistics. Statist. Prob. Letters, 7, 93-96.

36. Goldburger, A. S. (1962). Best Linear Unbiased Predictors in the Generalized Linear Regression Model. J. Amer. Statist. Assoc. 57, 369-375.

37. Govindarajulu, Z. (1963). On moments of order statistics and quasi ranges from normal populations. Ann. Math. Statist. 34, 633-651.

38. Gradshteyn, I.S. and Ryzhik, I.M.(1980).Tables of Integrals, Series, and Products, Corrected and Enlarged Edition. Academic Press,Inc.New York, NY

39. Grosswald, E. and Kotz, S. (1981). An Integrated Lack of Memory Property of the Exponential Distribution. Ann. Inst. Statist. Math. 33, A, 205-214

40. Gumbel E.J. (1961). The return period of order statistics.- *Ann. Inst. Statist. Math. 12*, n.3, 249-256.

41. Gupta, R. C. (1984). Relationships between Order Statistic and Record Values and Some characterization Results. J. Appl. Prob. 21, 425-430.

42. Inagaki N. (1980). The distribution of moving order statistics. In: Recent Developments in Statistical Inference and Data Analysis. Ed. K.Matusita. Amsterdam: North Holland, 137-142.

43. Johnson, N.L., Kotz, S. and Balakrishnan (1994). Continuous Univariate Distributions. Volume 1.Second Edition. John Wiley and Sons. Inc. New York.

44. Johnson, N.L., Kotz, S. and Balakrishnan (1994). Continuous Univariate Distributions. Volume 2. Second Edition. John Wiley and Sons. Inc. New York.

45. Joshi,P.C. (1971).Recurrence relations for the mixed moments of order statistics. Ann. Math.Statist.42,1096-1098.

46. Kabir, A. B. M and Ahsanullah, M. (1974). Estimation of the location and scale parameters of a power function distribution by linear function of order statistics. Comm. Statist. 3, 463-467.

47. Kagan, A. M., Linnik Y. V. and Rao, C. R. (1973). Characterization Problems in Mathematical Statistics, John Wiley, New York.

48. Kakosyan, A. V., Klebanov, L. B. and Melamed, J. A. (1984). Characterization of Distribution by the Method of Intensively Monotone Operators. Lecture Notes Math. 1088, Springer Verlag, New York, N.Y.

49. Kaminsky, K.S. and Nelson, P.I. (1975). Best linear unbiased prediction of order statistics in location and scale families. J. Amer.Statist. Assoc. 70, 145-150.

50. Karlin, S. (1966). A first course in stochastic processes. Academic Press, New York.

51. Klebanov, L. B. and Melamed, J. A. (1983). A method associated with characterizations of the exponential distribution. Ann. Inst. Statist. Math. 35,A, 105-114.

52. Kubat, P. and Epstein, B. (1980). Estimation of quantiles of location-scale distributions based on two or three order statistics. *Technometrics* **22**, 575-581.

53. Lawless, J. F. (1982). Statistical Models and Methods for Lifetime Data. John Wiley and Sons, New York, NY.

54. Lehman, E. H. (1963). Shapes, moments and estimators or the Weibull distribution. *IEEE Trans. Reliab.* **12**, 32-38.

55. Lieblein, J. (1953). On the exact evaluation of the variances and covariances of order statistics in samples from the extreme-value distribution. *Ann. Math. Statist.* **24**, 282-287.

56. Lieblein, J. (1962). On moments of order statistics from Weibull distribution. Ann. Math. Statist. 26,330-333.

57. Lurie, D. and Hartley, H. O. (1972). Machine generation of order statistics for Monte-Carlo Computations. Amer. Statistician **26**, 26-27.

58. Lau, Ka-sing and Rao, C. R.(1982). Integrated Cauchy functional equation and characterization of the exponential. Sankhya A, 44, 72-90.

59. Lloyd, E.H.(1952). Least squares estimation of location and scale parameters using order statistics. Biometrika 39, 88-95.

60. Mahmoud, M. W. and Ragad, A. (1975). On order statistics in samples drawn from the extreme value distribution. *Math* Operationsforschung u. Stat. **6**, 800-816.

61. Mann, N. R. (1970). Estimators and exact confidence bounds for Weibull parameters based on a few ordered observations. Technometrics **12**, 345-361.

62. Mann, N. R., Schaefer, R. E. and Singpurwala, N. D. (1974). Methods for Statistical Analysis of Reliability and Life Data. John Wiley and Sons, New York, NY.

63. Marsglia, G. and Tubilla, A. (1975). A note on the lack of memory property of the exponential distribution. Ann. Prob. 3, 352-354.

64. McCool, J. I. (1969). Unbiased maximum-likelihood estimation of a Weibull percentile when the shape parameter is known. IEEE Trans. Reliab. **18**, 78-79.

65. McCool, J. I. (1970b). Evaluating Weibull endurance data by the method of maximum likelihood. Trans. Amer. Soc. Lubric. Eng. **13**, 189-202.

66. Menon, M. V. (1963). Estimation of the shape and scale parameters of the Weibull distribution. Technometric*s* **5**, 175-182.

67. Nagaraja, H. N. (1986). Comparison of estimators and predictors from two parameter exponential distribution. Sankhya B, 48, 10-18.

68. Nagaraja, H. N. (1986). Some characterizations of continuous distributions based on regressions off adjacent order statistics and record values. Sankhya A 50, 70-73.

69. Nevzorov, V.B. and Ahsanullah, M. (1996).. Order statistics from exponential distribution. BEPOHTHOCB. (Russian). J..Math Science, 24-30.

70. Nevzorov, V.B. and Ahsanullah, M. (2000). Extremes and Records for concomitants of Order statistics and record values. J. Appl. Stat. Sc. (to appear)

71. Nevzorov, V.B. and Ahsanullah, M. Some distributions of induced records. (2000). Biometrical Journal 42, 8, 1069-1081.

72. Nevzorov, V.B. and Ahsanullah, M. (2001).. Distributions between uniform and exponential. J.Appl.Stat. Sc. V, 9-20. (eds. M. Ahsanullah, J. Kennyon and S. K. Sarkar).

73. Pfeifer,D.(1982). Characterizations of exponential distributions by independent non stationary increments. J. Appl. Prob. 19,127-135.

74. Puri, P.S. and Rubin, H, (1970). A characterization based on absolute difference to two i.i.d. random variables. Ann. Math. Statist. 41, 2113-2122.

75. Raqab, M.Z. and Nagaraja, H.N. (1995). On future predictors of future records. Metron, LIII n.1-2,185-202.

76. Raqab, M.Z. (1997). Modified maximum likelihood predictors of future order statistics from normal samples. Computational Statistics & Data Analysis.25, 91-106.

77. Rao, C.R. (1983). An extension of Denny's theorem and its application to characteizations of probability distributions. A Festchrift for Erich L. Lehman. Ed. Bickel et. al. Wordsworth International Group, Belmont, CA.

78. Rossberg, H.J. (1972). Characterization of distributions of certain functions of order statistics. Sankhya A, 34, 111-120.

79. Saleh, A.K. Md. E.(1976). Characterization of distributions using expected spacings between consecutive order statistics. Journal of Statistical Research, 10,1-13.

80. Saleh, A.K.Md. E., Scott, C, and Jenkins, D.B. (1975). Exact first and second order statistics from the truncated exponential distribution. Naval Res. Logist. Quart. 22, 65-77.

81. Sarhan, A.E. and Greenberg, B.G. (Eds) (1962). Contributions to order statistics. Wiley, New York, NY.

82. Sethuramann, J. (1965). On a characterization of the three limiting types of extremes. Sankhya, A. 27, 357-364.

83. Schucany, W. R. (1972). Order statistics in simulation. J. Statist. Comput. Simul. **1**, 281-286.

84. Sen, P. K. (1970). A note on order statistics for heterogeneous distributions. Ann. Math. Statist. 41, 2137-2139.

85. Siddiqui M.M. (1970). Order statistics of a sample and of an extended sample. In: Nonparametric Techniques in Statistical Inference, Ed. M.L. Puri. Cambridge University Press, 417-423.

86. Srikantan, K.S. (1962). Recurrence relations between PDF's of order statistics, and some applications. Ann. Math. Statist. 33, 169-177.

87. Weibull, W. (1951). A statistical distribution function of wide applicability. J. Appl. Mech. 18, 293-297.

88. Weissman, I. (1978). Estimation of parameters and large quantiles based on the kth largest observations. J. Amer. Statist. Assoc. **63**, 812-815.

89. Wesolowski, J. and Ahsanullah, M. (1997) .On characterizing distributions via linearity of regression for order statistics. Austral. J. Statist. 39(1), 69-78.

90. White, J.S. (1969). The moments of log - Weibull order statistics. Technometrics, 11, 373-386.

91. Young, D.H. (1970). The order statistics of the negative binomial distribution. Biometrika, 57, 181-186.

Chapter 2

EXTREMES

2.0 INTRODUCTION

Extreme value distributions arise in probability theory as limit distributions of maximum or minimum of n independent and identically distributed (i.i.d.) random variables for some normalizing constants. Suppose $X_1, X_2, \ldots, X_n$ are n independent and identically distributed random variables. If the largest order statistic $X_{n,n}$ with some normalizing constants has a non degenerate limiting distribution, then its distribution will converge to one of the following three types of limiting extreme value distributions as n $\to \infty$.

(1) Type 1: (Gumbel) $F(x) = \exp(-e^{-x})$, for all x,

(2) Type 2: (Frechet) $F(x) = \exp(-x^{-\delta})$, $x > 0$, $\delta > 0$

and

(3) Type 3: (Weibull) $F(x) = \exp(-(-x)\delta)$, $x < 0$, $\delta > 0$.

Since the smallest order statistic $X_{1,n} = Y_{n,n}$, where $Y = -X$, $X_{1,n}$ with some appropriate normalizing constants will also converse to one of the above three limiting distributions if we change X to -X in (1), (2) and (3).. Gumbel (1958) has given various applications of these distributions

Suppose $X_1, X_2, \ldots X_n$ are n i.i.d exponential random variables having the cumulative distribution function F(x) where $F(x) = 1-e^{-x}$. Then with a sequence of real numbers

$$a_n = \ln n \text{ and } b_n = 1, \text{we have } P(X_{n,n} \le a_n + b_n x) = P(X_{n,n} \le \ln n + x) = (1 - e^{-(\ln n + x)})^n$$

$$= (1 - \frac{e^{-x}}{n})^n \to e^{-e^{-x}} \text{ as } n \to \infty.$$

Thus the limiting distribution of $X_{n,n}$ with the constants a_n = ln n and b_n =1 when X's are distributed as exponential with unit mean is Type 1 extreme value distribution, The numbers a_n and b_n are known as normalizing constants. It can be shown that Type 1 (Gumbel distribution) extreme value distribution is the limiting distribution of $X_{n,n}$ when F(x) is normal, log normal, logistic, gamma etc. The type 2 and type 3 distributions can be transformed to Type 1 distribution by the transformations ln X and - ln X respectively. We will denote the Type 1 distribution as T_{10} and Type 2 and Type 3 distribution as $T_{2\delta}$ and $T_{3\delta}$ respectively. If the $X_{n,n}$ of n independent random variables from a distribution F with some normalizing constants has the limiting distribution T, then we will say that F belongs to the domain of attraction of T and write F$\in$ D(T). Similarly if the $X_{1,n}$ of n independent random variables from a distribution F with some normalizing constants has the limiting distribution H, then we will say that F belongs to the domain of attraction of H and write F$\in$ $\underline{D}(H)$.

The extreme value distributions were originally introduced by Fisher and Tippet (1928). These distributions have been used in the analysis of data concerning floods, extreme sea levels and air pollution problems. For details see Gumbel (1958), Horwitz (1980), Jenkinson (1955) and Roberts (1979 a,b).

Probability density function (pdf) of type 1 extreme value distribution.

The probability density function of type 1 extreme value distribution is given in figure 2.0.1.

The type 1 extreme value distribution is unimodal with mode at 0 and the points of inflection are at ± $\ln\left((3+\sqrt{5})/2\right)$. The pth percentile η_p, (0<p<1) of the curve can be calculated by the relation $\eta_p = -\ln(-\ln p)$.The median of X is -lnln2. The moment generating function $M_{10}(t)$, of the distribution for some t , 0<|t|<δ, is $M_{10}(t)$ $= \int_{-\infty}^{\infty} e^{tx} e^{-x} e^{-e^{-x}} dx = e^{t}\Gamma(1-t)$. The mean = γ^*, the Euler's constant and the variance $= \pi^2/6$.

$$f(x) := e^{-x} \cdot e^{-\left(e^{-x}\right)}$$

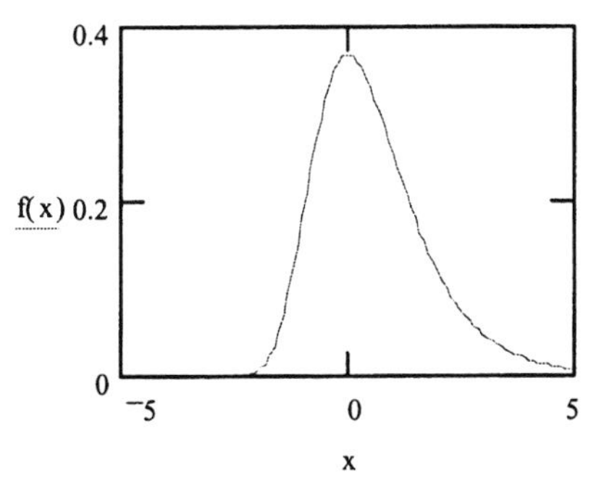

Figure 2.0.1

Probability density functions of Type 2

The probability density function of T_{21} is given in figure 2.0.2

$$f(x) := \left(\frac{1}{x^2}\right) \cdot e^{-\left(x^{-1}\right)}$$

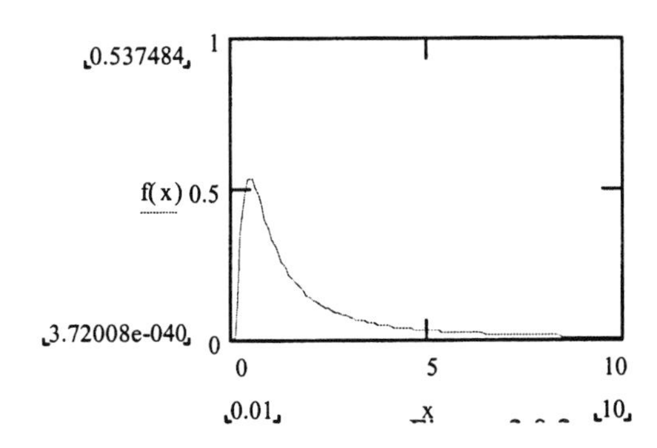

The mode of T_{21} is at x = 1/2. For $T_{2\delta}$ the mode is at $1/(\delta + 1)$, for $\delta > 1$, $E(X) = \Gamma(1 - \frac{1}{\delta})$ and for $\delta > 2$, $Var(X) = \Gamma(1 - \frac{2}{\delta}) - (\Gamma(1 - \frac{1}{\delta}))^2$.

The probability density function for type 3

The probability density function for type distribution for $\delta = 2$ is given in figure 2.0.3. Note for $\delta = 1$ T_{31} is the reverse exponential distribution.

$$f(x) := -2 \cdot x \cdot e^{-x^2}$$

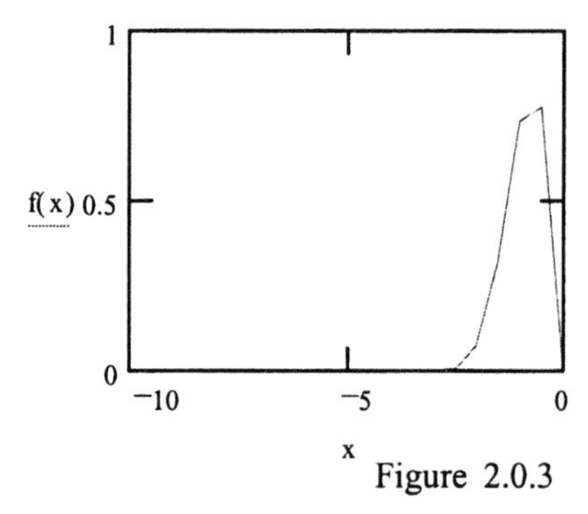

Figure 2.0.3

The mode of the Type 3 distribution is at $\left(\frac{\delta - 1}{\delta}\right)^{\frac{1}{\delta}}$. For type 3 distribution, E(X) = $\Gamma\left(1 + \frac{1}{\delta}\right)$ and Var(X) = $\Gamma(1 + \frac{2}{\delta}) - \left(\Gamma\left(1 + \frac{1}{\delta}\right)\right)^2$.

Table 2.0.1 gives the percentile points of T_{10}, T_{21} , T_{31} and T_{32} for some selected values of p.

Table 2.0.1. Percentile points of T_{10}, T_{21}, T_{31} and T_{32}

P	T_{10}	T_{21}	T_{31}	T_{32}
0.1	-0.83403	0.43429	-2.30259	-1.51743
0.2	-0.47589	0.62133	-1,60844	-1.26864
0.3	-0.18563	0.83058	-1.20397	-1.09726
0.4	0.08742	1.09136	- 0.91629	-0.95723
0.5	0.36651	1.44270	-0.69315	-0.83255
0.6	0.67173	1.95762	-0.51083	-0.71472
0.7	1.03093	2.80367	-0.35667	-9.59722
0.8	1.49994	4.48142	-0.22314	-0.47239
0.9	2.2504	9.49122	-0.10536	-0.324598

2.1 DOMAIN OF ATTRACTION

In this section we will study the domain of attraction of various distributions. The maximum order statistics $X_{n,n}$ of n independent and identically distributed random variable will considered. We will say that X_{nn} will belong to the domain of attraction of T(x) if the $\lim_{n\to\infty} P(X_{nn} \le a_n + b_n x) =$ T(x) for some normalizing constants a_n and b_n.

For example consider the uniform distribution with pdf f(x) = 1, 0<x<1. Then for t<0, $P(X_{nn} \le 1+t/n) = (1+t/n)^n \to e^t$. Thus X_{nn} from the uniform distribution belong to the domain of attraction of Type 3 distribution with $F(x) = e^x$, $-\infty<x<0$.
The following lemma will be helpful in proving the theorems of the domain of attraction.

Lemma 2.1.1

Let $\{X_n, n \ge 1\}$ be a sequence of independent and identically distributed random variables with distribution function F. Consider a sequence $(e_n, n \ge 1\}$ of real numbers. Then for any ξ, $0 \le \xi < \infty$, the following two statements are equivalent

(i) $\lim_{n\to\infty} n(\bar{F}(e_n)) = \xi$, $\bar{F} = 1 - F$

(ii) $\lim_{n\to\infty} P\left(X_{n,n} \le e_n\right) = e^{-\xi}$.

Proof.

Suppose (i) is true, then

$$\lim_{n\to\infty} P(X_{n,n} \le e_n) = \lim_{n\to\infty} F^n(e_n) = \lim_{n\to\infty} (1 - \bar{F}(e_n)))^n = \lim_{n\to\infty} (1-\xi/n + o(1))^n = e^{-\xi}.$$

Suppose (ii) is true, then

$$e^{-\xi} = \lim_{n\to\infty} P(X_{n,n} \le e_n) = \lim_{n\to\infty} F^n(e_n) = \lim_{n\to\infty} (1 - \bar{F}(e_n)))^n$$

Taking the logarithm of the above expression, we get

$$\lim_{n\to\infty} n \ln(1 - \bar{F}(e_n)) = -\xi \,.\; n\bar{F}(e_n))(1 + o(1)) \to \xi$$

Note The above theorem is true if $\xi = \infty$.

Domain of attraction of Type I distribution for X_{nn}.

The following theorem is due to Gnedenko (1943).

Theorem 2.1.1

Let $X_1, X_2, ..$ be a sequence of i.i.d random variables with distribution function F and $\alpha(F) = \inf\{x: F(x) > 0\}$ and $\beta(F) = \sup\{x: F(x) < 1\}$. Assume further that for a finite $\gamma < \beta(F)$, $\int_\gamma^{\beta(F)} \bar{F}(x)\,dx$ is finite for $\alpha(F)<\gamma<\beta(F)$. Then $F\in T_{10}$ iff $\lim_{t\to\alpha(F)} \frac{\bar{F}(t+xg(t))}{\bar{F}(t)} = e^{-x}$, $\bar{F}=1-F$, $g(t) = \frac{1}{\bar{F}(t)}\int_\gamma^{\beta(F)} \bar{F}(x)dx$, $\gamma< t < \beta(F)$ and for all real x.

Proof.

We choose the normalizing constants a_n and b_n of $X_{n,n}$ such that

$$a_n = \inf\{x : F(x) \le 1 - \frac{1}{n}\}$$

$b_n = g(a_n)$ and $a_n \to \beta(F)$ as $n \to \infty$. Suppose

$$\lim_{t\to\beta(F)} \frac{\bar{F}(t+xg(t))}{\bar{F}(t)} = e^{-x}, \text{ then}$$

By Lemma 2.1.1 we have $P(X_{n,n} \le a_n + b_n x) = e^{-e^{-x}}$.

Suppose $P(X_{n,n} \le a_n + b_n x) = e^{-e^{-x}}$, then we have by Lemma 2.1.1,

$$\lim_{n\to\infty} n\bar{F}(a_n + b_n x)) = e^{-x}.$$

$$e^{-x} = \lim_{n\to\infty} n\bar{F}(a_n+b_n x) = \lim_{n\to\infty} n\bar{F}(a_n)(\frac{\bar{F}(a_n+b_n x)}{\bar{F}(a_n)}) = \lim_{n\to\infty}(\frac{\bar{F}(a_n+b_n x)}{\bar{F}(a_n)}) = \lim_{t\to\beta(F)} \frac{\bar{F}(t+xg(t))}{\bar{F}(t)}.$$

The following Lemma (see Von Mises (1936)) gives a sufficient condition for the domain of attraction of Type 1 extreme value distribution for X_{nn}..

Lemma 2.1.2.

Suppose the distribution function F has a derivative on $[c_0, \beta(F)]$ for some c_0, $0<c_0<\beta(F)$, then if $\lim_{x\uparrow\beta(F)} \frac{f(x)}{\overline{F}(x)} = c, c>0$, then $F\in D(T_{10})$.

Example 2.1.1.

The exponential distribution $F(x) = 1-e^{-x}$ satisfies the sufficient condition, since $\lim_{x\to\infty} \frac{f(x)}{\overline{F}(x)} = 1$. For the logistic distribution $F(x) = \frac{1}{1+e^{-x}}$, $\lim_{x\to\infty} \frac{f(x)}{\overline{F}(x)}$ $= \lim_{x\to\infty} \frac{1}{1+e^{-x}} = 1$. Thus the logistic distribution satisfies the sufficient condition.

Example 2.1.2.

For the standard normal distribution with x > 0, (Abramowitz and Stegun (1968 p.932). $\overline{F}(x) = \frac{e^{-\frac{x^2}{2}}}{x\sqrt{2\pi}} h(x)$, where $h(x) = 1 - \frac{1}{x^2} + \frac{1.3}{x^4} + \cdots + \frac{(-1)^n 1.3.....(2n-1)}{x^{2n}} + R_n$ and

$R_n = (-1)^{n+1} 1.3....(2n+1) \int_x^\infty \frac{e^{-\frac{1}{2}u^2}}{\sqrt{2\pi}\, u^{2n+2}} du$ which is less in absolute value than the first neglected term.

It can be shown that $g(t) = 1/t + 0(t^3)$. Thus

$$\lim_{t\to\infty} \frac{\overline{F}(t+xg(t))}{\overline{F}(t)} = \lim_{t\to\infty} \frac{te^{\frac{t^2}{2}}}{\left(t+xg(t)\right)e^{\frac{1}{2}(t+xg(t))^2}} \frac{h(t+xg(t))}{h(t)} = \lim_{x\to\infty} \frac{e^{-xm(t,x)}}{t+xg(t)},$$

where $m(t,x) = g(t)\,(t+\frac{1}{2}xg(t)$. Since as $t\to\infty$, $m(t,x)\to 1$, we $\lim_{t\to\infty} \frac{\overline{F}(t+xg(t))}{\overline{F}(t)} = e^{-x}$.

Thus normal distribution belongs to the domain of attraction of Type I distribution.

Since $\lim_{x\to\infty} \frac{e^{-\frac{x^2}{2}}}{\sqrt{2\pi}\, x\overline{F}(x)} = \lim_{x\to\infty} h(x) = 1$, the standard normal distribution does not satisfy the Von Mises sufficient condition for the domain of attraction of the Type I distribution.

We can take $a_n = \frac{1}{b_n} - \frac{b_n}{2}(\ln\ln n + 4\pi)$ and $b_n = (2\ln\ln n)^{-1/2}$. However this choice of a_n and b_n is not unique. The rate of convergence of $P(X_{nn} \le a_n + b_n x)$ to $T_{10}(x)$ depends on the choices of a_n and b_n.

Domain of Attraction of Type 2 distribution for X_{nn}.

Theorem 2.1.2.

Let $X_1, X_2, ..$ be a sequence of i.i.d random variables with distribution function F and $\beta(F) = \sup\{x: F(x) < 1\}$. If $\beta(F) = \infty$, then $F \in T_{2\delta}$ if $\lim_{t\to\infty} \frac{\bar{F}(tx)}{\bar{F}(t)} = x^{-\delta}$ for x> 0 and some constant $\delta > 0$.

Proof.

Let $a_n = \inf\{x: \bar{F}(x) \le \frac{1}{n}\}$, then $a_n \to \infty$ as $n \to \infty$. Thus $\lim_{n\to\infty} n(\bar{F}(a_n x)) = \lim_{n\to\infty} n(\bar{F}(a_n)) \frac{\bar{F}(a_n x)}{\bar{F}(a_n)} =$ $x^{-\delta} \lim_{n\to\infty} n\bar{F}(a_n)$.

It is easy to show that $\lim_{n\to\infty} n\bar{F}(a_n) = 1$. Thus $\lim_{n\to\infty} n(\bar{F}(a_n x)) = x^{-\delta}$ and the proof of the Theorem follows from Lemma 2.1.1.

Example 2.1.3.

For the Pareto distribution with $\bar{F}(x) = \frac{1}{x^\delta}$, $\delta > 0$, $0 < x < \infty$, $\lim_{t\to\infty} \frac{\bar{F}(tx)}{\bar{F}(t)} = \frac{1}{x^\delta}$.

Thus the Pareto distribution belongs to domain of attraction of $T_{2\delta}$.

The following Theorem gives a necessary and sufficient condition for the domain of attraction of Type 2 distribution for X_{nn} when $\beta(F) < \infty$.

Theorem 2.1.3.

Let $X_1, X_2, ..$ be a sequence of i.i.d random variables with distribution function F and $\beta(F) = \sup\{x: F(x) < 1\}$. If $\beta(F) < \infty$, then $F \in T_{2\delta}$ if $\lim_{t\to\infty} \frac{\bar{F}(\beta(F) - \frac{1}{tx})}{\bar{F}(\beta(F) - \frac{1}{t})} = x^{-\delta}$ for $x > 0$ and some constant $\delta > 0$.

Proof.

Similar to Theorem 2.1.2.

Example 2.1.4.

The truncated Pareto distribution $f(x) = \frac{\delta}{x^{\delta+1}} \cdot \frac{1}{1-b^{\delta}}$, $1 \le x<b$, $b>1$,

$\lim_{t\to\infty} \frac{\overline{F}(\beta(F)-\frac{1}{tx})}{\overline{F}(\beta(F)-\frac{1}{t})} = \lim_{t\to\infty} \frac{\overline{F}(b-\frac{1}{tx})}{\overline{F}(b-\frac{1}{t})} = \lim_{t\to\infty} \frac{\left(b-\frac{1}{tx}\right)^{-\delta} - b^{-\delta}}{\left(b-\frac{1}{t}\right)^{-\delta} - b^{-\delta}} = x^{-1}$. Thus the truncated Pareto distribution belongs to the domain of attraction of T_{21} distribution.

The following Lemma (Von Mises (1936)) gives a sufficient condition for the domain of attraction of Type 2 extreme value distribution for X_{nn}.

Lemma 2.1.3.

Suppose the distribution function F is absolutely continuous in $[c_0, \alpha(F)]$ for some c_0, $0<c_0<\beta(F)$, then if $\lim_{x\uparrow\beta(F)} \frac{xf(x)}{\overline{F}(x)} = \delta, \delta > 0$, then $F \in D(T_{2\delta})$.

Proof.

Let $q(x) = \frac{xf(x)}{\overline{F}(x)}$, then $q(x) = -x\frac{d}{dx}\left(\ln \overline{F}(x)\right)$. Thus $\overline{F}(x) = k e^{-\int_{c\alpha\alpha}^{x} \frac{q(u)}{u} du}$, where k is a positive constant and $c_0 < \alpha < \beta(F)$.

Now $\lim_{t\to\infty} \frac{\overline{F}(tx)}{\overline{F}(t)} = \lim_{t\to\infty} e^{-\int^{tx} \frac{q(u)}{u} du} = \lim_{t\to\infty} e^{-\int_1^{x} \frac{q(tu)}{u} du} = e^{-\delta \ln x} = x^{-\delta}$

Example 2.1.5.

The truncated Pareto distribution $f(x) = \frac{\delta}{x^{\delta+1}} \cdot \frac{1}{1-b^{-\delta}}$, $1 \le x<b$, $b>1$, $\lim_{x\to\infty} \frac{xf(x)}{\overline{F}(x)} =$

$\lim_{x\to b} \frac{\delta x^{-\delta}}{x^{-\delta} - b^{-\delta}} = \infty$. Thus the truncated Pareto distribution does not satisfy the Von Mises sufficient condition. However it belongs to the domain of attraction of the Type 2 extreme value distribution, because $\lim_{t\to\infty}\frac{\bar{F}(\beta(F)-\frac{1}{tx})}{\bar{F}(\beta(F)-\frac{1}{t})}=x^{-\delta}$ for x > 0 and some constant $\delta > 0$.

The domain of attraction of type 3 distribution for X_{nn}.

The following theorem gives a necessary and sufficient condition for the domain of attraction of type 3 distribution for X_{nn}.

Theorem 2.1.4.

Let $X_1, X_2, ..$ be a sequence of i.i.d random variables with distribution function F and $\alpha(F) = \sup\{x: F(x) < 1\}$. If $\beta(F) < \infty$, then $F \in T_{3\delta}$ if $\lim_{t\to 0^+}\frac{\bar{F}(\beta(F)+tx)}{\bar{F}(\beta(F)-t)}=(-x)^{\delta}$ for $x < 0$ and some constant $\delta > 0$.

Proof. Similar to Theorem 2.1.2.

Suppose X is a negative exponential distribution truncated at $x= b > 0$. The pdf of X is $f(x)=\frac{e^{-x}}{F(b)}$, then for $x< 0$,

$$P\left(X_{nn}\le b+\frac{x(e^b-1)}{n}\right)=\left(\frac{1-e^{-(b+\frac{x(e^b-1)}{n})}}{1-e^{-b}}\right)^n \to e^x$$

as $n \to \infty$.

Thus the truncated exponential distribution belongs to domain of attraction of T_{31}.

Since $\lim_{t\to 0^+}\frac{\bar{F}(\beta(F)+tx)}{\bar{F}(\beta(F)-t)}=\lim_{t\to 0^+}\frac{e^{-(b+tx)}-e^{-b}}{e^{-(b-t)}-e^{-b}}=-x$, the truncated exponential distribution satisfies the necessary and sufficient condition for the domain of attraction of type 3 distribution for maximum..

The following Lemma gives Von Mises sufficient condition for the domain of attraction of type 3 distribution for X_{nn}.

Lemma 2.1.4.

Suppose the distribution function F is absolutely continuous in $[c_0, \beta(F)]$ for some c_0, $0<c_0<\beta(F) < \infty$, then if $\lim_{x\uparrow\beta(F)}\frac{(\beta(F)-x)f(x)}{\bar{F}(x)}=\delta, \delta > 0.$, then $F\in D(T_{3\delta})$.

Proof.

Similar to Lemma 2.1.3

Example.2.1.6.

Suppose X is a negative exponential distribution truncated at x= b >0, then the pdf of X is $f(x)=\frac{e^{-x}}{F(b)}$. Now $\lim_{x\uparrow\beta(F)}\frac{(\beta(F)-x)f(x)}{\overline{F}(x)}=\lim_{x\uparrow b}\frac{(b-x)e^{-x}}{e^{-x}-e^{-b}}=1$. Thus the truncated exponential distribution satisfies the Von Mises sufficient condition for the domain of attraction to Type 3 distribution.

A distribution that belongs to the domain of attraction of Type 2 distribution cannot have finite $\beta(F)$. A distribution that belongs to the domain of attraction of Type 3 distribution must have finite $\beta(F)$.. The normalizing constants of X_{nn} are not unique for any distribution. From the table it is evident that two different distributions (exponential and logistic) belong to he domain of attraction of the same distribution and have the same normalizing constants. The normalizing constants depends on F and the limiting distribution. It may happen that X_{nn} with any normalizing constants may not converge in distribution to a non degenerate limiting distribution but W_{nn} where W = u(X) , a function of X, may with some normalizing constants converge in distribution to one of the three limiting distribution. We can easily verify that the rv X whose pdf, f(x) = $\frac{1}{x(\ln x)^2}, x\geq e$ does not satisfy the necessary and sufficient conditions for the convergence in distribution of X_{nn} to any of the extreme value distributions. Suppose W= lnX, then $F_W(x)$ = 1-1/x for y>1. Thus with as $a_n=0$ and b_n = 1/n, $P(W_{nn\cdot}\leq x)\rightarrow T_{31}$ as $n\rightarrow\infty$

Following Pickands (1975), the following theorem gives a necessary and sufficient condition for the domain of attraction of X_{nn} from a continuous distribution.

Theorem 2.1.5.

For a continuous random variable the necessary and sufficient condition for X_{nn} to belong to the domain of attraction of the extreme value distribution of the maximum is

$$\lim_{c\to 0}\frac{F^{-1}(1-c)-F^{-1}(1-2c)}{F^{-1}(1-2c)-F^{-1}(1-4c)}=1 \text{ iff } F\in T_{10},$$

$$\lim_{c\to 0}\frac{F^{-1}(1-c)-F^{-1}(1-2c)}{F^{-1}(1-2c)-F^{-1}(1-4c)}=2^{1/\delta} \text{ iff } F\in T_{2\delta}$$

and

$$\lim_{c\to 0}\frac{F^{-1}(1-c)-F^{-1}(1-2c)}{F^{-1}(1-2c)-F^{-1}(1-4c)} = 2^{-1/\delta} \text{ iff } F\in T_{3\delta}.$$

Example 2.1.7.

For the exponential distribution, E(0,σ), with pdf f(x) = $\sigma^{-1}e^{-\sigma^{-1}x}$, x ≥ 0, $F^{-1}(x) = -\sigma^{-1}\ln(1-x)$ and $\lim_{c\to 0}\frac{F^{-1}(1-c)-F^{-1}(1-2c)}{F^{-1}(1-2c)-F^{-1}(1-4c)} = \lim_{c\to 0}\frac{-\ln\{1-(1-c)\}+\ln\{1-(1-2c)\}}{-\ln\{1-(1-2c)\}+\ln\{1-(1-4c)\}} = 1$.

Thus the domain of attraction of X_{nn} from the exponential distribution, E(0,σ), is T_{10}.

For the Pareto distribution, P(0,0,α) with pdf f(x) = $\alpha x^{-(\alpha+1)}, x>1, \alpha>0$, $F^{-1}(x) = (1-x)^{-1/\alpha}$ and $\lim_{c\to 0}\frac{F^{-1}(1-c)-F^{-1}(1-2c)}{F^{-1}(1-2c)-F^{-1}(1-4c)} = \lim_{c\to 0}\frac{c^{-1/\alpha}-(2c)^{-1/\alpha}}{(2c)^{-1/\alpha}-(4c)^{=1/\alpha}} = 2^{1/\alpha}$. Hence the domain of attraction of X_{nn} from the Pareto distribution, P(0,0,α) is $T_{2\alpha}$.

For the uniform distribution, U(-1/2 , 1/2), with pdf f(x) = $\frac{1}{2}$, $-\frac{1}{2}<x<\frac{1}{2}$, $F^{-1}(x) = 2x-1$. We have $\lim_{c\to 0}\frac{F^{-1}(1-c)-F^{-1}(1-2c)}{F^{-1}(1-2c)-F^{-1}(1-4c)} = \lim_{c\to 0}\frac{2(1-c)-1-2(1-2c)+1}{2(1-2c)-1-2(1-4c)+1} = 2^{-1}$.

Consequently the domain of attraction of X_{nn} from the uniform distribution, U(-1/2, 1/2) is T_{31}.

It may happen that X_{nn} from a continuous distribution does not belong to the domain of attraction of any one of the distribution. In that case X_{nn} has a degenerate limiting distribution. Suppose the rv X has the pdf f(x), where f(x) = $\frac{1}{x(\ln x)^2}, x\geq e$, then we have $F^{-1}(x) = e^{\frac{1}{1-x}}, 0<x<1$. and $\lim_{c\to 0}\frac{F^{-1}(1-c)-F^{-1}(1-2c)}{F^{-1}(1-2c)-F^{-1}(1-4c)} =$

$$\lim_{c\to 0}\frac{e^{\frac{1}{c}}-e^{\frac{1}{2c}}}{e^{\frac{1}{2c}}-e^{\frac{1}{4c}}} = \lim_{c\to 0}\frac{e^{\frac{1}{c}}-1}{1-e^{\frac{1}{2c}}} = \lim_{c\to 0}\frac{2e^{\frac{1}{c}}}{e^{\frac{1}{2c}}} = \lim_{c\to 0} 2e^{\frac{1}{2c}} = \infty$$

Thus the limit does not exit. Hence the rv X does not satisfy the necessary and sufficient condition as given in Theorem 2.1.5.

Theorems 2.1.1, 2.1.2, 2.1.3 and 2.1.4 are also true for discrete distributions. If the random variables X_j, j=1,2,..., n have discrete distribution with finite number of points of support, then X_{nn} can not converge to one of the extreme value distributions. Thus X_{nn} from binomial and discrete uniform distribution will converge to degenerate distributions.

The following Lemma gives a necessary condition for the convergence of $X_{n,n}$ to one of the extreme value distributions.

Lemma 2.1.5.

Suppose X is a discrete random variable with infinite number points in its support and taking values on non negative integers with $P(X= k) = p_k$. Then a necessary condition for the convergence of $P(X_{nn} \le a_n + b_n x)$ for a suitable sequence of a_n and b_n to one of the three extreme value distributions is $\lim_{k\to\infty} \frac{p_k}{P(X \ge k)} = 0$.

For the geometric distribution, $P(X= k)= p(1-p)^{k-1}$, $k \ge 1$, $0<p<1$, $\frac{p_k}{P(X \ge k)} = p$. Thus X_{nn} from the geometric distribution will have degenerate distribution as limiting distribution of X_{nn}..

Consider the distribution: $P(X= k) = \frac{1}{k(k+1)}$, $k= 1,2,\ldots$, then $P(X \ge k) = \frac{1}{k}$ and $\lim_{k\to\infty} \frac{p_k}{P(X \ge k)} = \lim_{k\to\infty} \frac{1}{k+1} = 0$. But $\lim_{t\to\infty} \frac{\bar{F}(tx)}{\bar{F}(t)} = x^{-1}$. Thus X belongs to the domain of attraction of T_{21}. The normalizing constants are $a_n = 0$ and $b_n = n$. However the condition $\lim_{k\to\infty} \frac{p_k}{P(X \ge k)} = 0$ is necessary but not sufficient.

Consider the discrete probability distribution with $P(X=k) = \frac{c}{k(\ln(k+1))^6}$, $k = 1,2,\ldots$ where $1/c = \sum_{k=1}^{\infty} \frac{1}{k(\ln(k+1))^6} \cong 9.3781$. Since $1 - \sum_{k=1}^{n} \frac{1}{k(\ln(k+1))^6} \propto \frac{1}{(\ln n)^5}$, $\frac{P(X=n)}{1-\sum_{k=1}^{n=1} P(X=k)} \to 0$ an $n\to\infty$, this probability distribution does not satisfy the necessary and sufficient conditions for the convergence of X_{nn} to one of the three extreme value distributions.

We can use the following lemma to calculate the normalizing constants for various distributions belonging to the domain of attractions of $T(x)$..

Lemma 2.1.6.

Suppose $P(X_{nn} \le a_n + b_n x) \to T(x)$ as $n \to \infty$, then

(i) $a_n = F^{-1}(1-\frac{1}{n}), b_n = F^{-1}(1-\frac{1}{ne}) - F^{-1}(1-\frac{1}{n})$ if $T(x) = T_{10}(x)$,

(ii) $a_n = 0, b_n = F^{-1}(1-\frac{1}{n})$ if $T(x) = T_{2\delta}(x)$,

(iii) $a_n = F^{-1}(1), b_n = F^{-1}(1) - F^{-1}(1-\frac{1}{n})$ if $T(x) = T_{3\delta}(x)$

We have seen that the normalizing constants are not unique. However we can use the following Lemma to select simplified normalizing constants.

Lemma 2.1.7.

Suppose a_n and b_n is a sequence of normalizing constants for X_{nn} for the convergence to the domain of attraction of any one of the extreme value distributions. If a_n^* and b_n^* is another sequence such that $\lim_{n\to\infty}\frac{a_n - b_n^*}{b_n} = 0$ and $\lim_{n\to\infty}\frac{b_n^*}{b_n} = 0$, then a_n^* and b_n^* can be substituted for as the normalizing constants a_n and b_n for X_{nn}.

Example 2.1.7.
We have seen (see Table 2.1.1) that for the Cauchy distribution with pdf $f(x) = \frac{1}{\pi(1+x^2)}$, $-\infty < x < \infty$, the normalizing constants as $a_n = 0$ and $b_n = \cot(\pi / n)$.

However we can take $a_n^* = 0$ and $b_n^* = \frac{n}{\pi}$.

Example 2.1.8.

For Type 1 extreme value distribution, $F(x) = e^{-e^{-x}}$, $x > 0$ and X_{nn} with appropriate normalizing constants converges in distribution to Type 1 extreme value distribution. Here $F^{-1}(x) = -\ln(-\ln x)$. We can take $a_n = F^{-1}(1-\frac{1}{n}) = \ln n$ and $b_n = F^{-1}(1-\frac{1}{ne}) - F^{-1}(1-\frac{1}{n}) = 1$.

For Type 2 extreme value distribution with $F(x) = e^{-x^{\alpha}}$, $x>0$, $\alpha > 0$, X_{nn} with appropriate normalizing constants converges in distribution to Type 2 extreme value distribution. Here $F^{-1}(x) = (-\ln x)^{1/\alpha}$. We have $a_n = 0$ and $b_n = F^{-1}(1-\frac{1}{n}) = \frac{1}{n^{1/\alpha}}$.

For Type 3 extreme value distribution, $F(x) = e^{-(-x)^{\alpha}}$, $x < 0$, $\alpha > 0$ and X_{nn} with appropriate normalizing constants converges in distribution to Type 3 extreme value distribution. Here e $F^{-1}(x) = -\ (-\ln x)^{1/\alpha}$. We have $a_n = 0$ and $b_n = F^{-1}(1) - F^{-1}(1-\frac{1}{n}) = \frac{1}{n^{1/\alpha}}$.

The following tables gives the normalizing constants for some well known distributions belonging the domain of attraction of the extreme value distributions.

Table 2.1.1 Normalizing Constants for X_{nn}

Distribution	f(x)	a_n	b_n	Domain
Beta	$cx^{\alpha-1}(1-x)^{\beta-1}$ $c = \frac{\Gamma(\alpha+\beta)}{\Gamma(\alpha)\Gamma(\beta)}$ $\alpha > 0, \beta > 0$ $0<x<1$	1	$\left(\frac{\beta}{nc}\right)^{1/\beta}$	$T_{3\beta}$
Cauchy	$\frac{1}{\pi(1+x^2)}$, $-\infty < x < \infty$	0	$\frac{n}{\pi}$	T_{21}
Exponential	$\sigma e^{-\sigma x}$, $0<x<\infty$, $\sigma > 0$	$\frac{1}{\sigma}\ln n$	$\frac{1}{\sigma}$	T_{10}
Gamma	$\frac{x^{\alpha-1}e^{-x}}{\Gamma(\alpha)}$, $0 < x < \infty$	$\ln n - \ln\Gamma(\alpha)$ $+(\alpha-1)\ln\ln n$	1	T_{10}
Distribution	f(x)	a_n	b_n	Domain
Laplace	$\frac{1}{2}e^{-\lvert x \rvert}$, $-\infty < x < \infty$	$\ln(n/2)$	1	T_{10}
Logistic	$\frac{e^{-x}}{(1+e^{-x})^2}$	$\ln n$	1	T_{10}
Lognormal	$\frac{1}{x\sqrt{2\pi}}e^{-\frac{1}{2}(\ln x)^2}$ $0<x<\infty$	e^{α_n}, $\alpha_n =$ $\frac{1}{\beta_n} - \frac{\beta_n D_n}{2}$, D_n $= \ln\ln n + \ln 4\pi$ $\beta_n = (2\ln n)^{-1/2}$	$(2\ln n)^{-1/2} e^{\alpha_n}$	T_{10}

Normal	$\frac{1}{\sqrt{2\pi}} e^{-\frac{1}{2}x^2}$, $-\infty<x<\infty$	$\frac{1}{\beta_n} - \frac{\beta_n D_n}{2}$, $D_n = \ln\ln n + \ln 4\pi$, $\beta_n=(2\ln n)^{-1/2}$	$(2\ln n)^{-1/2}$	T_{10}
Pareto	$\alpha\, x^{-(\alpha+1)}$, $x>1, \alpha>0$	0	$n^{1/\alpha}$	$T_{2\alpha}$
Power Function	$\alpha x^{\alpha-1}$, $0<x<1, \alpha>0$	1	$\frac{1}{n\alpha}$	T_{31}
Rayleigh	$\frac{2x}{\sigma^2} e^{-\frac{x^2}{\sigma^2}}$, $x \geq 0$	$\sigma(\ln n)^{\frac{1}{2}}$	$\frac{\sigma}{2}(\ln n)^{-\frac{1}{2}}$	T_{10}
Truncated Exponential	Ce^{-x}, $C=1/(1-e^{-e(F)})$, $0<x<\alpha(F)<\infty$	$\alpha(F)$	$\frac{e^{e(F)}-1}{n}$	T_{31}
T distribution	$\frac{k}{\left(1+\frac{x^2}{\upsilon}\right)^{(\upsilon+1)/2}}$, $k = \frac{\Gamma((\upsilon+1)/2)}{(\pi\upsilon)^{1/2}\Gamma(\upsilon/2)}$	0	$\left(\frac{kn}{\upsilon}\right)^{1/\upsilon}$	$T_{2\upsilon}$
Type 1	$e^{-x}e^{-e^{-x}}$	ln n	1	T_{10}
Distribution	f(x)	a_n	b_n	Domain
Type 2	$\alpha x^{-(\alpha+1)} e^{-x^{-\alpha}}$, $x>0, \alpha>0$	0	$n^{-1/\alpha}$	$T_{2\alpha}$
Type 3	$\alpha(-x)^{\alpha-1} . e^{-(-x)^\alpha}$, $x<0$, $\alpha>0$	0	$n^{-1/\alpha}$	$T_{3\alpha}$
Uniform	$1/\theta$, $0<x<\theta$	θ	θ/n	T_{31}
Weibull	$\alpha\, x^{\alpha-1} e^{-x^\alpha}$, $x>0, \alpha>0$	$(\ln n)^{1/\alpha}$	$\frac{(\ln n)^{\frac{1-\alpha}{\alpha}}}{\alpha}$	$T_{1\alpha}$

Let us consider X_{1n} of n i.i.d random variables. Suppose $P(X_{1n} \leq \underline{c}_n + d_n x) \rightarrow H(x)$ as $n \rightarrow \infty$, then the following three types of distributions are possible for H(x).

Type 1 distribution: $H_{10}(x) = 1- e^{-e^x}, -\infty < x < \infty$:;

Type 2 distribution: $H_{2\delta}(x) = 1- e^{-(-x)^{-\delta}}, x < 0, \delta > 0$.

Type 3 distribution : $H_{3\delta}(x)= 1- e^{-x^{\delta}}, x > 0, \delta > 0$.

It may happen that X_{nn} and X_{1n} may belong to different types of extreme value distributions. For example consider the exponential distribution, $f(x) = e^{-x}$, $x > 0$. The X_{nn} belongs to the domain of attraction of the type 1 distribution of the maximum , T_{10}. Since $P(X_{1n} > n^{-1}x) = e^{-x}$, X_{1n} belongs to the domain of attraction of Type 2 distribution of the minimum, H_{21} .It may happen that X_{nn} does not belong to any one of the three limiting distributions of the maximum but X_{1n} belong to the domain of attraction of one of the limiting distribution of the minimum Consider the rv X whose pdf, f(x) = $\frac{1}{x(\ln x)^2}, x \geq e$. We have seen that F does not satisfy the necessary and sufficient conditions for the convergence in distribution of X_{nn} to any of the extreme value distributions. However it can be shown that $P(X_{1n} > \alpha_n + \beta_n x) \to e^{-x}$ as $n \to \infty$ for $\alpha_n = e$ and $\beta_n = e^{-\frac{n-1}{n}} - e$. Thus the X_{1n} belongs to the domain of attraction of H_{21}.

If X is a symmetric random variable and X_{nn} belongs to the domain of attraction of $T_i(x)$, then X_{1n} will belong to the domain of attraction of the corresponding $H_i(x)$, i =1,2,3.

The following Lemma is needed to prove the necessary and sufficient conditions for the convergence of X_{1n} to one of the limiting distributions H(x)

Lemma 2.1.8.

Let $\{X_n, n \geq 1\}$ be a sequence of independent and identically distributed random variables with distribution function F. Consider a sequence $(e_n, n \geq 1 \}$ of real numbers. Then for any ξ, $0 \leq \xi < \infty$, the following two statements are equivalent

(iii) $\lim_{n \to \infty} n(F(e_n)) = \xi$

(iv) $\lim_{n \to \infty} P\left(X_{n,n} > e_n\right) = e^{-\xi}$.

Proof .

The proof of the Lemma follows from Lemma 2.1.1 by considering the fact $P(X_{1n} > e_n) = (1 - F(e_n))^n$

The following theorem gives a necessary and sufficient condition for the convergence of X_{1n} to $H_{10}(x)$.

Theorem 2.1.6.

Let X_1, X_2, .. be a sequence of i.i.d random variables with distribution function F. Assume further that E(X| X $\leq$ t is finite for some t $>\alpha(F)$ and h(t) =E(t-X|X$\leq$ t) .Then F$\in H_{10}$ iff $\lim_{t\to\alpha(F)} \frac{F(t+xh(t))}{F(t)} = e^x$ for all real x.

Proof. Similar to Theorem 2.1.1

Example 2.1.9.

Consider the logistic distribution with F(x) = $\frac{1}{1+e^{-x}}$, $-\infty < x < \infty$. Now

$h(t) = E(t-x|X\leq t) = t-(1+e^{-t})\int_{-\infty}^{t} \frac{xe^{-x}}{(1+e^{-x})^2}dx = (1+e^{-t})\ln(1+e^{t})$. It can easily be shown that h(t) $\to$1 as t $\to$- ∞. We have $\lim_{t\to\alpha(F)} \frac{F(t+xh(t))}{F(t)} = \lim_{t\to-\infty} \frac{1+e^{-t}}{1+e^{-(t+xh(t))}} = \lim_{t\to-\infty} \frac{e^{t+xh(t)}+e^{xh(t)}}{1+e^{t+xh(t)}} = e^x$.

Thus X_{1n} from logistic distribution belongs to the domain of H_{10}. Domain of Attraction of Type 2 distribution for X_{1n}.

Theorem 2.1.7.

Let X_1, X_2, .. be a sequence of i.i.d random variables with distribution function F then F$\in H_{2\delta}$ iff $\alpha(F) = -\infty$ and $\lim_{t\to\alpha(F)} \frac{F(tx)}{F(t)} = x^{\delta}$ for all x > 0.

Proof.

Suppose $H_{2\delta}(x)$ = 1- $e^{-(-x)^{-\delta}}$,$x<0, \delta>0$, then we have

$$\lim_{t\to\alpha(F)} \frac{F(tx)}{F(t)} = \lim_{t\to-\infty} \frac{1-e^{-(-tx)^{-\delta}}}{1-e^{-(-t)^{-\delta}}} = \lim_{t\to-\infty} \frac{\delta x(-tx)^{-(\delta+1)}e^{-(-tx)^{-\delta}}}{\delta(-t)^{-(\delta+1)}e^{-(-t)^{-\delta}}} = x^{-\delta}, \delta > 0.$$

Let $\lim_{t\to\alpha(F)} \frac{F(tx)}{F(t)} = x^{-\delta}, \delta > 0$. We can write Let $a_n = \inf\{x: \bar{F}(x) \leq \frac{1}{n}\}$, then $a_n \to -\infty$ as $n \to \infty$. Thus $\lim_{n\to-\infty} n(F(a_n x)) = \lim_{n\to-\infty} n(F(a_n))\frac{F(a_n x)}{F(a_n)} = x^{-\delta} \lim_{n\to-\infty} nF(a_n)$.

It is easy to show that $\lim_{n\to-\infty} nF(a_n) = 1$. Thus $\lim_{n\to-\infty} n(F(a_n x)) = x^{-\delta}$ and the proof of the Theorem follows from Lemma 2.1.8.

Example 2.1.10.

Consider the Cauchy distribution with F(x) = $\frac{1}{2}+\frac{arctg\,x}{\pi}$. Thus

$$\lim_{t\to\alpha(F)}\frac{F(tx)}{F(t)}=\lim_{t\to-\infty}\frac{\frac{1}{2}+\frac{arcg\,tx}{\pi}}{\frac{1}{2}+\frac{arctg\,t}{\pi}}=\lim_{t\to-\infty}\frac{x(1+t^2)}{1+(tx)^2}=x^{-1}.$$ Thus F belongs to the domain of attraction of H_{21}. Domain of attraction of Type 3 distribution for $X_{1,n}$.

Theorem 2.1.8.

Let X_1, X_2, .. be a sequence of i.i.d random variables with a distribution function F then $F\in H_{3\delta}$ iff $\alpha(F)$ is finite and $\lim_{t\to 0}\frac{F(\alpha(F)+tx)}{F(\alpha(F)+t)}=x^{\delta},\delta>0$ and for all x > 0.

Proof.

The proof is similar to Theorem 2.1.7.

Example 2.1.10.

Suppose X has the uniform distribution with F(x) = x, 0<x<1.Then

$\lim_{t\Rightarrow 0}\frac{F(tx)}{F(t)}=x$. Thus, $F\in H_{31}$. Following Pickands (1975), the following theorem gives a necessary and sufficient condition for the domain of attraction of $X_{1,,n}$ from a continuous distribution.

Theorem 2.1.9.

For a continuous random variable the necessary and sufficient condition for $X_{1,,n}$ to belong to the domain of attraction of the extreme value distribution of the minimum is

$$\lim_{c\to 0}\frac{F^{-1}(c)-F^{-1}(2c)}{F^{-1}(2c)-F^{-1}(4c)}=1 \text{ iff } F\in H_{10},$$

$$\lim_{c\to 0}\frac{F^{-1}(c)-F^{-1}(2c)}{F^{-1}(2c)-F^{-1}(4c)}=2^{1/\delta}\text{ iff } F\in H_{2\delta}$$

and

$$\lim_{c\to 0}\frac{F^{-1}(c)-F^{-1}(2c)}{F^{-1}(2c)-F^{-1}(4c)}=2^{-1/\delta}\text{ iff } F\in H_{3\delta}.$$

Example 2.1.11.

For the logistic distribution with F(x) = $\frac{1}{1+e^{-x}}, F^{-1}(x) = \ln x - \ln(1-x)$

$$\lim_{c\to 0}\frac{F^{-1}(c)-F^{-1}(2c)}{F^{-1}(2c)-F^{-1}(4c)} = \lim_{c\to 0}\frac{\ln c-\ln(1-c)-\ln 2c+\ln(1-2c)}{\ln 2c-\ln(1-2c)-\ln 4c+\ln(1-4c)} = 1.$$ Thus $F \in H_{10}$.

For the Cauchy distribution with F(x) = $\frac{1}{2}+\tan^{-1}(x)$. We have $F^{-1}(x) = \tan\pi(x-\frac{1}{2}) = -\frac{1}{\pi x}$ for small x. Thus

$$\lim_{c\to 0}\frac{F^{-1}(c)-F^{-1}(2c)}{F^{-1}(2c)-F^{-1}(4c)} = \frac{\frac{1}{2\pi c}-\frac{1}{\pi c}}{\frac{1}{4\pi c}-\frac{1}{2\pi c}} = 2.$$ Thus $F \in H_{21}$.

For the exponential distribution, E(0,σ), with pdf f(x) = $\sigma^{-1}e^{-\sigma^{-1}x}$, $x \geq 0$, $F^{-1}(x) = -\sigma^{-1}\ln(1-x)$ and $\lim_{c\to 0}\frac{F^{-1}(c)-F^{-1}(2c)}{F^{-1}(2c)-F^{-1}(4c)} = \lim_{c\to 0}\frac{-\ln\{1-c)+\ln\{1-2c)}{-\ln\{1-2c)+\ln\{1-4c)} = 2^{-1}$. Thus $F \in H_{31}$. We can use the following lemma to calculate the normalizing constants for various distributions belonging to the domain of attractions of $H_j(x)$, j =1,2,3.

Lemma 2.1.9.

Suppose P($X_{1,,n} \leq c_n + d_n x) \to H(x)$ as $n \to \infty$, then

(i) $c_n = F^{-1}(\frac{1}{n}), d_n = F^{-1}(\frac{1}{n}) - F^{-1}(\frac{1}{ne})$ if $H(x) = H_{10}(x)$,

(ii) $c_n = 0, b_n = |F^{-1}(\frac{1}{n})|$ if $H(x) = H_{2\delta}(x)$,

(v) $c_n = \alpha(F), b_n = F^{-1}(\frac{1}{n}) - \alpha(F)$ if $H(x) = H_{3\delta}(x)$

(vi)

We have seen (Lemma 2.1.6) that the normalizing constants are not unique for $X_{n,,n}$. The same is also true for the $X_{1,,n}$.

Example 2.1.12.

For the logistic distribution with F(x) = $\dfrac{1}{1+e^{-x}}$, X_{1n} with appropriate normalizing constants converge in distribution to Type 1(H_{10}) distribution. The normalizing constants are $c_n = F^{-1}(\frac{1}{n}) = \ln\left(\frac{1/n}{1-(1/n)}\right) = \ln(\frac{1}{n-1}) \cong -\ln n$ *and* $d_n = F^{-1}(\frac{1}{n}) - F^{-1}(\frac{1}{ne}) = 1$.

For Cauchy distribution with F(x) $= \frac{1}{2} + \frac{arctg\, x}{\pi}$, $X_{1,n}$ with appropriate normalizing constants converge in distribution to Type 2 (H_{21}) distribution. The normalizing constants are $c_n = 0$ and $d_n = |F^{-1}(\frac{1}{n})| = \tan\pi(\frac{1}{2} - \frac{1}{n}) = \cot(\frac{\pi}{n}) \cong \frac{n}{\pi}$.

For the uniform distribution with F(x) = x, 0<x<1, $X_{1,,n}$ with appropriate normalizing constants converge in distribution to Type 3 (H_{31}) distribution. The normalizing constants are $c_n = 0, b_n = F^{-1}(\frac{1}{n}) = \frac{1}{n}$.

Table 2.1.2 Normalizing Constants for $X_{1,,n}$

Distribution	f(x)	c_n	d_n	Domain
Beta	$c\, x^{\alpha-1}(1-x)^{\beta-1}$ $c = \dfrac{1}{B(\alpha,\beta)}$ $\alpha > 0, \beta > 0$ 0<x<1	0	$\left(\dfrac{c\,\alpha}{n}\right)^{1/\alpha}$ $c = \dfrac{\Gamma(\alpha)\Gamma(\beta)}{\Gamma(\alpha+\beta)}$	$H_{3\alpha}$
Cauchy	$\dfrac{1}{\pi(1+x^2)}$, $-\infty < x < \infty$	0	Cot(π/n)	H_{21}
Exponential	$\sigma e^{-\sigma x}$, 0<x<∞, σ > 0	0	$\dfrac{1}{n\sigma}$	H_{31}
Gamma	$\dfrac{x^{\alpha-1}e^{-x}}{\Gamma(\alpha)}$, $0 < x < \infty$	0	$\dfrac{\Gamma(\alpha)}{n}$	H_{31}
Laplace	$\frac{1}{2}e^{-\lvert x\rvert}$, $-\infty < x < \infty$	$\ln\left(\dfrac{n}{2}\right)$	1	H_{10}
Logistic	$\dfrac{e^{-x}}{(1+e^{-x})^2}$	-ln n	1	H_{10}

Lognormal	$\frac{1}{x\sqrt{2\pi}} e^{-\frac{1}{2}(\ln x)^2}$ $0<x<\infty$	$e^{-\alpha_n}$, $\alpha_n =$ $\frac{1}{b_n} - \frac{b_n D_n}{2}$, D_n $= \ln \ln n + \ln 4\pi$ $b_n = (2\ln n)^{-1/2}$	$(2\ln n)^{-1/2} e^{-\alpha_n}$	H_{10}
Normal	$\frac{1}{\sqrt{2\pi}} e^{-\frac{1}{2}x^2}$, $-\infty<x<\infty$	$\frac{c_n D_n}{2} - \frac{1}{c_n}$, D_n $= \ln \ln n + \ln 4\pi$ $b_n = (2\ln n)^{-1/2}$	$(2\ln n)^{-1/2}$	H_{10}
Distribution	f(x)	c_n	d_n	Domain
Pareto	$\alpha x^{-(\alpha+1)}$ $x > 1, \alpha > 0$	0	$\left(\frac{n}{n-1}\right)^{1/\alpha}$	H_{21}
Power Function	$\alpha x^{\alpha-1}$, $0 < x < 1, \alpha > 0$	0	$n^{-1/\alpha}$	H_{31}
Rayleigh	$\frac{2x}{\sigma^2} e^{-\frac{x^2}{\sigma^2}}$, $x \geq 0$	0	$\frac{\sigma}{\sqrt{n}}$	H_{32}
T distribution	$\frac{k}{\left(1+\frac{x^2}{\upsilon}\right)^{(\upsilon+1)/2}}$ $k =$ $\frac{\Gamma((\upsilon+1)/2)}{(\pi\upsilon)^{1/2}\Gamma(\upsilon/2)}$	0	$\left(\frac{kn}{\upsilon}\right)^{1/\upsilon}$	$H_{2\upsilon}$
Type 1 (for minimum)	$e^x e^{-e^x}$	-ln n	1	H_{10}
Type 2 (for minimum)	$\alpha(-x)^{-(\alpha+1)} e^{-(-x)^{-\alpha}}$ $x<0, \alpha>0$	0	$n^{1/\alpha}$	$H_{2\alpha}$
Type 3 (for minimum)	$\alpha x^{\alpha-1} . e^{-x^\alpha}$ $x>0, \alpha>0$	0	$n^{-1/\alpha}$	$H_{3\alpha}$
Uniform	$1/\theta$, $0<x<\theta$	0	θ/n	H_{31}
Weibull	$\alpha x^{\alpha-1} e^{-x^\alpha}$, $x>0, \alpha>0$	0	$n^{-1/\alpha}$	$H_{3\alpha}$

2.2 ASYMTOTIC DISTRIBUTION OF $X_{N-K+1,\,N}$

We will consider the case k is fixed. The asymptotic distribution of $X_{n-k+1,\,n}$ for fixed k as n tends to ∞ is given in the following Theorem.

Theorem 2.2.1.

Suppose $X_1, X_2, \ldots, X_n$ are n i.i.d. rv and $X_{n-k+1,,n}$ is the n-k+1 th order statistics. If for some normalizing constants a_n and b_n, $F^n(a_n + b_n x) \to T(x)$ in distribution as $n \to \infty$, then $P(X_{n-k+1\,n} \le a_n + b_n x) \to \sum_{j=0}^{k-1} T(x)(-\ln T(x))^j / j!$ in distribution as $n \to \infty$.for any fixed k and all x.

Proof.

Let us consider a sequence c_n, $n \geq 1$ such that as $n \to \infty$. $c_n \to c$, then $\lim_{n\to\infty}(1-\frac{c_n}{n})^n = e^{-c}$. Take $c_n = n(1-F(a_n + b_n x))$.

Now $P(X_{n-k+1\,n} \le a_n + b_n x) = \sum_{r=n-k+1}^{n} \binom{n}{r}(F(a_n + b_n x))^r (1-F(a_n + b_n x))^{n-r}$

$$= \sum_{r=0}^{k-1} \binom{n}{r}(c_n / n))^r (1-c_n / n))^{n-r}.$$

Thus we can consider $P(X_{n-k+1,n} \le a_n + b_n x)$ as a cumulative binomial distribution, B(n, c_n/n) evaluted at k-1. Since $n \to \infty$, $c_n \to c$ and T(x) = $\lim_{n\to\infty} F^n(a_n + b_n x) = \lim_{n\to\infty}(1-\frac{c_n}{n})^n = e^{-c}$, using Poisson approximation to the binomial distribution , we get $P(X_{n-k+1\,n} \le a_n + b_n x) \to \sum_{j=0}^{k-1} T(x)(-\ln T(x))^j / j!$ in distribution as $n \to \infty$.

Example 2.2.1.

Consider n i.i.d. random variables from the exponential distribution, $F(x) = 1- e^{-x}$, x>0.

Since with $a_n = \ln n$ and $b_n = 1$, $F^n(a_n + b_n x) \to e^{-e^{-x}}$ in distribution as $n \to \infty$.

We will have $P(X_{n-r+1\,n} \le a_n + b_n x) \to \sum_{j=0}^{k-1} \frac{e^{-jx}}{j!} e^{-e^{-x}}$ in distribution as $n \to \infty$. The asymptotic distribution of $X_{k,\,n}$ for fixed k as $n \to \infty$ is given in the following theorem.

Theorem 2.2.2.

Suppose $X_1, X_2, \ldots, X_n$ are n i.i.d. rv's and $X_{k,,n}$ is the k th order statistics. If for some normalizing constants a_n and b_n, $\overline{F}^n(a_n+b_nx)\to\overline{H}(x)$ in distribution as $n \to \infty$, then for finite k, $\lim_{n\to\infty} P(X_{k,n}>a_n+b_nx)\to\sum_{j=0}^{k-1}\overline{H}(x)\frac{(-\ln\overline{H}(x))^j}{j!}$ in distribution for any fixed k and all x.

Proof.

Similar to Theorem 2.2.1

Example 2.2.2.

Consider n i.i.d. random variables from the exponential distribution, F(x) = 1-e^{-x}, x>0. Since with $a_n=0$ and $b_n=1/n$, $\overline{F}^n(a_n+b_nx)\to\overline{H}(x)=e^{-x}$ as $n \to \infty$. But for all n, $\overline{F}^n(a_n+b_nx)=e^{-x}$ Hence we will have $P(X_{kn}>a_n+b_nx)==$ $\sum_0^{k-1}\frac{x^j}{j!}e^{-x}$, for all x > 0 and all n.

The following Theorem gives the limiting distribution of $X_{k,,n}$ for fixed k in terms of the exponentially distributed random variables.

Theorem 2.2.3.

Suppose $X_1, X_2, \ldots, X_n$ are n i.i.d. rv and $X_{n--k+1,n}$ is the n-k+1 th order statistics. If for some stablizing constants a_n and b_n, $F^n(a_n+b_nx)\to T(x)$ in distribution as $n \to \infty$, then $\lim_{n\to\infty} P(X_{n-k+1}\le a_n+b_n x) \underset{=}{d} W-(W_1+\frac{W_2}{2}+\ldots+\frac{W_{k-1}}{k-1})$, if X $\in D(T_{10})$, where W is distribted as T_{10} and $W_1, W_2. \ldots, W_{k-1}$ are indendent and identically distributed rv's with cdf F(x) = 1- e^{-x},

(i) $\lim_{n\to\infty} P(X_{n-k+1,,n}\le a_n+b_nx)\overset{d}{=}(W_1+W_2+\ldots..+W_k)^{-1/\delta}$, if X$\in D(T_{2\delta})$, where W_1, $W_2. \ldots, W_k$ are indendent and identically distributed rv's with cdf F(x) = 1- e^{-x},

(ii) $\lim_{n\to\infty} P(X_{n-k+1,n}\le a_n+b_nx)\overset{d}{=}-(W_1+W_2+\ldots..+W_k)^{1/\delta}$, if X$\in D(T_{3\delta})$, where W_1, $W_2. \ldots, W_k$ are independent and identically distributed rv's with cdf F(x) = 1- e^{-x}.

Proof.

The results follow from Theorem 2.2.1.

2.3 ESTIMATION OF THE PARAMETERS OF EXTREME VALUE DISTRIBUTION

Estimation of the index parameter

The three distributions, $T_{10}(x)$, $T_{2\delta}(x)$ and $T_{3\delta}(x)$ are members of the following family of distributions

$$G(x,\gamma) = \exp[-\{ 1+\gamma x)^{-1/\gamma}],$$
$$x >- 1/\gamma \text{ for } \gamma > 0, (T_{2\,1/\delta}(x)) = G((x-1)/\gamma,\gamma),$$
$$x <- 1/\gamma \text{ for } \gamma < 0,(T_{3\,1/\delta}(x) = G(-(x+1)/\gamma,\gamma)$$

and $G(x,0) = \exp(-e^{-x})$ for all x.

However we can take $G(x, 0) = \lim G(x,,\gamma)$ as γ tends to zero.
The above representation is known as Von Mises form.

Similarly the three distributions, $H_{10}(x)$, $H_{2\delta}(x)$ and $H_{3\delta}(x)$ are members of the following family of distributions

$$J(x,,\gamma) = 1- \exp[-\{ 1-\gamma x)^{-1/\gamma}],$$

$$x < 1/\gamma \text{ for } \gamma > 0, (H_{2\,1/\delta}(x)) = J((x+1)/\gamma,\gamma),$$
$$(x >1/\gamma \text{ for } \gamma < 0,(H_{3\,1/\delta}(x) = J(-(x-1)/\gamma,\gamma)$$

and

$$J(x,0) = 1-\exp(-e^{x}) \text{ for all x } (J(x,0) = H_{10}(x)).$$

The parameter γ is known as index parameter. We will consider first the estimation of the index parameter γ (= $1/\delta$) of T(x) . T(x). It can be shown that $F^{-1}(x) = \frac{1}{\gamma}[(-\ln x)^{-\gamma} - 1]$ and as x tends to ∞, for $U(x) = F^{-1}(1-\frac{1}{x})$,

$$\frac{U(xs)-U(xt)}{U(xu)-U(xv)} \approx \frac{s^{\gamma}-t^{\gamma}}{u^{\gamma}-v^{\gamma}} \text{ , for } \gamma \neq 0$$

and

$$\frac{U(xs)-U(xt)}{U(xu)-U(xv)} \approx \frac{\ln(s/t)}{\ln(u/v)}, \text{ for } \gamma = 0.$$

Taking s = 4m, t = u = 2m and v = m, we get as x tends to ∞,

$$\frac{U(xs)-U(xt)}{U(xu)-U(xv)} \approx 2^{\gamma}, \text{ for } \gamma \neq 0$$

and

$$\frac{U(xs)-U(xt)}{U(xu)-U(xv)} \approx 1, \text{ for } \gamma = 0.$$

The Pickand (1975) introduced the estimator $\hat{\gamma}(n,m)$ of γ as

$$\hat{\gamma}(m,n) = \frac{1}{\ln 2} \ln\left(\frac{X_{n-m+1n} - X_{n-2m+1n}}{X_{n-2m+1n} - X_{n-4m+1n}}\right)$$

for a selected m, m→ ∞ as n → ∞ but m/n → 0.This estimator is asymptotically consistent for γ.

Thus Hill (1975) introduced the estimator $\tilde{\gamma}$ (m,n) of γ as

$$\tilde{\gamma}\,(\mathrm{m,n}) = \frac{1}{m}\sum_{j=1}^{m} \ln X_{n-j+1n} - \ln X_{n-m},$$

for a selected m, m→ ∞ as n → ∞ but m/n → 0.This estimator is asymtotically consistent for γ when 1-F is regularly varying with index $-1/\gamma$.

Dekkers, Einmahl and de Hann (1989) presented an estimator, $\bar{\gamma}(m,n)$ of γ as

$$\bar{\gamma}(m,n) = \tilde{\gamma}(m.n) + 1 - \frac{1}{2}\left(1 - \frac{(\tilde{\gamma}(m.n))^2}{\tilde{k}(m,n)}\right)^{-1},$$

where $\tilde{k}(m.n) = \frac{1}{m}\sum_{j=1}^{m} (\ln X_{n-j+1n} - X_{n-mn})^2$ and for a selected m, m→ ∞ as n → ∞ but m/n → 0.

Estimation of Location and scale parameters

The standard representatives of three types of limiting distributions for suitable normalized extremes $X_{n-k+1,n}$ in the case when k is fixed have the form

$F_k(x) = \sum_{j=0}^{k-1} T(x)(-\ln T(x))^j / j!$, where T(x) is one of the functions $T_{10}(x)$, $T_{2,\delta}(x)$ and $T_{3,\delta}(x)$. Changing location and scale parameters we see that cdf $F_k((x-\mu)/\sigma)$ also can be limiting for $X_{n-k+1,n}$. Consider the moment characteristics of possible asymptotic distributions of extremal order statistics.

Let random variable T_k has the cdf $F_k((x-\mu)/\sigma)$, where μ is a location parameter and $\sigma>0$ is a scale parameter.

Consider the case of the type I of the extreme value distributions, when $T(x)=T_{1,0}(x)$. Due to Theorem 2.2.3 we have the following representation:

$$\frac{T_k - \mu}{\sigma} \stackrel{d}{=} W - (W_1 + \frac{W_2}{2} + \ldots + \frac{W_{k-1}}{k-1}) , \tag{2.3.1}$$

where W is distributed as $T_{10}(x)$ and $W_1, W_2, \ldots, W_{k-1}$ are independent and have the standard exponential distribution. Using (2.3.1) we obtain that $ET_k=\mu+\alpha_k\sigma$, $Var(T_k)= \sigma^2 V_{m,m}$, where α_1 coincides with the Euler's constant, and

$$\alpha_j = \alpha_1 - \sum_{i=1}^{j-1} i^{-1}, \ j \geq 2,$$

$$V_{11} = \pi^2 / 6$$

and

$$V_{jj} = V_{11} - \sum_{i=1}^{j-1} i^{-2}, \ j \geq 2.$$

If $T(x)=T_{2,\delta}(x)$, then Theorem 2.2.3 implies that

$$\frac{T_k - \mu}{\sigma} \stackrel{d}{=} -(W_1+W_2+\ldots+W_k)^{-1/\delta} \tag{2.3.2}$$

and hence

$$E(Tk) = \mu + -\sigma \frac{\Gamma(k - \frac{1}{\delta})}{\Gamma(k)} ,$$

$$Var(T_k) = \sigma^2 \{ \frac{\Gamma(k-\frac{2}{\delta})}{\Gamma(k)} - (\frac{\Gamma(k-\frac{1}{\delta})}{\Gamma(k)})^2 \},$$

$$Cov(T_k, T_n) = \sigma^2 \{ \frac{\Gamma(k - \frac{1}{\delta})}{\Gamma(k)} [\frac{\Gamma(n - \frac{2}{\delta})}{\Gamma(n - \frac{1}{\delta})} - \frac{\Gamma(n - \frac{1}{\delta})}{\Gamma(n)}] \}, \text{ for } k < n.$$

Analogous formulae can be obtained for the case when $T(x)=T_{3,\delta}$ and (due to Theorem 2.2.3)

$$\frac{T_k - \mu}{\sigma} \stackrel{d}{=} (W_1+W_2+\ldots+W_k)^{1/\delta}. \qquad (2.3.3)$$

Now we consider estimators of μ and σ based on m observed maximal order statistics.

The following Lemma gives estimators of μ and σ for Type I extreme value distribution.

Lemma 2.3.1.

Let $\hat{\mu}$ and $\hat{\sigma}$ be the minimum variance linear unbiased estimators of μ and σ based on the observed values $x_1, x_2, \ldots, x_m$ of $T_1, T_2, \ldots, T_m$ in the case of Type I extreme value distribution with location parameter μ and scale parameter σ. Then

$$\hat{\mu} = x_m - \alpha_m \hat{\sigma},$$

$$\hat{\sigma} = \frac{m}{m-1}(\overline{x}_m - x_m),$$

where

$$\overline{x}_m = (x_1 + x_2 + \ldots + x_m)/m,$$

The variances and the covariance of $\hat{\mu}$ and $\hat{\sigma}$ are given by

$$\mathrm{Var}(\hat{\mu}) = \sigma^2(\alpha_m^2(m-1)^{-1} + V_{mm}),$$

$$\mathrm{Var}(\hat{\sigma}) = \sigma^2(m-1)^{-1},$$

$$\mathrm{Cov}(\hat{\mu}, \hat{\sigma}) = -\sigma^2\alpha_m(m-1)^{-1}.$$

We consider here the type I extreme value distribution

It can be shown that

$$1'V^{-1} = (0,0,\ldots,V_{mm}^{-1}),\ \delta'V^{-1} = (1,1,\ldots,\alpha_m V_{mm}^{-1} - (m-1))$$

$$\delta'V^{-1}\delta = \alpha_m^2 V_{mm}^{-1} + m - 1,\ \delta'V^{-1}1 = \alpha_m V_{mm}^{-1},$$

where $1' = (1,1,\ldots,1)$, $\delta' = (\alpha_1, \alpha_2, \ldots, \alpha_m)$.

Using the above relations, we have

$$(X - \hat{\mu}1)'V^{-1}(X - \hat{\mu}1) = (X - \hat{\mu}1 - \hat{\sigma}\delta)'V^{-1}(X - \hat{\mu}1 - \hat{\sigma}\delta) + \hat{\sigma}^2\delta'V^{-1}\delta.$$

The ratio

$$W = \frac{\hat{\sigma}^2 \delta' V^{-1} \delta}{(X - \hat{\mu} 1)' V^{-1} (X - \hat{\mu} 1)} \quad (2.3.4)$$

can be considered as the proportion of variance of X explained by the regression of the type I extreme value distribution. Consequently W can be used to test the hypothesis that a distribution is of type I extreme value or belongs to the domain of attraction of the type I extreme value distribution. We know that if the rvs, X_i's, are from some specific distribution, then for suitable constants α_n and β_n, $(M_n - \alpha_n)/\beta_n \rightarrow E_1(0,1)$. It will be of interest to concentrate on a statistic which is invariant with respect to location and scale parameters. We took the following test statistic W* which is closely related to W and invariant with respect to location and scale parameter.

$$W^* = \frac{m}{m-1} \frac{(\bar{x}_m - x_m)^2}{\sum_{i=1}^{m} (x_i - \bar{x}_m)^2} \quad (2.3.5)$$

Further $m E(\bar{x}_m - x_m)^2 = (m-1) E \sum_{i=1}^{m} (x_i - \bar{x}_m)^2$. The exact distribution and the percentage points of W* are difficult to calculate. It is easy to see that W* is bounded by 1. We used simulation technique to calculate the percentage points of the distribution of W* when X_i's are from type I distribution.

Critical Region and the Power of the Test

The critical values and the power of the proposed statistic W* were studied by simulation. Twenty thousand samples(n) were generated from the type I extreme value distribution. For each n, n-m+1th (m = 2,...,20) order statistic was obtained. 5000 simulated samples were generated. The upper and lower 20%,10%,5%,2.5%,1% and 0.5% percentage points of these order were computed from these 5000 samples. In Table 2.3.1 we have reported the upper and lower 5% points ($\alpha = 0.05$) for m = 3(1) 20. Using these critical values, we tested the hypothesis that the observations are from Type II and Type III distributions. Powers of the tests using W^* for Type II and Type III distributions were identical for all α. We have reported the powers of Type II distributions based on one sided alternative with lower percentile points for $\alpha = 0.05$. It is found that the statistic W* is very powerful even for small value of m.. Table 2 gives the power of this test against Type II distribution for various values of δ .The power of the test against all alternatives increases with the increase of m. As δ decrease Type II approaches Type I and the power of the test decreases. Hasofer and Wang (1992) used a statistic W^{**}

Table 2.3.1. 5% points of W * of Type I Extreme Value Distribution

	m										
	3	4	5	6	7	8	9	10	12	16	20
Upper	0.9916	0.8658	0.6646	0.5160	0.4168	0.3414	0.2943	0.2572	0.2023	0.1372	0.1018
Lower	0.2703	0.1620	0.1183	0.0996	0.0821	0.0714	0.0630	0.0574	0.0475	0.0365	0.0298

Table 2.3.2. Power of W* against some selected alternatives

		m										
		3	4	5	6	7	8	9	10	12	16	20
	$\delta = 4$	0.4746	0.6210	0.7092	0.7976	0.8752	0.9126	0.9422	0.9666	0.9874	0.9988	1.0000
Type II	$\delta = 2$	0.2734	0.4212	0.4994	0.5904	0.6740	0.7382	0.7836	0.8326	0.9148	0.9842	0.9998
	$\delta = 1$	0.1466	0.2406	0.2956	0.3574	0.4170	0.4746	0.5178	0.5674	0.7214	0.9446	0.9994
	$\delta = .5$	0.0820	0.1330	0.1660	0.1928	0.2254	0.2530	0.2720	0.3036	0.4896	0.8662	0.9926
	$\eta = 4$	0.4886	0.6472	0.7542	0.8694	0.9482	0.9782	0.9930	0.9970	0.9994	1.0000	1.0000
	$\eta = 2$	0.3000	0.4778	0.5902	0.7208	0.8494	0.9126	0.9930	0.9802	0.9950	1.0000	1.0000
Pareto	$\eta = 1$	0.1958	0.3380	0.4332	0.5450	0.6752	0.7670	0.8404	0.8998	0.9606	0.9978	1.0000
	$\eta = .5$	0.1594	0.2728	0.3404	0.4206	0.5178	0.6020	0.6744	0.7520	0.8360	0.9696	0.9972
U(a,b)	$b = 1$	0.9444	0.1630	0.2248	0.3116	0.4450	0.5600	0.6590	0.7660	0.8914	0.9894	0.9996

similar to our W^* based on k largest order statistics for a test of domain of attraction of Type I extreme value distribution. These two statistics are based on completely different types of data. Hasofer and Wang (1992) gave the power of W^{**} using 50 upper order statistics out of a sample of 300 observations for various values of δ (= α in their paper). We also computed power of W* when records are from Pareto (member of Type II) and Uniform (member of Type III) distributions. Several members of the classical Pareto distribution ($F(x)=1- x^{\gamma}, \gamma > 0, x \geq 1$) were considered. Table 2.3.2 gives the power of the W^* test for various values of η (= $1/\gamma$) . We obtained for $\alpha = 0.05$ in the case of Pareto distribution the powers of W^* as 0.9978 and 0.9696 for m = 16 corresponding to $\eta = 1$ and $\eta = 1/2$, whereas the corresponding powers of W^{**} for $\alpha = 0.05$ using 50 upper order statistics out of a sample of 300 observations were given by Hasofer and Wang (1992) as 0.996 and 0.851 for $\eta = 1$ and $\eta = 1/2$. We found that the power of W^* increases as m increases . The power of W^* decreases as η decreases. For uniform distribution, $(F(x) = x/(b-a)$, $a < x < b)$,it was found that the test statistic W* does not depend is invariant with respect to a and b.

REFERENCES

1. Ahsanullah, M. (1996). Some inferences of the generalized extreme value distribution based on record values. Journal of Mathematical Sciences 78, 1-10.

2. Ahsanullah, M. (1995). *Record Statistics*, Nova Science Publishers, Inc. Commack, NY.

3. Ahsanullah, M. (1990). Estimation of the parameters of the Gumbel distribution based on m record values. Comp. Statist. Quarterly, 3, 231-239.

4. Ahsanullah, M. (1988). *Introduction to Record Statistics*, Ginn Press, Needham Heights, MA.

5. Ahsanullah, M.(1994).Records of the extreme value distribution. Pak.J.Stat. 10,1,147 Statistics.Vol.10 (1), 147-170.

6. Ahsanullah, M. (1994). Record values from univariate continuous distributions. Proceedings of the Extreme Value Theory and Applications., 1-12.

7. Ahsanullah, M. and Holland, B. (1994). On the use of record values to estimate the location and scale parameters of the generalized extreme value distribution. Sankhya, A, 56,480-499

8. Ahsanullah, M. and Bhoj, D.S.(1996). Record values of extreme value distributions and a test for domain of attraction of type I extreme value distribution . Sankhya 58, B,151-158.

9. Balakrishnan, N., Ahsanullah, M. and Chan, P.S. (1992). Relations for single and product moments of record values from Gumbel distribution. Stat.and Prob.Letters, 15, 223-227.

10. Balakrishnan, N. and Ahsanullah, M. and Chan, P.S. (1993). Recurrence relations for moments of record values from generalized extreme value distribution. Comm. Statist. Theory and Methods. 22(5),1471-1482.

11. Deken, J.G. (1978). Record values, scheduled maxima sequences. J. Appl. Prob. 15, 491- 496.

12. Dwass, M. (1964). External Processes. Ann. Math. Statist. 35, 1718-1725.

13. Feller, W. (1966). An Introduction to Probability Theory and its Applications. Vol. II, Wiley, New York.

14. Freudenberg, W. and Szynal, D. (1976). Limit laws for a random number of record values. Bull. Acad. Polon. Sci.Ser. Math. Astr. Phys. 24, 195-199.

15. Galambos,J. (1987). The Asymptotic Theory of Extreme Order Statistics. Robert E. Krieger Publishing Co. Malabar, Florida.

16. Gaver, D. P. and Jacobs, P.A. (1978). Non homogeneously paced random records and associated extremal processes. J. Appl. Prob. 15, 543-551.

17. Gnedenko, B. (1943). Sur la Distribution Limite du Terme Maximum d'une Serie Aletoise. Ann. Math. 44, 423-453.

18. Gumbel,E.J. (1958). Statistics of Extremes.Columbia Univ. Press, New York.

19. Horwitz, J. (1980). Extreme values from a non stationary stochastic process: an application to air quality analysis (with discussion) . Technometrics 22,469-482.

20. Hasofer, A. M. and Wang, Z. (1992). At test for Extreme Value Domain of Attraction. J. Amer. Stat. Assoc. 87, 171-177.

21. Jenkinson, A.F. (1955). The frequency distribution of the annual maximum (or minimum) values of meteerological elements. Quart. J. Meter. Soc. 87, 158-171.

22. Jellinek, H. H. G. (1958). The influence of imperfections on the strength of ice. *Proc. Phys. Soc.* London 71, 797 - 814.

23. Johnson, N. L. and Kotz, S. (1972). *Distributions in Statistics.* John Wiley and Sons, New York.

24. Judckaja, P. I. (1974). On the maximum of a Gaussian sequence. Theory Probab. Math. Statist. 2, 259-267.

25. Juncosa, M. L. (1949). On distribution of the minimum in a sequence of mutually independent random variables. *Duke Math. J.*16. 609-618.

26. Karlin, S. (1966). A first course in stochastic processes. Academic Press, New York.

27. Karr, A. (1976). Two extreme value processes arising in hydrology. J Appl. Prob.. 13, 190-94.

28. Kawata, T. (1951). Limit distributions of single order statistics. Rep. Stat. Appl. Res. JUSE 1, 4-9.

29. Kimbeall, B. F. (1946). Significant statistical estimation functions for the parameters of the distribution of maximum values. *Ann Math. Statist.* 17, 299-309.

30. Kimball, B. F. (1949). An approximation to the sampling variance of an estimated maximum value of given frequency based on fit of doubly exponential distribution of maximum values. Ann. Math. Statist. 20, 110-113.

31. Kotz, S. and Johnson, N. L. (1982). Encyclopedia of Statistical Sciences. John Wiley and Sons, New York, NY.

32. Kunio, O, T., Shimizu, M., Yamada, K. and Kimura, Y. (1974). An interpretation of the scatter of fatigue limit on the basis of the theory of extreme value. Trans. JSME 40, 2101-2109.

33. Kwerel, S. M. (1975a). Most stringent bounds on aggregated probabilities of partially specified dependent systems. J. Amer. Statist. Assoc. 70, 472-479.

34. Kwerel, S. M. (1975b). Bounds on the probability of the union and intersection of *m* events. Adv. Appl. Prob. *7,* 431-448.

35. Lamperti, J. (1964). On extreme order statistics. Ann. Math. Statist. 35, 1726-1737.

36. Lawless, J. F. (1978). Confidence interval estimation for the Weibull and extreme value distribution. Technometrics, 20, 355-364.

37. Leadbetter, M. R. (1975). Aspect of extreme value theory for stationary pocesses – A Survey. *In: Stochastic Processes and Related Topics*, **1.** (edited by Puri), New York.

38. .Leadbetter, M. R. and Lindgreen, G. and Rootzen, H. (1983). Extremes and Related Properties of Random Sequences and Processes. Springer-Verlag, New York.

39. Lehman, E. H. (1963). Shapes, moments and estimators or the Weibull distribution, IEEE Trans. Reliab. **12**, 32-38.

40. Levi, R. (1949). Calculs probabilistes de la securité des constructions. Annales des Ponts Et Chaussées 119, 493-539.

41. Lieblein, J. (1953). On the exact evaluation of the variances and covariances of order statistics in samples from the extreme-value distribution. Ann. Math. Statist. 24, 282-287.

42. Lieblein, J, and Zelen, M. (1956). Statistical investigation of the fatigue life of deep- groove ball bearings, J. Res. Nat. Bur.Stand. 57, 273-316.

43. Lindgren, G. (1971). Extreme values of stationary normal processes. Z. Wahrsch. verw. Geb. 17, 39-47.

44. Lindgren G. (1972b). Wave-length and amplitude for a stationary Gaussian process after a high maximum. Z. Wahrsch. verw. Geb. 23, 293-326.

45. Mahmound, M. W. and Ragad, A. (1975). On order statistics in samples drawn from the extreme value distribution. Math Operationsforschung u. Stat. **6,** 800-816.

46. Mallows, C. L. (1973). Bounds on distribution functions in terms of expectations of order statistics. Ann. Prob. 1, 297-303.

47. Mann, N. R. (1967a). Results on location and scale parameter estimation with application to the extreme value distribution. ARL 67-0023, Aerospace Research Laboratory, Wright-Patterson AFB, Ohio, AD 653575.

48. Mann, N. R. (1967 b). Tables for obtaining the best linear invariant estimates of parameters of the Weibull distribution. Technometrics 9, 629-645.

49. Mann, N. R. (1968b). Point and interval estimation procedures for the two-parameter Weibull and extreme-value distributions. Technometrics 10, 231-256.

50. Mann, N. R. (1970c). Estimators and exact confidence bounds for Weibull parameters based on a few ordered observations. Technometrics 12, 345-361.

51. Mansour, A. (1972). Methods of computing the probability of failure under extreme values of bending moment, J. Ship Res. *6, 113-123.*

52. Marcus, M.B. and Pisky, M. (1969). On the domain of attraction of exp[-exp (-*x*)]. J. Math. Anal Appl. 28, 440-449.

53. Mardia, K. V. (1964b).Asymptotic independence of bivariate extremes. Calcutta Stat. Assoc. Bull. 13, 172-178.

54. Marshall, A. W. and Olkin, I. (1983). Domains of attraction of multivariate extreme value distributions. Ann. Prob. 11, 168-177.

55. Mason, D. M. (1982). Laws of large numbers for sums of extreme values. Ann. Prob. 10, 754-764.

56. Mathar, R. (1984). The limit behaviour of the maximum of random variables with applications to outlier-resistance. J. Appl. Prob. 21, 646-650.

57. McCool, J. I. (1969). Unbiased maximum-likelihood estimation of a Weibull percentile when the shape parameter is known. IEEE Trans. Reliab. 18, 78-79.

58. McCool, J. I. (1970a). Inference on Weibull percentiles and shape parameter from maximum likelihood estimates. IEEE Trans. Reliab. 19, 2-9.

59. McCool, J. I. (1970b). Evaluating Weibull endurance data by the method of maximum likelihood. Trans. Amer. Soc. Lubric. Eng. 13, 189-202.

60. McCord, J. R. (1964). On asymptotic moments of extreme statistics. Ann. Math. Statist. 35, 1738-1745.

61. Mandel, J. (1959). The theory of extreme values. ASTM Bulletin No. 236, 29-30.

62. Maritz, J. S. and Munro, A.H. (1967). On the use of the generalized extreme value distribution in estimating extreme percentiles. Biometrics , 79-103.

63. Mejler, D.G. (1956). On the problem of the limit distribution for the maximal term of a variational series. L'vov Politechn. Inst. Naucn. Zp. 38, 90-109.

64. Menon, M. V. (1963). Estimation of the shape and scale parameters of the Weibull distribution. Technometrics 5, 175-182.

65. Metcalfe, A. G. and Smitz G. K. (1964). Effect of Length on the strength of glass fibres. Proc. Amer. Soc. Testing and Materials 64, 1075-1093.

66. Mexia, J. T. (1967). Studies on the extreme double exponential distribution. Sequential estimation and testing for the location parameter of Gumbel distribution. Revista da Faculdade de Ciencias, Universidade de Lisboa A 12, 5-14.

67. Michael, J. R. (1983). The stabilized probability plot. Biometrika 70, 11-17.

68. Mihram, G. A. (1969). Complete sample estimation techniques for reparameterized Weibull distributions. IEEE Trans. Reliab. 18, 190-195.

69. Millan, J. and Yevjevich, V. (1971). Probabilities of observed draughts. Colorado State Univ. Hydrology papers, N. 50, Fort Collins, Colorado.

70. Mises, R. Von (1936). La distribution de la plus grande de n valeurs. Revue Mathématique de l'Union Interbalkanique (Athens) 1, 141-160.

71. Misteth E. (1974). Dimensioning of structures for flood discharge according to the theory of probability. Acta Technica Academiae Scientiarum Hungaricae 76, 107-127.

72. Mittal, Y. and Ylvisaker, D. (1976). Strong laws for the maxima of stationary Gaussian processes. Ann. Prob. 4, 357-371.

73. Mogyorodi, J. (1967). On the limit distribution of the largest term in the order statistics of a sample of random size. Magyar Tud. Akad. Mat. Fiz. Oszt. Kozl. **17**, 75-83.

74. Monfort, M. A. J., Van (1970). On testing that the distribution of extremes is of type I when type II is the alternative. J. Hydrol. 11, 421-427.

75. Monfort, M. A. J., Van (1973). An asymmetric test on the type of the distribution of extremes. Mededelingen Landbouwhogeschool 73, 1-15.

76. Moore, A. H. and Harter, H. L. (1967). One-order-statistic conditional estimators of shape parameters of limited and Pareto distributions and scale parameters of type II asymptotic distributions of smallest and largest values. IEEE Trans. Reliab. 16, 100-103.

77. Mori, T. (1976). Limit laws for maxima and second maxima for strong mixing processes. Ann Prob. *4, 122-126.*

78. Moriguti, S. (1951). Extremal properties of extreme value distributions. Ann. Math. Statist. 22, 523-536.

79. Moses, F. (1974). Reliability of structural systems. J. of Structural Division, ASCE, 100, ST 9, 1813-1820.

80. Moyer, C. A., Bush, J. J., and Ruley, B. T. (1962). The Weibull distribution function for fatigue life. Research and Standards 2, 405-411.

81. Nair, K. A. (1976). Bivariate extreme value distributions. Comm. in Statist. 5, 575-581.

82. Nelson, W. and Thompson, V. C. (1971). Weibull probability papers. J. Qual. Technol. 3, 45- 50

83. Nevzorov, V.B (1987). Records. Theory. Probab. Appl. 32, 201-228.

84. Nevzorov, V.B (2000). Rrecorde: Mathematical Theory. Translations of Mathematical Monographs. Volume 194. America Mathematical Society.

85. Nevzorov, L.N. and Nevzorov, V.B (2000).ordered random variables. Acta Appl. Math 58, no. 1-3, 217-219.

86. Newell, G. F. (1964). Asymptotic extremes for *m*-dependent random variables. Ann. Math. Statist. 35, 1322-1325.

87. Obert, T. L. (1972). Brittle Fracture of Rock. Fracture: An Advanced Treatise, VII: Fracture of Nonmetals and Composites. Academic Press, New York.

88. Ochi, M. K. (1973). On prediction of extreme values. *J. Ship Res.* **17**, 29-37.

89. Passos Coelho, D. and Pinho G., T. (1963). Studies on extreme double exponential distribution The location parameter. Revista da Faculdade de Ciencias de Lisboa A10, 37-46

90. Pfeifer, D.(1981). Asymptotic expansions for the mean and variance of logarithmic inter-record times. Meth. Operat.Res. 39,113-121.

91. Pfeifer, D.(1982). Characterizations of exponential distributions by independent non stationary increments. J. Appl. Prob. 19,127-135.

92. Pfeifer, D.(1988).Limit Laws for inter-record times for non homogeneous Markov chains. J. Organizational Behav. Statist. 1, 69-74.

93. Pickands, J. III (1967). Sample sequences of maxima. Ann. Math. Statist. 38, 1570- 1574.

94. Pickands, J. III (1968). Moment convergence of sample extremes. Ann. Math. Statist. 39, 881- 889.

95. Pickands, J. III . III (1975). Statistical inference using extreme order statistics. Ann. Statist. 3, 119-131.

96. Rao, C.R. (1983). An extension of Denny's theorem and its application to characteizations of probability distributions. A Festchrift for Erich L. Lehman. Ed. Bickel et. al. Wordsworth International Group, Belmont, CA.

97. Reiss, R.D. (1989). Approximate Distributions of Order Statistics. Springer - Verlag, New York, NY.

98. Reiss, R. D. (1981). Uniform approximation to distributions of extreme order statistics. Adv. Appl. Prob. 13, 533-547.

99. Renyi, A. (1953). On the theory of order statistics. Acta Math. Acad. Sci. Hungar. 4, 191- 231.

100. Renyi, A. (1962). On outstanding values of a sequence of observations. In: Selected paper of A Rényi, 3. Akadémiai Kiadó Budapest, 50-65.

101. Renyi, A. (1962). Theorie des elements saillants d'une suit d'observations collq. combinatorial Meth. Prob. Theory, Aarhus University 104-115.

102. Resnick, S.I. (1973). Record values and maxima. Ann. Prob.1, 650-662.

103. Resnick, S.(1987). Extreme Values , Regular Variation and Point Processes. Springer Verlag, New York.

104. Resnick, S.I. (1971). Asymptotic location and recurrence properties of maxima of a sequence of random variables defined on a Markov chain. Zeitschrift fur Wahrsch. verw Geb. 18,197-217.

105. Resnick, S.I. (1973b). Record values and maxima. Ann. Prob. 1, 650-662.

106. Rice, S. O. (1939). The distribution of the maxima of a random curve. Amer. J. Math. 61, 409- 416.

107. Ringer, L. J. and Springkle E. III (1972). Estimation of the parameters of the Weibull distribution from multicensored samples. IEEE Trans. Reliab. 21, 46-51.

108. Roberts, E. M. (1979a). Review of statistics of extreme values with applications to air quality data, Part I: review. J. Air Poll. Control Assoc. 29, 632-637.

109. Roberts, E. M (1979b). Review of statistics of extreme values with applications to air quality data, Part II: Applications. J. Air Pollut. Control Assoc. 29, 733-740.

110. Rootzen, H. (1974). Some properties of convergence in distribution of sums and maxima of dependent random variables. Zeitschrift fur Wahrsch. verw. Geb. 29, 295-307.

111. Rustagi, J. S. (1957). On minimizing and maximizing a certain integral with statistical applications. Ann. Math. Statist. **28**, 309-328.

112. Salamingo, F.J. and Whitaker, L.D. (1986). On estimating population characteristics from record breaking observations. I. Parametric Results. Naval Res. Log. Quart. 25, 531-543.

113. Sarhan, A. E. and Greenberg, B. G. (1962). Contributions to Order Statistics. John Wiley and Sons, New York.

114. Shannon, C.E. (1948). A mathematical theory of communication (concluded). Bell. Syst. Tech. J. 27, 629-631.

115. Sethuramann, J. (1965). On a characterization of the three limiting types of extremes. Sankhya, A. 27, 357-364.

116. Schafer, D. (1974). Confidence bounds for the minimum fatigue life. Technometrics. 16, 113- 123.

117. Schucany, W. R. (1972). Order statistics in simulation. J. Statist. Comput. Simul. 1, 281-286.

118. Schueller, G. I. (1984). *Application of Extreme Values in Structural Engineering.* In: Statistical extremes and Applications. NATO ASI Series. D. Reidel Publishing Company.

119. Schupbach, M. and Husler, J. (1983). Simple estimators for the parameters of the extreme-value distribution based on censored data. Technometrics 25, 189-192.

120. Schuster, E. F. (1984). Classification of probability laws by tail behavior. J. Amer. Statist Assoc. 9, 936-939.

121. Sen, P. K. (1959). On the moments of the sample quantiles. Calcutta Statist. Ass. Bull. 9, 1-19.

122. Sen, P. K. (1961). A note on the large sample behavior of extreme sample values from distributions with finite endpoints. *Bull Calcutta Statist. Assoc.* **10**, 106-115.

123. Sen, P. K. (1968). Asymptotic normality of sample quantiles for *m*-dependent process. Ann. Math. Statist. 39, 1724-1730.

124. Sen, P. K., Bhattacharyya, B. B. and Suh, M. W. (1973). Limiting behaviour of the extremum of certain sample functions. Ann. Statist. 1, 297-311.

125. Seneta, E. (1976). *Regularly Varying Functions.* Lecture Notes in Mathematics 508, Springer-Verlag, Heidelberg.

126. Serfozo, R. (1982). Functional limit theorems for extreme values of arrays of independent random variables. Ann. Prob. 10, 172-177.

127. Sethuraman, J. (1965). On a characterization of the three limiting types of the extreme. Sankhya A 27, 357-364.

128. Shepp, L. A. (1979). The joint density of the maximum and its location for a Wiener process with drift. J. Appl. Prob.16, 423-427

129. Sibuya, M. (1960). Bivariate extreme statistics. *Ann. Inst. Stat. Math.* **11**, 195-210.

130. Singpurwalla, N. D. (1972b). Extreme values for a lognormal law with applications to air pollution problems. *Technometrics* 14, 703-711.

131. Sirvanci, M. and YANG, G. (1984). Estimation of the Weibull parameters under type I censoring. J. Amer. Statist. Assoc. 79, 183-187.

132. Smith, R. L. (1981). Asymptotic distribution for the failure of fibrous materials under series-parallel structure and equal load-sharing. J. App. Mech. 103 75-82.

133. Sneyers, R. (1984). Extremes in meteorology. In: *Statistical Extremes and Applications.* NATO ASI Series. D. Reidel Publishing Company.

134. Stam, A. J. (1973). Regular variation of the tail of a subordinated probability distribution. Adv. Appl. Prob. 5, 308-327.

135. Suzuki, E. (1961). A new procedure of statistical inference on extreme values. *Pap.* Meterol. Geophys. 12, 1-17.

136. Sweeting, T. J. (1985). On domains of uniform local attraction in extreme value theory. Ann. Prob.. 13, 196-205.

137. Teugels, J. L. (1981). Limit theorems on order statistics. Ann. Prob. 9, 868-880.

138. Thoman, D. R., Bain, L. J and Antle, C.E. (1969). Inferences on the parameters of the Weibull distribution. *Technometrics* 11, 445-460.

139. Thom, H. C. S. (1967). Asymptotic Extreme Value Distributions Applied to Wind and Waves.NATO Seminar on extreme value problems, Faro, Portugal.

140. Thom, H. C. S. (1971). Asymptotic extreme value distributions of wave height in open ocean. J Mar. Res. 29, 19-27.

141. Thom, H. C. S. (1973). Extreme wave height distributions over oceans. *J. Waterw. Harb.* Coast. Div., ASCE, 99 WW3, 355-374.

142. Thomas, D. R. and WILSON, W. M. (1972). Linear order statistic estimation for the two-parameter Weibull and extreme value distributions from type II progressively censored samples. Technometrics 14, 679-691.

143. Tiago De Oliveira, J. (1957). Estimators and tests for continuous populations with l location and dispersion parameters. Rev. Fac. Cienc. Lisboa A6, 121-146.

144. Tiago De Oliveira, J. (1958). Extremal distributions. Rev. Fac. Cienc. Lisboa **A7**, 215- 227.

145. Tiago De Oliveira, J. (1961). The asymptotical independence of the sample mean and the extremes. Rev. Fac. Cienc. Lisboa **A8**, 299-310.

146. Tiago De Oliveira, J (1984). *Statistical Extremes and Applications*. NATO ASI Series. D.Reidel Publishing Company, Lisbon.

147. Tippett, L. H. C. (1925). On the extreme individuals and the range of samples taken from a normal population. Biometrika **17**, 364-387.

148. Todorovic, P. and Shen H. W. (1976). *Some Remarks on the Statistical Theory of Extreme Values Stoch. Approach.* Water Resour. **2** (edited by H. W. Sen). Univ. of Colorado Press, Fort Collins, Colorado.

149. Turkman, K. F. and Walker, A. M. (1983). Limit laws for the maxima of a class of quasi-stationary sequences. J. Appl. Prob.. 20, 814-821.

150. Walsh, J.E. (1969a). Asymptotic independence between largest and smallest of a set of independent observations. Ann. Inst. Statist. Math., Tokyo, 21, 287-289.

151. Walsh, J.E. (1969b). Approximate distributions for largest and for smallest of a set of independent observations. S. Afr. Statist. J. 3, 83-89.

152. Watts, V. Rootzen, H. and Leadbetter, M.R. (1982). On limiting distribution of Intermediate order statistics from stationary sequences. Ann. Prob. 10, 653-662.

153. Weibull, W. (1951). A statistical distribution function of wide applicability. J. Appl. Mech. 18, 293-297.

154. Weibull, W. (1977). References on the Weibull distribution. FTL A-report A 20: 23. Forvarets Teletekniska Laboratorium, Stockholm.

155. Weinstein, S. B. (1973). Theory and application of some classical and generalized asymptotic distributions of extreme values. IEEE Trans. Inf. Theor. 19, 148-154.

156. Weiss, L. (1969). The joint asymptotic distribution of the k-smallest sample spacings. J. Appl. Prob.. 6, 442-448.

157. Weissman, I. (1978). Estimation of parameters and large quantiles based on the kth largest observations. J. Amer. Statist. Assoc. 63, 812-815.

158. Weissman, I. (1980). Estimation of tail parameters under type I censoring. Comm. Statist. Theor. Meth. A9, 1165-1175.

159. White, J.S. (1969). The moments of log - Weibull order statistics. Technometrics, 11, 373-386.

160. Wilks, S. S. (1962). *Mathematical Statistics.* John Wiley and Sons, New York.

161. Yang, S. S. (1977). General distribution theory of the concomitants of order statistics. Ann. Statist. *5, 996-1002.*

162. Yang, G, S. (1981). Linear functions of concomitants of order statistics with applications to non- parametric estimations of a regression function. J. Amer. Statist. Assoc. 76, 658-662.

Chapter 3

ORDER STATISTICS BASED ON CUMULATIVE SUMS

In this chapter we study asymptotic properties of order statistics based on cumulative sums of random variables.

Let $X_1, X_2,\dots$ be a sequence of random variables, $S_0 = 0$ and

$$S_n=X_1+\dots+X_n,\ n=1,2,\dots.$$

Let now

$$S_{0,n}\le S_{1,n}\le\dots\le S_{n,n}$$

be the ordered sums $S_0, S_1,\dots,S_n$. In particular,

$$S_{0,n}= \min\{S_0,S_1,\dots,S_n\}$$

and

$$S_{n,n}= \max\{S_0,S_1,\dots,S_n\}.$$

A great number of different theorems are known for maximal sums $S_{n,n}$. Due to the evident equality

$$\min\{0, X_1, X_1+X_2,\dots, X_1+\dots+X_n\}=-\max\{0, Y_1, Y_1+Y_2,\dots, Y_1+\dots+Y_n\},$$

where $Y_k= -X_k$, all these results can be reformulated for minimal order statistics $S_{0,n}$.

Three main types of methods to investigate maximal sums $S_{n,n}$ will be discussed below.

The first approach is based on the various forms of the Donsker- Prohorov invariance principle, which provides that asymptotic distributions of a wide class of functionals of the sums $S_0, S_1,\ldots, S_n$ coincide with the distributions of the corresponding functionals of some processes (Wiener, stable, Brownian bridge, etc). It is very important that asymptotic distributions of these functionals are the same under different conditions on the initial random variables $X_1, X_2,\ldots$. Hence, it suffices to find the limit distribution of a certain functional for the simplest case, say for X's taking two values or for X's having the standard normal distributions. Due to the invariance principle, the functional of the corresponding process has the same distribution. Moreover, the distribution obtained in the simplest case is obliged to remain asymptotic for the functional of $S_1, S_2,\ldots$ under more general conditions on X's.

The second approach is based on the Pollaczek- Spitzer identity. This identity expresses characteristic functions of order statistics $S_{n,n}$ via characteristic functions of random variables $S_n^+=\max\{0,S_n\}$:

$$\sum_{n=0}^{\infty} E\exp(itS_{n,n})z^n=\exp\{\sum_{n=1}^{\infty} E\exp(itS_n^+)z^n/n!\},\ |z|<1,\ -\infty<t<\infty.$$

One more approach uses direct probabilistic methods and the classical limit theorems for sums of independent random variables.

It turns out that one can apply all these methods to study asymptotic distributions of other elements of the variational series based on cumulative sums.

The structure of this chapter is the following: at first we will demonstrate how the mentioned methods work in the case of extremes $S_{0,n}$ and $S_{n,n}$ and then we will get some results for any order statistics $S_{k,n}$, $1<k<n$.

3.1. Distributions of Extreme Order Statistics

Invariance principle

We begin our consideration with the classical i.i.d. case. Let $X, X_1, X_2,\ldots$ be a sequence of independent identically distributed random variables,

$$S_0=0\ ,\ S_n=X_1+\ldots+X_n,\ S_{0,n}=\min\{S_0,S_1,\ldots,S_n\}$$

and

$$S_{n,n}=\max\{S_0,S_1,\ldots,S_n\},\ n=1,2,\ldots.$$

Let us propose additionally that $EX=0$ and $VarX=\sigma^2$, $0<\sigma<\infty$.

For any $n=1,2,\ldots$ we construct broken lines

$X_n(t)=(S_{[nt]}+(nt-[nt])X_{[nt]+1})/\sigma n^{1/2}$, $0\le t\le 1$.

It follows from the invariance principle (for details see, for example, Billingsley (1968)) that

$$X_n(t) \xrightarrow{D} W(t),\ 0\le t\le 1.$$

where $W(t)$ is a Wiener process. It means that for any continuous functional $\varphi(x(t))$ defined on the set $C[0,1]$ of continuous functions $x(t)$, $0\le t\le 1$, $P\{\varphi(X_n(t))<x\}\to P\{\varphi(W(t))<x\}$, as $n\to\infty$.

Amongst these functionals there are *sup-* and *inf-* functionals and we get the following useful relations:

$$P\{\sup_{0\le t\le 1} X_n(t)<x\}\to P\{\sup_{0\le t\le 1} W(t)<x\}$$

and

$$P\{\inf_{0\le t\le 1} X_n(t)<x\}\to P\{\inf_{0\le t\le 1} W(t)<x\},\ n\to\infty.$$

Moreover, analogous relations are valid for joint distributions of functionals, say,

$$P\{\sup_{0\le t\le 1} X_n(t)<x,\ \inf_{0\le t\le 1} X_n(t)<y,\ X_n(1)<z\}\to$$

$$P\{\sup_{0\le t\le 1} W(t)<x,\ \inf_{0\le t\le 1} W(t)<y,\ W(1)<z\}.$$

Observing now that

$$\sup_{0\le t\le 1} X_n(t)=S_{n,n},\ \inf_{0\le t\le 1} X_n(t)=S_{0,n}$$

and

$$X_n(1)=S_n,$$

we obtain that

$$P\{S_{n,n}<x\sigma\sqrt{n},\ S_{0,n}<y\sigma\sqrt{n},\ S_n<z\sigma\sqrt{n}\}\to$$

$$P\{\sup_{0\le t\le 1} W(t)<x,\ \inf_{0\le t\le 1} W(t)<y,\ W(1)<z\}\text{ as } n\to\infty.$$

Consider now independent random variables $X_1,X_2,\ldots$ with zero means and finite variances

σ_k^2= Var (X_k), k=1,2,….

Let

B_n^2=Var $(S_n)=\sigma_1^2+\ldots+\sigma_n^2$, n=1,2,…,

and let now $X_n(t)$ be a broken line, generated by points $(B_k^2/B_n^2,\ S_k/B_n)$. Note that the previous i.i.d. construction of $X_n(t)$ has used points $(k/n,\ S_k/n^{1/2})$. Prohorov (1956) proved that in this case the invariance principle is also valid and

$$X_n(t) \xrightarrow{D} W(t),\ 0\le t\le 1.$$

if a sequence S_n/B_n converges to the standard normal law, as $n\to\infty$. Note that sums S_n/B_n have asymptotic normal distribution if, for instance, the Lindeberg condition holds, i.e.

$$\frac{1}{B_n^2}\sum_{k=1}^{n}\int_{|x|>\varepsilon B_n} x^2 dF_k(x)\to 0,\ n\to\infty,$$

for any fixed positive ε, where F_k is a distribution function of the random variable X_k, k=1,2,….
Therefore if a sequence $X_1,X_2,\ldots$ satisfies the Lindeberg condition and $EX_k=0$, k=1,2,…, then

$$P\{S_{n,n}<xB_n,\ S_{0,n}<yB_n,\ S_n<zB_n\}\to$$

$$P\{\sup_{0\le t\le 1} W(t)<x,\ \inf_{0\le t\le 1} W(t)<y,\ W(1)<z\},\ \text{as } n\to\infty.$$

We observe that asymptotic distributions of random variables $S_{0,n}/B_n$ and $S_{n,n}/B_n$ for a wide set of sequences $X_1, X_2, \ldots$ do not depend on distributions of X's and coincide with distributions of the corresponding functionals of Wiener processes.

Analogous result is valid for stable processes (see, Heyde (1969)). Let independent identically distributed random variables $X_1, X_2,\ldots$ have such a distribution that S_n/b_n converges to a stable law with a characteristic function

$$f(t)=\exp\{-\lambda|t|^{\alpha}(1+i\beta t/|t|\ \mathrm{tg}(\pi\alpha/2))\},$$

where $\lambda>0$, $0<\alpha<2$ and $\beta=0$, if $\alpha=1$, $|\beta|<1$, if $0<\alpha<1$ and $|\beta|\le 1$, if $1<\alpha\le 2$.

Let also Y(t) be a stable process with independent stationary increments corresponding to the characteristic function f(t), Y(0)=0. Then

$$P\{S_{n,n}<xb_n, S_{0,n}<yb_n, S_n<zb_n\}\to$$

$$P\{\sup_{0\le t\le 1} Y(t)<x, \inf_{0\le t\le 1} Y(t)<y, Y(1)<z\}, \text{as } n\to\infty.$$

It is rather easy to find the distributions of *sup*- and *inf*- functionals for Wiener processes. Really, Wiener paths are continuous with probability 1 and if the event

$$\{\sup_{0\le t\le 1} W(t)\ge x\}$$

holds for some x>0 then there exists such a τ, $0<\tau\le 1$, that $W(\tau)=x$.

Since $W(1)-W(\tau)$ and $W(\tau)$ are independent and

$$P\{W(1)-W(\tau)\ge 0\}=P\{W(1)-W(\tau)\le 0\}$$

we get that

$$P\{W(1)\ge x\}=P\{W(1)-W(\tau)\ge 0\}=$$

$$P\{W(1)-W(\tau)\le 0\}=P\{\sup_{0\le t\le 1} W(t)\ge x, W(1)\le x\}.$$

Now for x>0 we have equalities

$$P\{\sup_{0\le t\le 1} W(t)\ge x\}=$$

$$P\{\sup_{0\le t\le 1} W(t)\ge x, W(1)\ge x\} + P\{\sup_{0\le t\le 1} W(t)\ge x, W(1)\le x\}=$$

$$P\{W(1)\ge x\}+ P\{\sup_{0\le t\le 1} W(t)\ge x, W(1)\le x\}=2P\{W(1)\ge x\}.$$

Since

$$P\{W(1)\ge x\}=1-\Phi(x),$$

where Φ is the distribution function of the standard normal law, we obtain that

$$P\{\sup_{0\le t\le 1} W(t)<x\}=2\Phi(x)-1, x\ge 0.$$

Evidently,

$$P\{\sup_{0\le t\le 1} W(t)<x\}=0, \text{ if } x<0,$$

and finally we have the following formula:

$$G(x)= P\{\sup_{0\le t\le 1} W(t)<x\}=\max\{0, 2\Phi(x)-1\}, -\infty<x<\infty.$$

In fact, we have used the so-called reflection principle: each path $W(t,\omega)$, $0\le t\le 1$, of a Wiener process such that $W(\tau,\omega)=x$ for some τ $(0<\tau<1)$ has its equiprobability twin path $W(t, \omega^*)$ such that

$$W(t, \omega^*)= W(t, \omega) \text{ for } 0<t\le\tau, \text{ and}$$

$$W(t, \omega^*)- W(\tau, \omega^*)=-(W(t, \omega)- W(\tau, \omega)) \text{ for } \tau<t\le 1.$$

Applying the same approach one can find that if $y<x$ and $x>0$ then

$$\Phi(y)=P\{W(1)<y\}=$$

$$P\{\sup_{0\le t\le 1} W(t)<x, W(1) <y\} +P\{\sup_{0\le t\le 1} W(t)\ge x, W(1) <y\}=$$

$$P\{\sup_{0\le t\le 1} W(t)<x, W(1) <y\}+ P\{\sup_{0\le t\le 1} W(t)\ge x, W(1) >2x-y\}.$$

Since $2x-y>x$, we get that

$$P\{\sup_{0\le t\le 1} W(t)\ge x, W(1) >2x-y\}=P\{W(1)>2x-y\}=1-\Phi(2x-y)$$

and by combining the latest equalities we come to the following relation:

$$P\{\sup_{0\le t\le 1} W(t)<x, W(1) <y\}=\Phi(y)+\Phi(2x-y)-1.$$

If $y>x\ge 0$, then

$$P\{\sup_{0\le t\le 1} W(t)<x, W(1) <y\}= P\{\sup_{0\le t\le 1} W(t)<x\}=2\Phi(x)-1.$$

Observing that $-W(t)$ also is a Wiener process one can see that

$$\inf_{0\le t\le 1} W(t) \stackrel{d}{=} -\sup_{0\le t\le 1} W(t).$$

This relation enables us to write that

$$P\{\inf_{0\le t\le 1} W(t)<x\}=1-G(-x)=\min\{2\Phi(x),1\}.$$

More complicated method also based on the reflection principle gives the joint distribution of

$$\inf_{0\le t\le 1} W(t)$$

and

$$\sup_{0\le t\le 1} W(t).$$

It appears (see , for example, Billingsley (1968) or Feller (1971)) that

$$P\{x<\inf_{0\le t\le 1} W(t) \le \sup_{0\le t\le 1} W(t)<y\}=$$

$$\sum_{k=-\infty}^{\infty} (-1)^k(\Phi(y+k(y-x))- \Phi(x+k(y-x)))$$

for $x<0<y$. It follows from this equality that

$$P\{\sup_{0\le t\le 1} |W(t)|<x\}$$

$$= P\{-x<\inf_{0\le t\le 1} W(t) \le \sup_{0\le t\le 1} W(t)<x\}$$

$$= \sum_{k=-\infty}^{\infty} (-1)^k(\Phi((2k+1)x))- \Phi((2k-1)x)) \text{ for } x>0.$$

The latest formula has another useful form:

$$P\{\sup_{0\le t\le 1} |W(t)|<x\}$$

$$= 1-\frac{4}{\pi}\sum_{k=1}^{\infty}\frac{(-1)^k}{2k+1}\exp\{-\pi^2(2k+1)^2/8x^2\}.$$

All formulae, which we have got here for Wiener processes, give us expressions for asymptotic distributions of suitable normalized extreme order statistics $S_{0,n}$ and $S_{n,n}$ and even for random variables

$$\max\{|S_0|,|S_1|,\ldots,|S_n|\}$$

in the situation when

$$EX_k = 0,\ k=1,2,\ldots,$$

and there exist such normalizing constants B_n, n=1,2,..., that sums S_n/B_n are asymptotically normal.

Sometimes we need to know distributions of maximal and minimal order statistics based on sums S_k, k=1,2,...,n, conditioned by the fixed value of the last sum S_n. In this case we can use the invariance principle with processes of Brownian bridge as limit. A Brownian bridge

$$W^a(t),\ 0\le t\le 1,$$

is a Wiener process W(t) conditioned by the event W(1)=a. Consider a Brownian bridge $W^0(t)$, which represents a Gaussian process satisfying the following conditions:
$W^0(0)= W^0(1)$, $EW^0(t)=0$

and

$$EW^0(s)W^0(t)=s(1-t),\ 0\le s\le t\le 1.$$

Note that the following useful representation connecting a Brownian bridge with a Wiener process is valid:

$$W^0(t) \overset{d}{=} W(t)-tW(1),\ 0\le t\le 1.$$

The structure of Brownian bridges enables us to use the following limit relation to find distributions of functionals $\varphi(W^0(t))$:

$$P\{\varphi(W^0(t))<x\}= \lim P\{\varphi(W(t))<x|0<W(t)<\varepsilon\},\ \text{as } \varepsilon\to 0.$$

Recall that

$$P\{\sup_{0\le t\le 1} W(t)<x,\ W(1)<y\}=\Phi(y)+\Phi(2x-y)-1.$$

For any $0<\varepsilon<x$, we get now that

$$P\{\sup_{0\le t\le 1} W(t)<x \mid 0<W(1)<\varepsilon\}$$

$$= P\{\sup_{0\le t\le 1} W(t)<x,\ 0<W(1)<\varepsilon\}/P\{0<W(1)<\varepsilon\}$$

$$= (\Phi(\varepsilon)+\Phi(2x-\varepsilon)-\Phi(0)-\Phi(2x))/\ (\Phi(\varepsilon)-\Phi(0))$$

and letting ε to zero we obtain that

$$P\{\sup_{0\le t\le 1} W^0(t)<x\}=1-\exp(-2x^2),\ x>0.$$

The similar arguments give the equality

$$P\{\sup_{0\le t\le 1} W^a(t)<x\}=1-\exp(-2x(x-a)),\ x>0,\ x>a.$$

for a Brownian bridge $W^a(t)$, corresponding to any value a..

For our collection we present two more formulae for a Brownian bridge $W^0(t)$:

$$P\{x<\inf_{0\le t\le 1} W^0(t)\le \sup_{0\le t\le 1} W^0(t)<y\}=$$

$$\sum_{k=-\infty}^{\infty} (\exp\{-2k^2(y-x)^2\}-\exp\{-2(y+k(y-x))^2\}),\ x<0<y,$$

and

$$P\{\sup_{0\le t\le 1} |W^0(t)|<x\}=1+2\sum_{k=1}^{\infty} (-1)^k\exp\{-2k^2x^2\},\ x>0.$$

Below we will also give some explicit expressions for distributions of the supremum and the infimum of stable processes.

Pollaczek- Spitzer's identity

We will prove the so-called Pollaczek- Spitzer identity and then it will be used to find asymptotic distributions of $S_{n,n}$ in the case when X's have a stable distribution.

Theorem 3.1.1.

For any $|z|<1$ and $-\infty<t<\infty$ the following equality holds:

$$\sum_{n=0}^{\infty} E\{\exp(itS_{n,n})\}z^n=\exp\{\sum_{n=1}^{\infty} E\exp\{it S_n^+\}z^n/n\}, \qquad (3.1.1)$$

where

$$S_n^+=\max(0,S_n).$$

Proof. We will apply the approach of Heinrich (1985) to prove this theorem. We need some auxiliary results.
Denote

$$S_{(k,n)}=\max(0,S_n\text{-}S_k).$$

Lemma 3.1.1.

The equality

$$P\{S_n^+<x,\ \max_{1\le j\le n-1} S_j\ge x\}=\sum_{k=1}^{n-1} P\{\max_{1\le j\le k} S_j+S_{(k,m)}<x,\ S_{(k,m)}\ge x\} \qquad (3.1.2)$$

holds for any $x>0$.

Evidently, for any $1\le k\le n-1$,

$$P\{S_n<x,\ \max_{1\le j\le n-1} S_j\ge x\}=\sum_{k=1}^{n-1} P\{\max_{0\le j\le k-1} S_{n-j}<x,\ S_{n-k}\ge x\}. \qquad (3.1.3)$$

Since random variables $X_1,X_2,\ldots,X_n$ are independent and identically distributed, the following relations hold

$$(X_1,X_2,\ldots,X_k,X_{k+1},\ldots,X_n)\overset{d}{=}(X_{n-k+1},\ldots,X_n,X_1,\ldots,X_{n-k})$$

and

$$(S_1,S_2,\ldots,S_k,\ S_n\text{-}S_k)\overset{d}{=}(X_{n-k+1},\ldots,X_{n-k+1}+\ldots+X_n,X_1+\ldots+X_{n-k}).$$

Then,

$$P\{\max_{0\le j\le k-1} S_{n-j}<x,\ S_{n-k}\ge x\}=$$

$$P\{\max\{S_{n-k+1},\ldots,S_n\}<x,\ X_1+\ldots+X_{n-k}\ge x\}=$$

$$P\{\max_{1\le j\le k} S_j+S_{(k,n)}<x,\ S_{(k,n)}\ge x\}$$

and (3.1.2) follows from (3.1.3).

Lemma 3.1.2.

For any $n\ge 1$,

$$nP\{S_{n,n}<x\}=\sum_{k=0}^{n-1} P\{S_{(k,k)}+S_{(k,n)}<x\}. \qquad (3.1.4)$$

It is evident, that (3.1.4) is valid for $x\le 0$. Consider $x>0$. We will prove (3.1.4) by induction. It is clear that this relation holds for $n=1$. Now let us suppose that (3.1.4) is true for $n=1,2,\ldots,m$ and show that then it holds for $n=m+1$. Observing that

$$(X_1,X_2,\ldots,X_m,X_{m+1})\overset{d}{=}(X_{m+1},X_1,\ldots,X_m),$$

one has for positive x the following equalities:

$$\begin{aligned}&mP\{S_{(m+1,m+1)}<x\}\\&=mP\{\max(X_1,X_1+X_2,\ldots,X_1+\ldots+X_{m+1})<x\}\\&=mP\{\max(X_2,X_2+X_3,\ldots,X_2+\ldots+X_{m+1})+X_1<x\}\\&=mP\{\max(X_1,X_1+X_2,\ldots,X_1+\ldots+X_m)+X_{m+1}<x\}.\end{aligned} \qquad (3.1.5)$$

By the induction assertion and from independence of $(X_1,\ldots,X_m)$ and X_{m+1}, one gets that

$$mP\{\max(X_1,X_1+X_2,\ldots,X_1+\ldots+X_m)+X_{m+1}<x\}=\sum_{k=0}^{m-1} P\{S_{k,k}+S_{(k,m)}+X_{m+1}<x\}. \qquad (3.1.6)$$

Now we can continue equalities as follows:

$$\sum_{k=0}^{m-1} P\{S_{k,k}+S_{(k,m)}+X_{m+1}<x\}$$

$$= \sum_{k=0}^{m-1} P\{\max(0, X_2, X_2+X_3, \ldots, X_2+\ldots+X_{k+1}) + \max(0, S_{m+1}-S_{k+1}) + X_1 < x\}$$

$$= \sum_{k=0}^{m-1} P\{\max(X_1, X_1+X_2, \ldots, X_1+\ldots+X_{k+1}) + \max(0, S_{m+1}-S_{k+1}) < x\}$$

$$= \sum_{k=0}^{m-1} P\{\max_{1\le j\le k+1} S_j + S_{(k+1,m+1)} < x\}. \tag{3.1.7}$$

On combining (3.1.6) and (3.1.7) we get that

$$(m+1)P\{S_{m+1,m+1} < x\} =$$

$$P\{S_{m+1,m+1} < x\} + \sum_{k=0}^{m-1} P\{\max_{1\le j\le k+1} S_j + S_{(k+1,m+1)} < x\} =$$

$$P\{S_{m+1,m+1} < x\} + \sum_{k=1}^{m} P\{\max_{1\le j\le k} S_j + S_{(k,m+1)} < x\}. \tag{3.1.8}$$

To complete the proof we have to show that

$$P\{S_{m+1,m+1} < x\} + \sum_{k=1}^{m} P\{\max_{1\le j\le k} S_j + S_{(k,m+1)} < x\} = \sum_{k=0}^{m} P\{S_{k,k} + S_{(k,m+1)} < x\}. \tag{3.1.9}$$

Note that

$$P\{S_{m+1,m+1} < x\}$$

$$= P\{\max(0, S_{m+1}) < x, \max_{1\le j\le m} S_j < x\}$$

$$= P\{\max(0, S_{m+1}) < x\} - P\{\max(0, S_{m+1}) < x, \max_{1\le j\le m} S_j \ge x\}$$

and

$$P\{\max_{1\le j\le k} S_j + S_{(k,m+1)} < x\} = P\{\max_{1\le j\le k} S_j + S_{(k,m+1)} < x, S_{(k,m+1)} \ge x\} + P\{S_{k,k} + S_{(k,m+1)} < x\}. \tag{3.1.10}$$

Now from (3.1.10) and the assertion of lemma 3.1.1 we get (3.1.9). It completes the proof of lemma 3.1.2.

Observing that $S_{(k,k)}$ and $S_{(k,n)}$ are independent we obtain from (3.1.4) that

$$nE\exp(itS_{n,n})$$

$$=\sum_{k=0}^{n-1} E\exp(itS_{k,k})E\exp(itS_{(k,n)})$$

$$=\sum_{k=0}^{n-1} E\exp(itS_{k,k})E\exp(it\,S^{+}_{n-k}) \tag{3.1.11}$$

for any n=1,2,.... Equalities (3.1.11) imply that

$$\sum_{n=0}^{\infty} nE\exp(itS_{n,n})z^{n-1}$$

$$=\sum_{n=1}^{\infty} nE\exp(itS_{n,n})z^{n-1}$$

$$=\sum_{n=1}^{\infty}\sum_{k=0}^{n-1} E\exp(itS_{k,k})E\exp(it\,S^{+}_{n-k})z^{n-1}$$

$$=\sum_{k=0}^{\infty} E\exp(itS_{k,k})\sum_{n=k+1}^{\infty} E\exp(it\,S^{+}_{n-k})z^{n-1}$$

$$=\sum_{k=0}^{\infty} E\exp(itS_{k,k})z^{k}\sum_{n=1}^{\infty} E\exp(it\,S^{+}_{n})z^{n-1}. \tag{3.1.12}$$

Denote

$$A(z)=\sum_{k=0}^{\infty} E\exp(itS_{k,k})z^{k}$$

and

$$B(z) = \sum_{n=1}^{\infty} \frac{1}{n} E\exp(itS_n^+)z^n.$$

Then (3.1.11) can be rewritten as

$$A'(z)=A(z)B'(z).$$

Since A(0)=1 and B(0)=0 we get that

$$A(z)=\exp\{B(z)\}.$$

The theorem is proved.

Remark 3.1.1.

Theorem 3.1.1 was proved for independent identically distributed random variables. Analyzing the proof one can see that really it is also valid for sequences of exchangeable X's and even for sequences of cyclically exchangeable random variables. Note that $X_1, X_2, \ldots$ are cyclically exchangeable if for any n=1,2,..., k=1,...,n and $x_1, x_2, \ldots \in R^1$

$$P\{X_1<x_1,\ldots, X_{n-k}<x_{n-k}, X_{n-k+1}<x_{n-k+1},\ldots, X_n<x_n\}$$

$$= P\{X_{k+1}<x_1,\ldots, X_n<x_{n-k}, X_1<x_{n-k+1},\ldots, X_k<x_n\}.$$

Now we will show how Theorem 3.1.1 can be used to find limit distributions of $S_{n,n}$ for the case when X's have a stable distribution. We will apply Darling's method (Darling (1956)) which was developed by Heyde (1969) and Nevzorov (1979). Darling has studied asymptotic distribution of maximal order statistics $S_{n,n}$ for symmetric stable laws.

Heyde and Nevzorov have investigated more general situations. For the sake of simplicity we will study symmetric stable laws only.

Let $X_1, X_2, \ldots$ be independent identically distributed random variables with a characteristic function

$$f_\gamma(t)=E\exp(itX_k)= \exp(-|t|^\gamma),\ 0<\gamma\le 2.$$

Note that for any n=1,2,... there exist density functions p_n of sums S_n and

$$p_n(x)=\frac{2}{\pi}\int_{-\infty}^{\infty} \exp(-itx-n|t|^\gamma)dt$$

$$= \frac{1}{\pi}\int_0^\infty \cot x \exp(-n\, t^\gamma)dt. \quad (3.1.13)$$

We will apply the Pollaczek- Spitzer identity in terms of Laplace transforms:

$$\sum_{n=0}^{\infty} E\{\exp(-sS_{n,n})\}z^n = \exp\{\sum_{n=1}^{\infty} E\exp\{-s\, S_n^+\}z^n/n\},\ s\geq 0,\ |z|<1. \quad (3.1.14)$$

Recalling that S_n also are symmetric random variables we get from (3.1.13) that

$$\sum_{n=0}^{\infty} E\{\exp(-sS_{n,n})\}z^n$$

$$= P\{S_n<0\}+\frac{1}{\pi}\int_0^\infty \exp(-sx)\int_0^\infty \cos tx \exp(-n\, t^\gamma)dtdx$$

$$= ½+\frac{1}{\pi}\int_0^\infty \exp(-n\, t^\gamma)\int_0^\infty \exp(-sx)\cos tx dx\, dt.$$

Since

$$\int_0^\infty \exp(-sx)\cos tx dx = s/(s^2+t^2),$$

we obtain that

$$E\exp(-s\, S_n^+)=1/2+\frac{1}{\pi}\int_0^\infty \exp(-ns^\gamma t^\gamma)/(1+t^2)\, dt. \quad (3.1.15)$$

Equalities (3.1.14) and (3.1.15) imply that

$$\sum_{n=0}^{\infty} E\{\exp(-sS_{n,n})\}z^n = \exp\{-\frac{1}{2}\ln(1-z)-\frac{1}{\pi}\int_0^\infty (\ln(1-z\exp -s^\gamma t^\gamma))/(1+t^2)\, dt\}$$

and

$$(1-z)\sum_{n=0}^{\infty} E\{\exp(-sS_{n,n})\}\, z^n = \exp\{-\frac{1}{\pi}\int_0^\infty \ln\frac{1-z\exp(-t^\gamma s^\gamma)}{1-z}\frac{dt}{1+t^2}\} \quad (3.1.16)$$

for $0 \le z < 1$ and $s \ge 0$.

It follows from (3.1.16) that

$$\lim_{z \uparrow 1} (1-z) \sum_{n=0}^{\infty} E\{\exp(-s(1-z)^{1/\gamma} S_{n,n})\} z^n$$

$$= \lim_{z \uparrow 1} \exp\{-\frac{1}{\pi} \int_0^{\infty} \ln \frac{1 - z\exp(-t^{\gamma}(1-z)s^{\gamma})}{1-z} \frac{dt}{1+t^2}\}$$

$$= \exp\{-\frac{1}{\pi} \int_0^{\infty} \ln(1+t^{\gamma}s^{\gamma}) \frac{dt}{1+t^2}\}. \qquad (3.1.17)$$

From the invariance principle for stable laws it is known that there exists a limit distribution function G_{γ} for $S_{n,n}/n^{1/\gamma}$ and

$$G_{\gamma}(x) \equiv \lim_{n \to \infty} P\{n^{-1/\gamma} S_{n,n} < x\} = P\{\sup_{0 \le t \le 1} Y_{\gamma}(t) < x\},$$

where $Y_{\gamma}(t)$ ($Y_{\gamma}(0)=0$) is a stable process with independent stationary increments corresponding to the characteristic function $f_{\gamma}(t) = \exp(-|t|^{\gamma})$.
Denote

$$\psi_{\gamma}(s) = \int_0^{\infty} \exp(-sx) d\, G_{\gamma}(x) = \lim_{n \to \infty} E\exp(-s\, S_{n,n}/n^{1/\gamma}).$$

For any $0<z<1$ introduce an integervalued random variable $N=N(z)$ which is independent with $X_1, X_2, \ldots$ and have the distribution

$$P\{N(z)=n\} = (1-z)z^n,\ n=0,1,\ldots.$$

Note that

$$\lim_{z \uparrow 1} P\{(1-z)N_z \le x\} = 1 - \lim_{z \uparrow 1} P\{(1-z)N_z > x\}$$

$$= 1 - \lim_{z \uparrow 1} P\{N_z > [x/(1-z)]\} = 1 - \lim_{z \uparrow 1} z^{1+[x/(1-z)]}$$

$$= 1 - \exp\{-x\},\ x>0.$$

Since

$$(1\text{-}z)\sum_{n=0}^{\infty} E\{\exp(-sS_{n,n})\}z^n = E\exp\{-S_{N,N}\},$$

$$P\{N\to\infty\}=1$$

and the limit distribution of (1-z)N is exponential, as $z\to 1$, we get now that

$$\lim_{z\uparrow 1}(1\text{-}z)\sum_{n=0}^{\infty} E\{\exp(-s(1\text{-}z)^{1/\gamma} S_{n,n})\}\, z^n$$

$$= \lim_{z\uparrow 1} E\{\exp(-s((1\text{-}z)N)^{1/\gamma} S_{N,N}\, N^{-1/\gamma})\}$$

$$= \int_0^{\infty} \exp(-x)\, \psi_\gamma\, (sx^{1/\gamma})dx. \tag{3.1.18}$$

Equalities (3.1.17) and (3.1.18) imply that

$$\int_0^{\infty} \exp(-x)\, \psi_\gamma\, (sx^{1/\gamma})dx = h_\gamma(s), \tag{3.1.19}$$

where

$$h_\gamma(s) = \exp\{-\frac{1}{\pi}\int_0^{\infty} \ln\,(1+t^\gamma s^\gamma)\, \frac{dt}{1+t^2}\}.$$

Thus we have got an integral equation for the Laplace transform of the limit distribution function G_γ. Darling has suggested to use the Mellin transform for extracting G_γ from (3.1.19). Let

$$\Im_\gamma(u) = \int_0^{\infty} s^{u-1} dG_\gamma(s)$$

and

$$H_\gamma(u) = \int_0^{\infty} s^{u-1} h_\gamma(s)ds$$

be correspondingly the Mellin- Stieltjes transform for distribution function G_γ and the Mellin transform for $h_\gamma(s)$. It follows from (3.1.19) that

$$H_\gamma(u)= \int_0^\infty s^{u-1}h_\gamma(s)ds$$

$$= \int_0^\infty s^{u-1}\int_0^\infty \exp(-x)\,(\int_0^\infty \exp(-sx^{1/\gamma}v)dG_\gamma(v))dxds$$

$$= \int_0^\infty dG_\gamma(v)\int_0^\infty \exp(-x)\,(\int_0^\infty s^{u-1}\exp(-sx^{1/\gamma}v)ds)dv$$

$$= \Gamma(u)\int_0^\infty dG_\gamma(v)\int_0^\infty \exp(-x)\,(x^{1/\gamma}v)^{-u}dx$$

$$= \Gamma(u)\,\Gamma(1-u/\gamma)\int_0^\infty v^{-u}dG_\gamma(v) = \Gamma(u)\,\Gamma(1-u/\gamma)\Im_\gamma(1-u)$$

and

$$\Im_\gamma(u)= H_\gamma(1-u)/\,\Gamma(1-u)\,\Gamma(1-1/\gamma+ u/\gamma). \tag{3.1.20}$$

One can see that $\Im_\gamma(u)$ is defined for such u that Re (1-u) is sufficiently close to zero. From (3.1.20) we get that

$$g_\gamma(x)= G_\gamma'(x)$$

exists and

$$g_\gamma(x) = \frac{1}{2\pi i}\int_{c-i\infty}^{c+i\infty} x^{-s}\Im_\gamma(s)ds$$

$$= \frac{1}{2\pi i}\int_{c-i\infty}^{c+i\infty}\frac{x^{-s}}{\Gamma(1-s)\Gamma(1-\frac{1-s}{\gamma})}\int_0^\infty z^{-s}\exp\{-\frac{1}{\pi}\int_0^\infty \log(1+z^\gamma y^\gamma)\,dy\}\,dzds.$$

It turns out that the RHS of the above formula can be simplified for $\gamma=1$ and $\gamma=2$.

Let $\gamma=2$. Then $E\exp(itX)=\exp\{-t^2\}$, i.e. X_1, X_2,... have the normal distributions with $EX_i=0$ and $DX_i=2$, i=1,2,.... In this case

$h_2(s)=\exp\{-I_2(s)/\pi\}$,

where

$$I_2(s) = \int_0^\infty \ln(1+t^2s^2)/(1+t^2)dt.$$

Since $I_2(0)=0$ and

$$I_2'(s)=\frac{2s}{s^2-1}\int_0^\infty \left(\frac{1}{1+t^2}-\frac{1}{1+s^2t^2}\right)dt=\pi/(1+s),$$

we get that

$I_2(s) = \pi\ln(1+s)$ and $h_2(s)=(1+s)^{-1}$.

Then

$$H_2(u) = \int_0^\infty s^{u-1}(1+s)^{-1}ds = \int_0^1 (1-x)^{u-1}x^{-u}dx = B(u,1-u)=\Gamma(u)\Gamma(1-u).$$

Recalling that

$$\Gamma(u)=2^{u-1}\Gamma(u/2)\,\Gamma(1/2+u/2)\,/\pi^{1/2}$$

one obtains that

$$\Im_\gamma(u) = \int_0^\infty s^{u-1}g_2(s)ds$$

$$= H_2(1-u)/\,\Gamma(1-u)\,\Gamma(1/2+u/2)\backslash$$

$$= \Gamma(u)/\,\Gamma(1/2+u/2)$$

$$= 2^{u-1}\Gamma(u/2)/\pi^{1/2}$$

$$= \frac{1}{\sqrt{\pi}}\int_0^\infty s^{u-1}\exp\{-s^2/4\}ds$$

and hence

$$g_2(x)=\frac{1}{\sqrt{\pi}}\exp\{-x^2/4\},\ x>0.$$

It implies that the limit distribution function for $S_{n,n}/n^{1/2}$ in the case, when X's have the standard normal distribution, is of the form

$$G(x) = \max\{0, 2\Phi(x)-1\}$$

and coincides with the expression for the distribution of the supremum of a Wiener process, which has been got earlier.

More complicated calculations have enabled Darling (1956) to obtain the following form of the density function $g_1(x)$ of the asymptotic distribution for $S_{n,n}/n$ in the case when

$$F(x)=\frac{1}{2}+\frac{arctg\ x}{\pi}.$$

It turns out that

$$g_1(x)=\frac{1}{\pi x^{1/2}(1+x^2)^{3/4}}\exp\{-\frac{1}{\pi}\int_0^x\frac{\log v}{1+v^2}dv\},\ x>0. \tag{3.1.21}$$

Nevzorov (1979) considered more general Cauchy distributions with

$$F(x)=\frac{1}{2}+\frac{arctg(x-a)}{\pi}$$

and he got that the limit density function $g_{1,a}$ for random variables $S_{n,n}/n$ is of the form

$$g_{1,a}(x)=\frac{x^{\delta-1}}{\pi(1+(x-a)^2)^{(1+\delta)/2}}(1+a^2)^{h(x)}\exp\{-\frac{1}{\pi}\int_0^x\frac{\ln v}{1+(v-a)^2}dv\}, \tag{3.1.22}$$

where

$$\delta=\frac{1}{2}+\frac{arctg\ a}{\pi},$$

and

$h(x) = (arctg\ (x-a)+arctg\ a)/2\pi$. Note that $g_{1,0}(x)$ coincides with $g_1(x)$.

Direct probabilistic methods

In some situations one can use direct probabilistic methods and the classical limit theorems for sums of independent random variables to find the asymptotic distributions of maximal order statistics.

There are some useful inequalities for $S_{n,n}$.

For example, for symmetric independent random variables $X_1, X_2, \ldots$ the following Levy's inequality holds for any x:

$$P\{S_{n,n} \geq x\} \leq 2P\{S_n \geq x\} . \tag{3.1.23}$$

For non-symmetric random variables we can apply an analogous inequality

$$P\{\max_{0 \leq k \leq n} (S_k - \mu_{1/2}(S_k - S_n)) \geq x\} \leq 2P\{S_n \geq x\} , \tag{3.1.24}$$

where $\mu_{1/2}(Y)$ denotes the median of a random variable Y. One more similar result requires the existence of the second moments of $X_1, X_2, \ldots$.

Kolmogorov inequality.

Let $X_1, X_2, \ldots$ be independent random variables and

$EX_k = 0$, $EX_k^2 < \infty$ for $k=1,2,\ldots$. Then for any x,

$$P\{S_{n,n} \geq x\} \leq 2P\{S_n \geq x - (2\sum_{k=1}^{n} EX_k^2)^{1/2} \}. \tag{3.1.25}$$

We will prove Petrov's inequality, which generalizes Levy's inequality. Let

$\mu_q(Y)$, $0<q<1$,

denote the quantile of order q of a random variable Y. It means that

$P\{Y \leq \mu_q(Y)\} \geq q$

and

$P\{Y \geq \mu_q(Y)\} \geq 1-q$.

Note that if $\mu_q(Y)$ is the quantile of order q of Y, then $-\mu_q(Y)$ coincides with $\mu_{1-q}(-Y)$ since

$P\{-Y\geq-\mu_q(Y)\}= P\{Y\leq\mu_q(Y)\} \geq q=1-(1-q)$

and

$P\{-Y\leq-\mu_q(Y)\} \geq 1-q.$

Theorem 3.1.2.

(Petrov (1975a)). For any n =1,2,... and 0<q<1, the following inequality holds:

$$P\{\max_{0\leq k\leq n} (S_k-\mu_q(S_k-S_n)) \geq x\}\leq\frac{1}{q}P\{S_n\geq x\} . \tag{3.1.26}$$

Proof.

Note that

$P\{S_n\geq-\mu_q(-S_n)\}=P\{S_n\geq\mu_{1-q}(S_n)\} \geq q$

and hence (3.1.26) holds for any

$x<-\mu_q(-S_n)= \mu_{1-q}(S_n).$

Moreover, if $x \geq-\mu_q(-S_n)$ then

$$P\{\max_{0\leq k\leq n} (S_k-\mu_q(S_k-S_n)) \geq x\}=P\{\max_{1\leq k\leq n} (S_k-\mu_q(S_k-S_n)) \geq x\} .$$

It means that instead of (3.1.26) it is sufficient to show that

$$P\{\max_{1\leq k\leq n} (S_k-\mu_q(S_k-S_n)) \geq x\}\leq\frac{1}{q}P\{S_n\geq x\} .$$

Let us introduce the following random variables and events:

$R_n=\max_{1\leq k\leq n} (S_k-\mu_q(S_k-S_n)), n=1,2,...,$

$A_1=\{S_1-\mu_q(S_1-S_n) \geq x \},$

$A_k=\{R_{k-1}<x, S_k-\mu_q(S_k-S_n) \geq x \}, k=2,...,n,$

$B_k=\{S_n-S_k-\mu_{1-q}(S_n-S_k) \geq 0\}, k=1,2,...,n.$

Note that

$$P\{A_iA_j\}=0$$

for any $i \neq j$ and A_k and B_k are independent for any $k=1,2,\ldots$.
One can see that

$$\{R_n \geq x\}=\bigcup_{k=1}^{n} A_k$$

and

$$P\{R_n \geq x\}=\sum_{k=1}^{n} P\{A_k\}.$$

Note also that

$$P\{B_k\}=P\{S_n-S_k \geq \mu_{1-q}(S_n-S_k)\} \geq q,\ k=1,2,\ldots,n.$$

It follows from the definition of A_k and B_k that

$$\bigcup_{k=1}^{n} A_k B_k \subset \{S_n \geq x\}$$

and then

$$P\{R_n \geq x\}=\sum_{k=1}^{n} P\{A_k\} \leq \frac{1}{q}\sum_{k=1}^{n} P\{A_k\}P\{B_k\} \leq \frac{1}{q}P\{S_n \geq x\}.$$

The theorem is proved.

Remark 3.1.2.

Inequalities (3.1.23) and (3.1.24) are the partial cases of the theorem 3.1.2 under $q=1/2$. The statement of theorem 3.1.2 enables us to obtain the following results (Nevzorov (1977)).

Theorem 3.1.3.

Let $X, X_1, X_2,\ldots$ be independent identically distributed random variables and let

$$EX=0$$

and

$\mathrm{Var}(X)=\sigma^2,\ 0<\sigma^2<\infty.$

Then

$$\sup_x |P\{\max_{0\le k\le n} (S_k+\mu_q(S_n-S_k)) < x\sigma n^{1/2}\}-G_q(x)|\to 0, \tag{3.1.27}$$

as $n\to\infty$, where

$$G_q(x)=\begin{cases} 0, & if \quad x<\Phi^{-1}(q) \\ \dfrac{\Phi(x)-q}{1-q}, & if \quad x\ge\Phi^{-1}(q) \end{cases},\ 0<q<1,$$

Φ is the distribution function of the standard normal law and Φ^{-1} is the inverse of Φ.

The next theorem is a generalization of Theorem 3.1.3. Consider a sequence of independent random variables $X_1, X_2,\ldots$ with zero means and finite variances $\sigma_1^2, \sigma_2^2,\ldots$. Denote

$$B_n^2=\sum_{k=1}^{n}\sigma_k^2$$

and for any $\varepsilon>0$ introduce the Lindeberg fraction

$$\Lambda_n(\varepsilon)=\frac{1}{B_n^2}\sum_{k=1}^{n}\int_{|x|>\varepsilon B_n} x^2 dP\{X_k<x\}.$$

Theorem 3.1.4.

If a sequence of independent random variables $X_1,X_2,\ldots$ satisfies the Lindeberg condition, i.e.

$\Lambda_n(\varepsilon)\to 0,\ n\to\infty,$

for any positive ε, then for any $0<q<1$,

$$\sup_x |P\{\max_{0\le k\le n} (S_k+\mu_q(S_n-S_k)) < xB_n\}-G_q(x)|\to 0,\ n\to\infty. \tag{3.1.28}$$

Since theorem 3.1.3 follows from (3.1.28) we will prove theorem 3.1.4 only.

Proof.

Denote

$$T_n = \max_{0 \le k \le n} (S_k + \mu_q(S_n - S_k)).$$

Recalling the statement of theorem 3.1.2 we get that

$$P\{T_n \ge x\} = P\{\max_{0 \le k \le n} (S_k - \mu_{1-q}(S_k - S_n)) \ge x\} \le \frac{1}{q} P\{S_n \ge x\}. \tag{3.1.29}$$

For any positive C introduce random variables

$$X_k^* = \begin{cases} X_k, & if \ |X_k| \le C \\ 0, & if \ |X_k| > C \end{cases}.$$

Let

$$S_0^* = 0,\ S_n^* = \sum_{k=1}^{n} X_k^*$$

and

$$T_n^* = \max_{0 \le k \le n} (S_k^* + \mu_q(S_n^* - S_k^*)).$$

It is easily seen that

$$|P\{S_n^* \ge x\} - P\{S_n \ge x\}| \le \sum_{k=1}^{n} P\{|X_k| > C\} \tag{3.1.30}$$

and

$$|P\{T_n^* \ge x\} - P\{T_n \ge x\}| \le \sum_{k=1}^{n} P\{|X_k| > C\}. \tag{3.1.31}$$

From the definition of quantiles it follows that

$$P\{ S_n^* - S_k^* > \mu_q(S_n^* - S_k^*)\} = 1 - P\{ S_n^* - S_k^* \le \mu_q(S_n^* - S_k^*)\} \le 1 - q. \tag{3.1.32}$$

Now, for

$X \geq \mu_q(S_n^*)$,

one gets that

$$P\{T_n^* \geq x\} = \sum_{k=1}^{n} P\{T_{k-1}^* < x, S_k^* + \mu_q(S_n^* - S_k^*) \geq x\} =$$

$$\sum_{k=1}^{n} P\{T_{k-1}^* < x, x+2C \geq S_k^* + \mu_q(S_n^* - S_k^*) \geq x\}. \qquad (3.1.33)$$

To verify that (3.1.33) is true we need to note that

$\mu_q(S_n^* - S_k^*) - \mu_q(S_n^* - S_{k-1}^*) \leq C$,

and hence, if the event

$\{S_{k-1}^* + \mu_q(S_n^* - S_{k-1}^*) < x\} \supset \{T_{k-1}^* < x\}$

happens, then

$S_k^* + \mu_q(S_n^* - S_k^*) =$

$S_{k-1}^* + \mu_q(S_n^* - S_{k-1}^*) + X_k^* + (\mu_q(S_n^* - S_k^*) - \mu_q(S_n^* - S_{k-1}^*)) \leq x + 2C.$

Relations (3.1.32) and (3.1.33) imply that for $x \geq \mu_q(S_n^*)$,

$P\{T_n^* \geq x\} \geq$

$$\frac{1}{1-q} \sum_{k=1}^{n} P\{T_{k-1}^* < x, x \leq S_k^* + \mu_q(S_n^* - S_k^*) \leq x+2C\} P\{S_n^* - S_k^* > \mu_q(S_n^* - S_k^*)\} \geq$$

$$\frac{1}{1-q} \sum_{k=1}^{n} P\{T_{k-1}^* < x, x \leq S_k^* + \mu_q(S_n^* - S_k^*) \leq x+2C, S_n^* > x+2C\} =$$

$$\frac{1}{1-q} P\{S_n^* > x+2C\}. \qquad (3.1.34)$$

If

$x<\mu_q(S_n^*)$,

then

$$P\{ T_n^* \geq x\}=1. \tag{3.1.35}$$

From inequalities (3.1.30), (3.1.31) and (3.1.34) we obtain that

$$P\{T_n \geq x\} \geq \frac{1}{1-q} P\{ S_n > x+2C\} - \frac{2-q}{1-q} \sum_{k=1}^{n} P\{|X_k|>C\} \tag{3.1.36}$$

for $x \geq \mu_q(S_n^*)$.

On combining inequalities (3.1.29) and (3.1.36) we get now that

$$\sup_{x \geq \mu_q(S_n^*)} |P\{T_n \geq x\} - \frac{1}{1-q} P\{ S_n \geq x\}| \leq \frac{1}{1-q} P\{x \leq S_n \leq x+2C\} + \frac{2-q}{1-q} \sum_{k=1}^{n} P\{|X_k|>C\}. \tag{3.1.37}$$

Since the Lindeberg condition holds,

$P\{ S_n \geq xB_n\} \to 1-\Phi(x)$,

and

$|P\{xB_n \leq S_n \leq xB_n+2C\} - \Phi(x+2C/B_n) + \Phi(x)| \to 0$,

as $n \to \infty$. Note also that

$$\sum_{k=1}^{n} P\{|X_k|>C\} \leq \frac{1}{C^2} \sum_{k=1}^{n} \int_{|x|>C} x^2 dP\{X_k<x\} = \Lambda_n(C/B_n).$$

Choosing

$C=C(n)=\varepsilon B_n$,

we can obtain the following estimate with absolute constant A:

$$\Phi(x+2C/B_n)-\Phi(x)+\sum_{k=1}^{n} P\{|X_k|>C\}\leq A(\varepsilon+\Lambda_n(\varepsilon)),$$

which is true for any positive ε. Since

$$\Lambda_n(\varepsilon)\to 0,\ n\to\infty,$$

for any fixed ε>0, we can get now that

$$|P\{T_n\geq xB_n\}-(1-\Phi(x))/(1-q)|\to 0,\ n\to\infty,$$

if

$$x\geq\mu_q(S_n^*)/B_n.$$

To complete the proof we need to show that

$$\mu_q(S_n^*)/B_n\to\Phi^{-1}(q),$$

as $n\to\infty$. Using the central limit theorem, which holds because the sequence $X_1,X_2,\ldots$ satisfies the Lindeberg condition, we easily get that

$$\mu_q(S_n^*)/B_n=\mu_q(S_n)/B_n+o(1)$$

and

$$\Phi(\mu_q(S_n^*)/B_n)\to q,\ n\to\infty.$$

These relations give us the desired statement.

Theorem 3.1.4 is proved.

Remark 3.1.3.

Note that

$$|\max_{0\leq k\leq n}(S_k+a(k,n))-\max_{0\leq k\leq n}(S_k+b(k,n))|\leq\max_{0\leq k\leq n}|b(k,n)-a(k,n)|.$$

Hence it follows that if we take

$$a(k,n)=\mu_q(S_n-S_k)+o(B_n),\ n\to\infty,$$

as the centering constants, then $G_q(x)$ remains the limit distribution function for

$$\max_{0\le k\le n} (S_k+a(k,n))/B_n$$

as well as for

$$\max_{0\le k\le n}(S_k+\mu_q(S_n-S_k))/B_n,$$

Consider, for example, conditions of Theorem 3.1.3, which is a partial case of Theorem 3.1.4. Under these conditions, from the central limit theorem we have that

$$|\Phi(\mu_q(S_n-S_k)/\sigma(n-k)^{1/2}) -q|=o(1),$$

as $n-k\to\infty$, and

$$\mu_q(S_n-S_k)=x_q\sigma(n-k)^{1/2}+o(n^{1/2}),n\to\infty, \qquad (3.1.38).$$

where

$$x_q=\Phi^{-1}(q).$$

From relation (3.1.38) and the Remark 3.1.3 (in our case $B_n=\sigma\, n^{1/2}$) we get the following result.

Theorem 3.1.5.

Let $X, X_1, X_2,\ldots$ be independent identically distributed random variables and let

$$EX=0$$

and

$$Var(X)=\sigma^2,\ 0<\sigma^2<\infty.$$

Then

$$\sup_x|P\{\max_{0\le k\le n} (S_k+x_q\sigma(n-k)^{1/2})<x\sigma n^{1/2}\}-G_q(x)|\to 0, \qquad (3.1.39)$$

where

$$x_q=\Phi^{-1}(q).$$

Remark 3.1.4.

We can rewrite (3.1.39) as follows:

$$\sup_x |P\{\max_{0\le k\le n} (\frac{S_k}{\sigma\sqrt{n}} + a\sqrt{1-\frac{k}{n}}) < x\} - \widehat{G}_a(x)| \to 0, \tag{3.1.40}$$

where

$$\widehat{G}_a(x) = \begin{cases} 0, & if \quad x < a \\ \dfrac{\Phi(x)-\Phi(a)}{1-\Phi(a)}, & if \quad x \ge a \end{cases} \tag{3.1.41}$$

and $-\infty<a<\infty$.

Remark 3.1.5.

From (3.1.40) and the invariance principle it follows that

$$P\{\sup_{0\le t\le 1} (W(t) + a\sqrt{1-t}) < x\} = \widehat{G}_a(x).$$

The family of distribution functions (3.1.41) includes the function

$$G(x) = \max\{0, 2\Phi(x)-1\},$$

which is limit for the suitably normalized order statistics $S_{n,n}$ in the case, when $EX_i=0$ and variances DX_i, $i=1,2,\ldots$, are finite. Taking $a=-\infty$ in (41), one gets the distribution function of the standard normal law $\Phi(x)$ as $\widehat{G}_{-\infty}(x)$. Below we will investigate such situations, when the asymptotic distribution function for maximal order statistics $S_{n,n}$ coincides with $\Phi(x)$.

Let now $X, X_1, X_2, \ldots$ be independent identically distributed random variables with $EX=a$. It is known (see, for example, Feller (1971), ch.XII), that if $a<0$ then sums S_n tend to $-\infty$, as $n\to\infty$, and

$$P\{\max(0,S_1,S_2,\ldots) < \infty\} = 1.$$

In this case the asymptotic distribution of $S_{n,n}$ coincides with the distribution of

$$M = \max\{0,S_1,S_2,\ldots\}$$

and

$$M(x)=P\{M<x\}$$

depends on a distribution function F of X. There are some methods to find M(x), each of them can be applied only for special types of distribution functions F. We mention here the following result given in Feller (1971), ch.XII.

Example 3.1.1.

Let X=Z+Y, where Z and Y are independent random variables with distribution functions

$$A(x)=\max\{0,1-\exp(-x/\lambda)\}$$

and B(x) correspondingly. Then

$$F(x)=P\{X<x\}=\int_{-\infty}^{\min(0,x)} A(x-u)dB(u).$$

Denote $\mu=EY$. Then $EX=\mu+\lambda$. It turns out that if $\mu+\lambda<0$ then

$$M(x)=P\{M<x\}=1-(1-\lambda\alpha)\exp\{-\alpha x\},\ x>0,$$

where α is the only positive root of the equation

$$\int_{-\infty}^{\infty} \exp(\alpha y)dF(y)=1.$$

Consider now the case a>0. In this situation the distribution of $S_{n,n}$ is close to the distribution of S_n. In fact, for any sequence of positive ε_n, n=1,2,...,

$$0\le P\{S_n<x\}-P\{S_{n,n}<x\}=P\{S_n<x,\ S_{n,n}\ge x\}=$$

$$P\{S_n<x-\varepsilon_n,\ S_{n,n}\ge x\}+P\{x-\varepsilon_n\le S_n<x,\ S_{n,n}\ge x\}\le$$

$$P\{S_{n,n}-S_n\ge\varepsilon_n\}+P\{x-\varepsilon_n\le S_n<x\}\le$$

$$P\{\max(0,-X_n,\ldots,-(X_n+\ldots+X_1))\ge\varepsilon_n\}+Q(S_n,\varepsilon_n), \tag{3.1.42}$$

where

$$Q(X,\varepsilon)=\sup_x P\{x\le X<x+\varepsilon\}$$

denotes the concentration function of a random variable X.

Let $Y_k=-X_k$. Then $EY_k=-a<0$ and

$$0\le\max\{0,-X_n,\dots,-X_n-\dots-X_1\}=\max\{0,Y_n,\dots,Y_1+\dots+Y_n\}.$$

We see that $\max\{0,Y_n,\dots,Y_1+\dots+Y_n\}$ has the same distribution as

$\max\{0,Y_1,\dots,Y_1+\dots+Y_n\}$, and $\max\{0,Y_1,\dots,Y_1+\dots+Y_n\}\le \mathcal{M}'$,

where

$$\mathcal{M}'=\max\{0,Y_1,Y_1+Y_2,\dots\}.$$

Note that $\mathcal{M}'$ (see the case a<0) is a proper random variable and

$$P\{\mathcal{M}'\ge\varepsilon_n\}\to 0,$$

if

$$\varepsilon_n\to\infty.$$

Now consider the second term on the RHS of (3.1.42). Petrov (1970) has proved that

$$Q(S_n,\varepsilon)\le L(\varepsilon+1)n^{-1/2}$$

for sums of independent identically distributed random variables, where L does not depend on ε and n.
Taking a sequence of positive ε_n in such a manner that

$$\varepsilon_n\to\infty$$

and

$$\varepsilon_n=o(n^{1/2}),\ n\to\infty,$$

we get from (3.1.42) that

$$\sup_x|P\{S_n<x\}-P\{S_{n,n}<x\}|\to 0,$$

as $n\to\infty$. It means that if $X_1,X_2,\dots$ are independent identically distributed random variables with positive expectations and there exist such constants a_n and b_n that

$$\sup_x | P\{S_n-a_n<xb_n\}-T(x)| \to 0, \tag{3.1.43}$$

for a certain limit distribution function T, then

$$\sup_x | P\{S_{n,n}-a_n<xb_n\}-T(x)| \to 0, \; n\to\infty. \tag{3.1.44}$$

As it is known from the central limit theorem, under the additional condition

$0<\sigma^2=DX<\infty$,

(3.1.43) can be rewritten as

$$\sup_x | P\{S_n-an<x\sigma n^{1/2}\}-\Phi(x)| \to 0.$$

The corresponding changing in (3.1.44) implies the following result.

Theorem 3.1.6

Let X, $X_1, X_2, \ldots$ be a sequence of independent identically distributed random variables, EX=a>0 and $0<\sigma^2=DX<\infty$. Then,

$$\sup_x | P\{S_{n,n}-an<x\sigma n^{1/2}\}-\Phi(x)| \to 0, \tag{3.1.45}$$

as $n\to\infty$.

Analogously, if X, X_1, $X_2, \ldots$ are independent identically distributed random variables, which belong to the domain of attraction of a stable distribution function G_α, $1<\alpha\leq 2$, that is

$$\sup_x | P\{S_n-a_n<xb_n\}- G_\alpha(x)| \to 0,$$

for suitably chosen constants a_n and b_n, and EX>0, then

$$\sup_x | P\{S_{n,n}-a_n<xb_n\}- G_\alpha(x)| \to 0, \; n\to\infty.$$

The similar arguments can be used for any sequence of random variables $X_1, X_2, \ldots$. If there exists such a sequence $\varepsilon_n>0$, n=1,2,…, that

$$\max\{0,-X_1,\ldots,-X_1-\ldots-X_n\}/\varepsilon_n \xrightarrow{p} 0$$

and

$$Q(S_n, \varepsilon_n) \to 0,\ n\to\infty,$$

then the following relation holds:

$$\sup_x |P\{S_n<x\}-P\{S_{n,n}<x\}| \to 0,\ n\to\infty. \tag{3.1.46}$$

For example, Nevzorov (1971a) has applied the analogous way to prove the following result.

Theorem 3.1.7.

Let $X_1,X_2,\ldots$ be a sequence of independent random variables with means $a_1,a_2,\ldots$ and finite variances σ_1^2, $\sigma_2^2,\ldots$. Denote

$$A_n=ES_n$$

and

$$B_n^2=\sigma_1^2+\ldots+\sigma_n^2,\ n=1,2,\ldots.$$

If the following conditions hold :

$$\sup_x |P\{S_n-A_n<xB_n\}-\Phi(x)\}| \to 0,\ n\to\infty, \tag{3.1.47}$$

$$n^{1/2} \min_{1\le k\le n} a_k \to \infty,\ n\to\infty, \tag{3.1.48}$$

$$dk\le B_k^2,\ \sigma_k^2\le D,\ k=1,2,\ldots, \tag{3.1.49}$$

where $d\le D$ are some positive constants, then

$$\sup_x |P\{S_{n,n}-A_n<xB_n\}-\Phi(x)\}| \to 0,\ n\to\infty. \tag{3.1.50}$$

Remark 3.1.6.

Theorem 3.1.6, which was formulated for identically distributed X's, follows from theorem 3.1.7. In (3.1.45) the centering constants A_n have the form

A_n=an,=1,2,....

The following counterexample shows that we can not change (3.1.48) , which presents a restriction on individual terms, by a more weak restriction (analogous to A_n=an) on moment characteristics of sums

$$\liminf A_n/n>0,\ n\to\infty. \tag{3.1.51}$$

Example.3.1.2.

Let $Y_1,Y_2,\ldots$ be independent random variables having the standard normal distribution. We construct a sequence $X_1,X_2,\ldots$ as follows:

$X_m=Y_m-m/2$,

if

$m=2^k$, k=1,2,...,

and

$X_m=Y_m+6$

for $m\neq 2^k$.

Since all X's are independent, have normal distributions and unit variances, (3.1.47) and (3.1.49) are valid with $B_n=n^{1/2}$ and d=D=1. It is easy to find that $A_n \geq n$ for any n=1,2,... . Really,

$$A_1=6>1,\ A_{2^n}=6(2^k-k)-\sum_{l=1}^{k}2^{l-1}=5\cdot 2^k-6k+1\geq 2^{k+1},\ k=1,2,...,$$

and if

$2^k\leq m<2^{k+1}$,

then

$A_m \geq A_{2^n} \geq 2^{k+1} > m.$

Now we will check that (3.1.50) fails in this situation. For any n=1,2,... we have equalities

$P\{ S_n - A_n < xB_n\} = \Phi(x).$

Since $S_{n,n} \geq S_{n-1}$, we get that

$P\{ S_{n,n} - A_n < xB_n\} \leq P\{ S_{n-1} - A_n < xB_n\} =$

$$P\left\{\frac{S_{n-1} - A_{n-1}}{B_{n-1}} < x\frac{B_n}{B_{n-1}} + \frac{a_n}{B_{n-1}}\right\} = \Phi\left(x\frac{B_n}{B_{n-1}} + \frac{a_n}{B_{n-1}}\right).$$

Let $n_k = 2^k$, k=1,2,.... Then,

$$P\{S_{n_k, n_k} - A_{n_k} < xB_{n_k}\} \leq \Phi\left(x\frac{B_{n_k}}{B_{n_k-1}} + \frac{a_{n_k}}{B_{n_k-1}}\right) =$$

$$\Phi\left(x\frac{2^{k/2}}{\sqrt{2^k - 1}} - \frac{2^{k-1}}{\sqrt{2^k - 1}}\right). \tag{3.1.52}$$

We get from (3.1.52) that for any fixed x, the RHS of (3.1.52) converges to zero, as $k \to \infty$, and (3.1.50) is not true.

Let us come back to sequences of i.i.d. random variables $X_1, X_2, \ldots$ with zero means and finite variances . From the results given above we can see that the form of asymptotic distributions of maximal sums

$$R_n = \max_{0 \leq k \leq n} (S_k + a(k,n))/\sigma n^{1/2}$$

depends essentially on centering constants a(k,n):
If

$a(k,n) = 0$

(or even if $a(k,n) = o(n^{1/2})$), then

$\lim P\{R_n<x\}=\max\{0,2\Phi(x)-1\}$, $n\to\infty$; (3.1.53)

a) If

$a(k,n)=dk$, $k=1,2,\dots,n$,

and d is a positive constant, then

$\lim P\{R_n<x+a(n,n)/\sigma n^{1/2}\}=\Phi(x)$, $n\to\infty$; (3.1.54)

(from theorem 3.1.7 one can see that (3.1.54) also holds for such a(k,n) that

$n^{1/2}\min_{1\le k\le n}(a(k,n)-a(k-1,n))\to\infty$, $n\to\infty$);

b) If

$a(k,n)= a\sigma\,(n-k)^{1/2}$,

then

$\lim P\{R_n<x\}=\widehat{G}_a(x)$, $n\to\infty$; (3.1.55)

where $\widehat{G}_a$ is defined in (3.1.40).

Besides (3.1.53)-(3.1.55) there is one more situation, in a certain sense intermediate between (3.1.53) and (3.1.54), when the limit distribution for maximal sum can be expressed in an explicit form (see, for example, Erdos and Kac (1946) and Chung (1948)):

c) If

$a(k,n)=d\sigma k/n^{1/2}+o(n^{1/2})$,

where d is a positive constant, then

$$\lim P\{R_n<x\}=H_d(x)=\int_{d^2/2}^{\infty}\frac{xd\exp(xd)}{2\pi^{1/2}\,t^{3/2}}\ \exp\{-\frac{x^2d^2}{4t}-t\}\,dt,\ n\to\infty. \quad (3.1.56)$$

It follows from (3.1.56) and the invariance principle that

$H_d(x)= P\{\sup_{0\le t\le 1}(W(t)+dt)<x\}$ (3.1.57)

for any positive d.

Other types of normalized constants for sums

In (3.1.53)-(3.1.57) as well as in (3.1.21) and (3.1.22) we listed explicit expressions (for different centering constants a(k,n)) of asymptotic distributions of maximal sums

$$\max_{0\le k\le n} (S_k+a(k,n))$$

normed by constants b_n , n=1,2,..., which depend on the number n of considered sums. In fact, $b_n=\sigma n^{1/2}$ in (3.1.53)-(3.1.57) and $b_n=n$ in (3.1.21) and (3.1.22). There are some results giving limit distributions of

$$\max_{0\le k\le n} S_k/b_k,$$

where each sum S_k is normed by its own normalizing constant. Darling and Erdos (1956) have obtained the first result of such a type.

Theorem 3.1.8.

Let X, $X_1,X_2,\ldots$ be a sequence of independent identically distributed random variables,
$EX=0$, $Var(X)==\sigma^2$, $0<\sigma<\infty$,

and

$E|X|^3<\infty$.

Then

$$P\{\max_{0\le k\le n} S_k/k^{1/2}<a_n+b_n/a_n+x/b_n\}\to\exp(-e^{-x}/2\pi^{1/2}),\ n\to\infty, \tag{3.1.58}$$

where

$a_n=(2\ln\ln n)^{1/2}$

and

$b_n=\frac{1}{2}\ln\ln\ln n.$

Recently this result has been extended (Bertoin (1998)) in respect of stable laws of index $\alpha\in(0,1)\cup(1,2)$. Bertoin supposes that

$$1-F(x)\sim cx^{-\alpha}$$

and

$$F(-x)=O(x^{-\alpha}),$$

as $x\to\infty$, c being a positive constant and $\alpha\in(0,1)\cup(1,2)$. The additional restriction EX=0 is assumed for $\alpha\in(1,2)$. Then the "stable" version of the Darling-Erdos theorem is given in the following form:

$$P\{\max_{0\le k\le n} S_k/k^{1/\alpha}<x(\ln n)^{1/\alpha}\}\to\exp(-cx^{-\alpha}), x>0,\ n\to\infty. \tag{3.1.59}$$

In fact , (3.1.59) means that in the given situation random variables

$$\max_{0\le k\le n} S_k/k^{1/\alpha}$$

and

$$n^{-1/\alpha}\max_{0\le k\le n} X_k$$

have the same asymptotic distributions. One can see that (3.1.58) and (3.1.59) cover all values $\alpha\in(0,2)$, except $\alpha=1$. The case $\alpha=1$ means the investigation of the limit distribution of random variables

$$V_n=\max_{0\le k\le n} S_k/k$$

for Cauchy distribution. Some results for V_n in a more general situation have been obtained in Nevzorov (1987). Below we will give a more detailed exposition of results for V_n, that is for maximum values in a sequence of sample means, but now , to complete Bertoin's equality (3.1.59) we present one partial case (Cauchy distribution) of Nevzorov's result.

It turns out that if

$$F(x)=\frac{1}{2}+\frac{1}{\pi}\ \text{arctg}\ (x-a),\ -\infty<a<\infty, \tag{3.1.60}$$

then

$$P\{V_n<x\}\sim n^{F(x)-1}/\Gamma(F(x)),\ -\infty<x<\infty, \tag{3.1.61}$$

where $\Gamma(s)$ is the gamma function, and

$$P\{V_n/\ln n<x\}\to \exp(-1/\pi x),\ x>0,\ n\to\infty. \tag{3.1.62}$$

It is evident that

$$1-F(x) \sim 1/\pi x,\ x\to\infty,$$

for the stable distribution (3.1.60), and one can see now that (3.1.62) corresponds to (3.1.59), with $\alpha=1$ and $c=1/\pi$.

As it is known, the fraction S_k/k has the same distribution as X_k for $F(x)$ given in (3.1.60). Of course, random variables S_k/k, $k=1,2,\ldots$, are dependent. It is interesting to compare the asymptotic distributions of

$$V_n= \max_{0\le k\le n} S_k/k$$

and

$$M(n)=\max\{0, X_1,\ldots,X_n\}$$

for Cauchy distributions (3.1.60). It is known in this case (see, for example, Galambos (1985)) that

$$P\{M(n)/n<x\}\to \exp(-1/\pi x),\ x>0,\ n\to\infty. \tag{3.1.63}$$

It follows from (3.1.62) and (3.1.63) that random variables $V_n/\ln n$ and $M(n)/n$ have the same asymptotic distribution and it does not depend on the location parameter a. Note that the analogous distribution for $S_{n,n}/n$, where

$$S_{n,n} =\max_{0\le k\le n} S_k ,$$

given in (3.1.21) and (3.1.22) does depend on a.

Now we will show how to find the limit distribution of V_n in the general case. Looking ahead we can observe that all limit distributions for normalized random variables

$$\max_{1\le k\le n} S_k/k ,$$

which we are going to get, have positive supports. Hence , the asymptotic distribution are the same for

$$\max_{1\le k\le n} S_k/k$$

and

$$\max_{0\le k\le n} S_k/k .$$

Therefore, for the sake of simplicity in the sequel we will propose that

$$V_n = \max_{1\le k\le n} S_k/k.$$

First of all we shall give relations making it possible to link the random variables V_n with variables whose asymptotic behavior has been fairly well studied. To this end , along with $X_1, X_2, \ldots$ introduce the random variables

$$Y_k(x)=X_k-x,\ -\infty<x<\infty,\ k=1,2,\ldots,$$

and sums

$$Z_n(x)=Y_1(x)+\ldots+Y_n(x).$$

Then we have

$$P\{ V_n\le x\}=P\{ \max_{1\le k\le n} Z_k(x) \le 0\}=P\{T(x)>n\}, \tag{3.1.64}$$

where

$$T(x)=\min\{k: Z_k(x)>0\}$$

is the first time when the random walk generated by a sequence of independent identically distributed random variables $Y_1(x), Y_2(x), \ldots$ reaches the positive axis. For each fixed x we are dealing with random variables T(x), which have been investigated by many authors (see, for example, Doney (1982), Feller (1971), Rogozin (1971), Spitzer (1974)). Using results of the listed papers and viewing x as a parameter we can derive some useful statements about the asymptotic behavior of V_n.

Theorem 3.1.9.

Let $X_1, X_2, \ldots$ be a sequence of independent identically distributed random variables. Then

$$H(x)=\lim_{n\to\infty} P\{V_n\le x\}=\exp\left(-\sum_{n=1}^{\infty}\frac{1}{n}P\{S_n>xn\}\right).$$

Remark 3.1.7.

Note that the series

$$\sum_{n=1}^{\infty}\frac{1}{n}P\{S_n>xn\}$$

converges for all $x>a$, if $E|X_1|<\infty$ and $EX_1\le a$. This result follows for instance from the results of Spitzer (1956) or Baum and Katz (1965) (see also Petrov (1975b), ch.9).

Proof of Theorem 3.1.9.

Denote

$$\tau_n=P\{S_1\le 0, \ldots, S_{n-1}\le 0, S_n>0\},\ n=1,2,\ldots,$$

and

$$\tau(s)=\sum_{n=1}^{\infty}\tau_n s^n,\ 0\le s\le 1.$$

Then (see Feller (1971), ch. XII, theorem 1),

$$\exp\left(-\sum_{n=1}^{\infty}\frac{s^n}{n}P\{S_n>0\}\right)=1-\tau(s). \tag{3.1.65}$$

Let

$$\tau_x(s)=\sum_{n=1}^{\infty}P\{T(x)=n\}s^n.$$

We see that $\tau_x(s)$ is the analog of $\tau(s)$ for random sums

$Z_n(x)=Y_1(x)+\ldots+Y_n(x)=S_n-nx,$

that is

$$\tau_x(s)=\sum_{n=1}^{\infty} P\{Z_1(x)\leq 0, \ldots, Z_{n-1}(x)\leq 0, Z_n(x)>0\}=\sum_{n=1}^{\infty} P\{V_{n-1}\leq x, V_n>x\}s^n. \qquad (3.1.66)$$

It follows from (3.1.65) that

$$1-\tau_x(s)=\exp(-\sum_{n=1}^{\infty}\frac{s^n}{n} P\{Z_n>0\})=\exp(-\sum_{n=1}^{\infty}\frac{s^n}{n} P\{S_n>nx\}). \qquad (3.1.67)$$

Since

$$1-\tau_x(1)=P\{T(x)=\infty\}=\lim_{n\to\infty} P\{V_n\leq x\},$$

letting s to one in (3.1.67), we obtain the statement of the theorem.

Remark 3.1.8.

To prove (3.1.62) for distribution function (3.1.60) we can use the following equality, which is close to (3.1.65):

$$\exp(\sum_{n=1}^{\infty}\frac{s^n}{n} P\{S_n\leq 0\})=1+\sum_{n=1}^{\infty} P\{S_1\leq 0, \ldots, S_{n-1}\leq 0, S_n\leq 0\}s^n. \qquad (3.1.68).$$

To prove (3.1.68) we must recall relation (3.1.14):

$$\sum_{n=0}^{\infty} E\{\exp(-sS_{n,n})\}z^n=\exp\{\sum_{n=1}^{\infty} E\exp\{-s S_n^+\}z^n/n\},\ s\geq 0,\ |z|<1.$$

Evidently, the latest equality, if s converges to ∞, transforms as follows:

$$\sum_{n=0}^{\infty} P\{S_{n,n}=0\}z^n=\exp\{\sum_{n=1}^{\infty} P\{S_n^+=0\}z^n/n\},\ |z|<1.$$

Using the evident equalities

$$P\{S_{0,0}=0\}=1,$$

$$P\{S_{n,n}=0\}= P\{S_1\leq 0, \ldots, S_{n-1}\leq 0, S_n\leq 0\}$$

and

$P\{S_n^+=0\}=P\{S_n\le 0\}$, n=1,2,...,

we come to (3.1.68).
In fact, if we rewrite (3.1.68) for random sums Z_n, we get

$$1+\sum_{n=1}^{\infty} P\{V_n\le x\}s^n=\exp\left(\sum_{n=1}^{\infty}\frac{s^n}{n}P\{Z_n\le 0\}\right)=\exp\left(\sum_{n=1}^{\infty}\frac{s^n}{n}P\{S_n\le nx\}\right). \tag{3.1.69}$$

For Cauchy distribution (3.1.60) one obtains that

$$P\{S_n\le nx\}=P\{X_1\le x\}=F(x)=\frac{1}{2}+\frac{1}{\pi}\ \mathrm{arctg}\ (x-a).$$

It implies that

$$1+\sum_{n=1}^{\infty} P\{V_n\le x\}s^n=(1-s)^{-F(x)}. \tag{3.1.70}$$

On expanding the RHS of (3.1.70) in powers of s, we find that

$$P\{V_n\le x\}=(1-\gamma)(1-\gamma/2)\ldots(1-\gamma/n)=\exp\left\{\sum_{k=1}^{n}\ln(1-\gamma/k)\right\}, \tag{3.1.71}$$

where

$$\gamma=1-F(x)\sim 1/x\pi,\ x\to\infty.$$

From (3.1.71), taking

$$x=y\ln n,\ y>0,$$

we obtain the desired relation

$$P\{V_n/\ln n<y\}\to\exp(-1/\pi y),\ x>0,$$

as $n\to\infty$.
Consider now the following example, which illustrates the statement of Theorem 3.1.9.

Example 3.1.3.

Let

$$F(x)=\Phi(x)=\int_{-\infty}^{x} \varphi(t)dt ,$$

where

$$\varphi(t)=\frac{1}{\sqrt{2\pi}} \exp(-t^2/2),$$

that is the random variables $X_1,X_2,\ldots$ have the standard normal distribution. It turns out that in this case the rate of decreasing of the right tail of the limit distribution function

$$H(x)= \exp(-\sum_{n=1}^{\infty}\frac{1}{n}P\{S_n>xn\})$$

for large x is determined only by the first summand in the sum

$$\Sigma=\sum_{n=1}^{\infty}\frac{1}{n} P\{S_n>xn\}.$$

Evidently,

$$\Sigma=1-\Phi(x)+(1-\Phi(x\sqrt{2}))/2+\sum_{n=3}^{\infty}\frac{1}{n}(1-\Phi(x\sqrt{n})).$$

Since for any positive x,

$$1-\Phi(x) \leq\varphi(x)/x, \qquad (3.1.72)$$

the following inequalities are valid if $x\geq1$:

$$0 \leq\sum_{n=3}^{\infty}\frac{1}{n}(1-\Phi(x\sqrt{n}))\leq \int_{2}^{\infty}\frac{1}{t} (1-\Phi(x\sqrt{t}))dt\backslash$$

$$=2 \int_{x\sqrt{2}}^{\infty}\frac{1}{v}(1-\Phi(v))dv$$

$$\leq2 \int_{x\sqrt{2}}^{\infty}\frac{1}{v^2}\varphi(v)dv$$

$$\leq \frac{1}{x^2} \int_{x\sqrt{2}}^{\infty} \varphi(v)dv \leq 1-\Phi(x\sqrt{2}). \tag{3.1.73}$$

From (3.1.73) we obtain the relation

$$1-\Phi(x) \leq \Sigma \leq 1-\Phi(x)+\frac{3}{2}(1-\Phi(x\sqrt{2})),$$

which leads to the following inequalities:

$$1-\Phi(x)-\frac{1}{2}(1-\Phi(x))^2 \leq 1-H(x)=1-\exp(-\Sigma) \leq 1-\Phi(x)+\frac{3}{2}(1-\Phi(x\sqrt{2})), \tag{3.1.74}$$

if $x \geq 1$. Note that

$$1-\Phi(x\sqrt{2}) = \int_{x\sqrt{2}}^{\infty} \varphi(v)dv = \sqrt{2} \int_{x}^{\infty} \varphi(t\sqrt{2})dt$$

$$= \sqrt{2} \int_{x}^{\infty} \exp(-t^2/2)\varphi(t)dt$$

$$\leq \exp(-x^2/2)\sqrt{2} \int_{x}^{\infty} \varphi(t)dt$$

$$= \exp(-x^2/2)\sqrt{2}(1-\Phi(x)), \tag{3.1.75}$$

for any positive x. Due to (3.1.72), (3.1.74) and (3.1.75) we get the final estimate:

$$\left| \frac{1-H(x)}{1-\Phi(x)} - 1 \right| \leq \frac{3}{\sqrt{2}} \exp\{-x^2/2\}. \tag{3.1.76}$$

From (3.1.76) we find that

$$1-H(x) \sim 1-\Phi(x), \ x \to \infty.$$

Remark 3.1.9.

We have mentioned above (see relation (3.1.61)) that
$P\{V_n<x\} \sim n^{F(x)-1}/\Gamma(F(x)), -\infty<x<\infty,$

if

$$F(x)=\frac{1}{2}+\frac{1}{\pi}\text{ arctg }(x-a),\ -\infty<a<\infty.$$

This result has been obtained in Nevzorov (1987). We will formulate (without proving) one more result for stable distributions from the same paper.

Assertion. Let $X_1,X_2,\ldots$ be independent strictly stable random variables with an exponent $\alpha<1$ and a distribution function F. Denote

$$\delta=P\{X_1<0\}$$

and

$$C(x)=\sum_{n=1}^{\infty}\frac{1}{n}(F(0)-F(xn^{1-1/\alpha})),\ -\infty<x<\infty.$$

From the boundedness of densities of stable distributions, it follows that

$$-\infty<C(x)<\infty.$$

It turns out that if $0<\delta<1$, then

$$V_n(x)\sim\exp\{-C(x)\}/\,\Gamma(\delta)n^{1-\delta},\ n\to\infty, \text{ for each fixed x.}$$

Convergence rates to limit distributions of maximal order statistics

There is a rich history of results, which give estimates of convergence rate in limit theorems for $S_{n,n}$. Almost all of such results consider one of the following two principal situations:

a) Distribution functions of maximal sums of centered X's converge to the distribution function $G(x)=\max\{0, 2\Phi(x)-1\}$;
b) Distribution functions of maximal sums of shifted to the right random variables X_1, $X_2,\ldots$ converge to the standard normal distribution function $\Phi(x)$.

Let $X_1,X_2,\ldots$ be independent random variables with zero expectations, variances $\sigma_1^2,\sigma_2^2,\ldots$ and let moments

$$\gamma_{p,k}=E|X_k|^p,\ k=1,2,\ldots,$$

be finite under some p, where $2<p\leq 3$. Denote

$$B_n^2=\sigma_1^2+\ldots+\sigma_n^2,$$

$$\Gamma_{p,n}=\gamma_{p,1}+\ldots+\gamma_{p,n}$$

and introduce

$$L_{n,p}=\Gamma_{p,n}/B_n^p,$$

which is called the Lyapunov fraction of order p.

Let

$$\Delta_n(x)=|P\{S_{n,n}<xB_n\}-G(x)|.$$

For identically distributed X's, Erdos and Kac (1946) have showed that

$$\Delta_n(x)\to 0,$$

as $n\to\infty$. Two years later, in more general situation, when

$$\sigma_k^2=\sigma^2>0,\ k=1,2,\ldots,$$

and

$$\gamma_{3,k}=O(1),\ k\to\infty,$$

Chung (1948) obtained that

$$\sup_x \Delta_n(x)=O(n^{-1/26}\log n),\ n\to\infty. \tag{3.1.77}$$

He has used the multidimensional central limit theorem to prove (3.1.77).

Prohorov (1956) got an estimate of convergence rate in the invariance principle, from which he proved that

$$\sup_x \Delta_n(x)=O((L_{n,3})^{1/4}\log^2 L_{n,3}),\ n\to\infty, \tag{3.1.78}$$

if

$$\gamma_{3,k}<\infty$$

for any $k=1,2,\ldots$.

Sawyer (1968) improved (3.1.78), applying a truncation of X's on the suitable levels. He has obtained that for any p>2,

$$\sup_x \Delta_n(x) \leq 300(L_{n,p})^{1/(p+1)}, \qquad (3.1.79)$$

if

$\gamma_{p,k} < \infty$, k=1,2,....

Nevzorov (1973) increased the exponent on the RHS of (3.1.79) and got the first nonuniform estimate for $\Delta_n(x)$:

$$\Delta_n(x) \leq C\max\{L_{n,p}, (L_{n,p})^{1/p}\}/(1+|x|^p), \qquad (3.1.80)$$

C being an absolute constant, $2<p\leq 3$.
Meanwhile, Nagaev (1970) has found that for identically distributed X's with finite third moments the following estimate is valid :

$$\sup_x \Delta_n(x) \leq C(\gamma_3/\sigma^3)^2 n^{-1/2}, \qquad (3.1.81)$$

where

$\gamma_3 = E|X|^3$

and
$\sigma^2 = Var(X)$.

As it is known, the optimal estimates in the central limit theorem for sums S_n of independent identically distributed X's with finite moments $E|X|^3$ has the order $n^{-1/2}$. Hence, we can expect that (3.1.81) is optimal in order of n. To justify this fact, we can recall equality (3.1.68):

$$\exp\left(\sum_{n=1}^{\infty} \frac{s^n}{n} P\{S_n \leq 0\}\right) = 1 + \sum_{n=1}^{\infty} P\{S_1 \leq 0, \ldots, S_{n-1} \leq 0, S_n \leq 0\} s^n.$$

If we consider independent symmetric random variables $X_1, X_2, \ldots$, with a continuous distribution function F, then the sums S_n also are symmetric and

$$P\{S_n \leq 0\} = 1/2$$

for any n=1,2,....We get in this case that the LHS of (3.1.68) coincides with

$(1-s)^{-1/2}$

and

$P\{S_1\le 0, \ldots, S_{n-1}\le 0, S_n\le 0\}=(2n)!/(n!)^2 2^{2n}$, n=1,2,....

Applying Stirling's approximation we obtain that

$P\{S_{n,n}\le 0\}=P\{S_1\le 0, \ldots, S_{n-1}\le 0, S_n\le 0\}=$

$$\frac{1}{\sqrt{\pi n}}\left(1-\frac{1}{12n}+O(n^{-2})\right), \tag{3.1.82}$$

as $n\to\infty$. Since G(0)=0, we can state that the order $n^{-1/2}$ can not be improved in (3.1.81).

The classical central limit theorem for sums of nonidentically distributed random variables provides an estimate, in which the Lyapunov fraction $L_{n,p}$ has the exponent one. If we consider the more interesting and natural case $L_{n,p}<1$, then the objective exponent of $L_{n,p}$ on the RHS of (3.1.80) is $1/p<1/2$. The question arises whether it is possible to obtain an estimate for $\Delta_n(x)$ analogous to the corresponding result for sums of independent nonidentically distributed random variables and, if not, what is the best estimate for $\Delta_n(x)$ in terms of the Lyapunov fraction. The answer was given by Arak and Nevzorov (1973). For any $2<p\le 3$ they have constructed sequences of independent symmetric random variables $X_1, X_2,\ldots$ for which the relation

$$\sup_x \Delta_n(x)=o((L_{n,p})^{1/p}),\ n\to\infty, \tag{3.1.83}$$

does not hold. It follows from (3.1.80) and (3.1.83) that

$$\sup_x \Delta_n(x) \le C(L_{n,p})^{1/p}, \tag{3.1.84}$$

where C is an absolute constant, is the optimal estimate in terms of the Lyapanov fraction of order p. Moreover, even for symmetric X's this estimate can not be improved.

Now we see that there is a discordance between estimates for the i.i.d. case and the non-i.i.d. case. Take for simplicity p=3. Then in the situation , when $X_1,X_2,\ldots$ have the same distribution with zero expectations, variances $Var(X_k)=\sigma^2$ and finite moments

$E|X_k|^3=\gamma_3$, k=1,2,...,

(3.1.84) converts to the following inequality:

$$\sup_x \Delta_n(x) \leq C(\gamma_3/\sigma^3)^{1/3} n^{-1/6}. \qquad (3.1.85)$$

On comparing (3.1.85) with (3.1.81) one can see that the optimal (in some sense) estimate for non-i.i.d. random variables becomes bad for i.i.d. X's. Note that in the classical limit theorem for sums, the Lyapunov inequality (optimal for non-i.i.d. random variables) includes, as a partial case, the Berry Esseen estimate for identically distributed random variables. The explanation of such an illusory contradiction is simple: $S_{n,n}$, unlike S_n, is not symmetric with respect to random variables $X_1, X_2, \ldots, X_n$, and good estimates must take into account the different contribution of X's in the generation of $S_{n,n}$. Arak (1974) got the following result.

Theorem 3.1.10.

Let X_1, X_2,... be independent random variables with zero expectations, variances $\sigma_1^2, \sigma_2^2, \ldots$ and finite moments

$$\gamma_{3,k} = E|X_k|^3, \; k=1,2,\ldots.$$

Then for any n=1,2,... and x>0,

$$\Delta_n(x) \leq C \sum_{k=1}^{n} \frac{xB_n + \rho_k}{xB_n + \rho_k + B_k} \frac{\gamma_{3,k}}{(1+x)(B_k^2 + B_n^2 x^2) B_{k-1,n}}, \qquad (3.1.86)$$

where C is an absolute constant,

$$B_{k-1,n}^2 = B_n^2 - B_k^2 = \sum_{j=k}^{n} \sigma_j^2$$

and

$$\rho_k = \max_{1 \leq j \leq k} \frac{\gamma_{3j}}{\sigma_j^2}, \; k=1,2,\ldots.$$

There are some interesting nontrivial corollaries of this theorem (see Arak (1974)).

Corollary.

For any x>0,

$$\Delta_n(x) \leq C\frac{\rho_n}{B_n(1+|x|^3)}, \tag{3.1.87}$$

and if additionally $X_1, X_2, \ldots$ are identically distributed, then

$$\Delta_n(x) \leq C(\gamma_3/\sigma^3)/n^{1/2}(1+|x|^3). \tag{3.1.88}$$

Uniform estimate (3.1.88) improves (3.1.81).

Remark 3.1.10.

It turns out that (3.1.84) is also a consequence of Theorem 3.1.10. We add one more useful result for normal distributions (Nevzorov (1971b)).

Theorem 3.1.11.

Let $X_1, X_2, \ldots$ be independent random variables and let X_k have normal $N(0, \sigma_k^2)$ distribution, $k=1,2,\ldots$. If

$$\sigma_k^2 \leq D^2, \; k=1,2,\ldots,$$

and

$$B_n^2 \geq d^2 n, \; n=1,2,\ldots,$$

$0<d\leq D<\infty$ being positive constants, then for any $x\geq 0$ the following inequality holds:

$$0 \leq P\{S_{n,n}<xB_n\} - G(x) \leq \frac{3D}{d\sqrt{n}}\exp\{-x^2/2\}, \; n=1,2,\ldots. \tag{3.1.89}$$

Remark 3.1.11.

If in theorem 3.1.11 we additionally suppose that $X_1, X_2, \ldots$ have the standard normal distribution then (3.1.89) can be rewritten as

$$0 \leq P\{S_{n,n}<x\sqrt{n}\} - G(x) \leq \frac{3}{\sqrt{n}}\exp\{-x^2/2\}, \; n=1,2,\ldots.$$

Now we consider the second principal case when the random variables $X_1, X_2, \ldots$ have positive expectations. In theorems 3.1.6 and 3.1.7 we have obtained relations of the type

$$\sup_x |P\{S_{n,n}-A_n<xB_n\}-\Phi(x)\}|\to 0,\ n\to\infty$$

in the situations when X's have finite variances. Now we will study the closeness of the distribution functions of $S_{n,n}$ and the distribution function $\Phi(x)$ of the standard normal law under some additional moment conditions on X's.

Rogozin (1966) have considered independent identically distributed random variables X, X_1,X_2,...., such that

$$EX=a>0,\ Var(X)=\sigma^2>0$$

and

$$\gamma_3=E|X-a|^3<\infty.$$

He has got that

$$0\le P\{S_n<x\}-P\{S_{n,n}<x\}\le C_1 n^{-1/2},. \tag{3.1.90}$$

and hence

$$\sup_x |P\{S_{n,n}-a_n<x\sigma n^{1/2}\}-\Phi(x)\}| \le C_2\, n^{-1/2}. \tag{3.1.91}$$

where constants C_1 and C_2 depend on a, σ, γ_3 and $E|X|$.

The order in n of the right hand sides of (3.1.90) and (3.1.91) is optimal. It follows from the following result obtained by Nevzorov (1971b).

Theorem 3.1.12

Let X, X_1, X_2,.... be independent identically distributed random variables with a density function p(x), such that

$$EX=a>0,\ 0<Var(X)=\sigma^2<\infty$$
and
$$\max_x p(x) \le C,$$

C being a positive constant. Then for any fixed x and a sufficiently large n the following inequality is valid:

$$P\{S_n-an<x\sigma n^{1/2}\}-P\{S_{n,n}-an<x\sigma n^{1/2}\}\geq\frac{P^2\{X<0\}}{1-P\{X<0\}}\frac{\exp(-x^2/2)}{10C\sigma\sqrt{\pi n}}. \tag{3.1.92}$$

Remark 3.1.12.

Under a weaker condition

$$\gamma_p=E|X-a|^p<\infty,$$

where $2<p<3$, the method, suggested by Rogozin, can be also used to show that in this situation the analogous to (3.1.90) statement has the form

$$0\leq P\{S_n<x\}-P\{S_{n,n}<x\}\leq C_3 n^{1-p/2},$$

and (3.1.91) can be rewritten as

$$\sup_x|P\{S_{n,n}-a_n<x\sigma n^{1/2}\}-\Phi(x)|\leq C_4 n^{1-p/2}, \tag{3.1.93}$$

where constants C_3 and C_4 depend on a, σ, γ_p and $E|X|$.

Nevzorov (1971a) applied another method to show that if i.i.d. X's have a positive expectation a and

$$\gamma_p=E|X-a|^p<\infty$$

for some $1<p\leq 2$, then

$$0\leq P\{S_n<x\}-P\{S_{n,n}<x\}\leq \tilde{C}n^{(1-p)/2p},$$

where $\tilde{C}$ does not depend on n.

Below we will show that the method, which was used by Rogozin to prove (3.1.91), can be applied in a more general situation.

We consider independent random variables $X_1, X_2,\ldots$ and denote

$$a_k=EX_k,\ \sigma_k^2=\mathrm{Var}(X_k),\ \gamma_{3,k}=E|X_k-a_k|^3,\ k=1,2,\ldots,$$

$$A(n)=ES_n=a_1+\ldots+a_n,$$

$$B_n^2=\mathrm{Var}(S_n)=\sigma_1^2+\ldots+\sigma_n^2,\ n=1,2,\ldots.$$

Theorem 3.1.13.

(Nevzorov (1971a)). Let $X_1, X_2, \ldots$ be independent random variables and there exist such positive constants $a \le A$, d and M, that

$$a \le a_k \le A, k=1,2,\ldots \tag{3.1.94}$$

$$B_n^2 \ge dn, n=1,2,\ldots \tag{3.1.95}$$

and

$$\gamma_{3,k} \le M, k=1,2,\ldots. \tag{3.1.96}$$

Then

$$\sup_x |P\{S_{n,n}-A_n<xB_n\}-\Phi(x)| \le C_5 n^{-1/2}, \tag{3.1.97}$$

where C_5 depends on a,A,d and M only.

Proof of Theorem 3.1.13.

One can see that

$$0 \le P\{S_n<x\}-P\{S_{n,n}<x\}$$

$$=P\{S_{n,n}\ge x, S_n<x\}=\sum_{k=1}^{n-1} P\{S_0<x, S_1<x,\ldots, S_{k-1}<x, S_k\ge x, S_n<x\}$$

$$\le \sum_{k=1}^{n-1} P\{S_0<x, S_1<x,\ldots, S_{k-1}<x, S_k\ge x\}P\{S_n-S_k<0\} \le T_1+T_2, \tag{3.1.98}$$

where

$$T_1=\max_{0\le k\le[n/2]} P\{S_n-S_k<0\} \sum_{k=1}^{[n/2]} P\{S_0<x, S_1<x,\ldots, S_{k-1}<x, S_k\ge x\}$$

and

$$T_2=\sum_{k=[n/2]+1}^{n-1} P\{S_0<x, S_1<x,\ldots, S_{k-1}<x, S_k\ge x\}P\{S_n-S_k<0\}.$$

Note, that under conditions of the theorem,

$$\sigma_k^2 \le (\gamma_{3,k})^{2/3} \le M^{2/3}, \ k=1,2,\ldots. \tag{3.1.99}$$

Applying Chebychev's inequality we obtain that

$$P\{S_n-S_k<0\}=P\{S_n-S_k-(A_n-A_k)<-(A_n-A_k)\} \le Var(S_n-S_k)/(A_n-A_k)^2 \le M^{2/3}/(n-k)a^2$$

and then

$$T_1=\max_{0\le k\le[n/2]} P\{S_n-S_k<0\} \le \max_{0\le k\le[n/2]} M^{2/3}/(n-k)a^2 \le 2\, M^{2/3}/\, a^2 n. \tag{3.1.100}$$

It suffices now to estimate T_2 . We will use the non-uniform estimate of Bikyalis (1966) for independent random variables:

$$|P\{\sum_{k=1}^{n} (X_k-a_k)<xB_n\}-\Phi(x)| \le CL_{n,3}/(1+|x|^3), \tag{3.1.101}$$

where C is an absolute constant and

$$L_{n,3}=\sum_{k=1}^{n} \gamma_{3,k}/B_n^3$$

is the Lyapunov fraction of order 3.

Since

$$P\{S_n-S_k<0\}=P\{\sum_{m=k+1}^{n} (X_m-a_k)<-\sum_{m=k+1}^{n} a_m\}=P\{\frac{\sum_{m=k+1}^{n}(X_m-a_m)}{\sqrt{\sum_{m=k+1}^{n}\sigma_m^2}}<-\frac{\sum_{m=k+1}^{n}a_m}{\sqrt{\sum_{m=k+1}^{n}\sigma_m^2}}\},$$

on applying (3.1.101) with

$$x=x_0=-\sum_{m=k+1}^{n} a_m/(\sum_{m=k+1}^{n} \sigma_m^2)^{1/2},$$

we get that

$$|P\{S_n-S_k<0\}-\Phi(x_0)|\leq C(\sum_{m=k+1}^{n}\gamma_{3,k})(\sum_{m=k+1}^{n}\sigma_m^2)^{-3/2}/(1+|x_0|^3)$$

$$\leq C|x_0|^{-3}(\sum_{m=k+1}^{n}\gamma_{3,k})(\sum_{m=k+1}^{n}\sigma_m^2)^{-3/2}=C(\sum_{m=k+1}^{n}\gamma_{3,k})/(\sum_{m=k+1}^{n}a_m)^3.$$

Using inequalities (3.1.94) and (3.1.96) we obtain now that

$$|P\{S_n-S_k<0\}-\Phi(x_0)|\leq CM/a^3(n-k)^2,\ k=1,2,\ldots,n-1. \quad (3.1.102)$$

It is easy to see that

$$\Phi(-x)\leq \exp(-x^2/2)/(2\pi)^{1/2}x\leq(2/\pi)^{1/2}e^{-1}x^{-3}, \quad (3.1.103)$$

for any positive x. Since x_0 is negative, it follows from (3.1.94), (3.1.99) and (3.1.103) that

$$\Phi(x_0)\leq (2/\pi)^{1/2}e^{-1}(\sum_{m=k+1}^{n}\sigma_m^2)^{3/2}/(\sum_{m=k+1}^{n}a_m)^3\leq(2/\pi)^{1/2}e^{-1}M/(n-k)^{3/2}a^3,\ k=1,2,\ldots,n-1. \quad (3.1.104)$$

From (3.1.102) and (3.1.104) we obtain that

$$P\{S_n-S_k<0\}\leq C_6/(n-k)^{3/2},$$

where constant C_6 depends on a and M only, and then

$$T_2\leq C_6\max_{[n/2]+1\leq m\leq n-1}P\{S_{m-1,m-1}<x,\ S_m\geq x\}\sum_{m=[n/2]+1}^{n-1}1/(n-m)^{3/2}\leq$$

$$C_6\max_{[n/2]+1\leq m\leq n-1}P\{S_{m-1,m-1}<x,\ S_m\geq x\}\sum_{m=1}^{\infty}m^{-3/2}=$$

$$C_7\max_{[n/2]+1\leq m\leq n-1}P\{S_{m-1,m-1}<x,\ S_m\geq x\}, \quad (3.1.105)$$

where C_7 is also a constant, which depends on a and M only.

The next stage of the proof is to estimate the expression

$$\max_{[n/2]+1\leq m\leq n-1}P\{S_{m-1,m-1}<x,\ S_m\geq x\}.$$

We have

$P\{S_{m-1,m-1}<x,\ S_m\geq x\}\leq P\{S_{m-1}<x,\ S_m\geq x\}=$

$P\{S_{m-1}<x, X_m>0\}-P\{\ S_m<x, X_m>0\}=$

$$\int_0^\infty (P\{S_{m-1}<x\}-P\{\ S_{m-1}<x-y\})dP\{X_m<y\}. \qquad (3.1.106)$$

To estimate the RHS of (3.1.106) we apply the Lyapunov inequality for the differences

$\Delta_{m-1}=P\{S_{m-1}<x\}-P\{\ S_{m-1}<x-y\}$

and get that

$$\Delta_{m-1}\leq\Phi\left(\frac{x-\sum_{j=1}^{m-1}a_j}{B_{m-1}}\right)-\Phi\left(\frac{x-y-\sum_{j=1}^{m-1}a_j}{B_{l-1}}\right)+2CL_{m-1,3}, \qquad (3.1.107)$$

C being an absolute constant . Since

$0\leq\Phi(x_1)-\ \Phi(x_2)\ \leq(2\pi)^{-1/2}(x_1-x_2),$

if $x_1>x_2$, we get from (3.1.107) that

$\Delta_{m-1}\leq(2\pi)^{-1/2}y/B_{m-1}+2CL_{m-1,3}.$ (3.1.108)

From (3.1.106) and (3.1.108) one obtains the following inequality:

$$P\{S_{m-1,m-1}<x,\ S_m\geq x\}\leq 2CL_{m-1,3}\ P\{X_m>0\}+(2\pi)^{-1/2}(B_{m-1})^{-1}\int_0^\infty ydP\{X_m<y\}. \qquad (3.1.109)$$

Note that

$$\int_0^\infty ydP\{X_m<y\}\leq\int_{-\infty}^\infty |y|dP\{X_m<y\}\leq\int_{-\infty}^\infty |y-a_m|dP\{X_m<y\}+a_m\leq(\gamma_{3,m})^{1/3}+a_m. \qquad (3.1.110)$$

On combining (3.1.109) with (3.1.110) and inequalities for moments (3.1.94)-(3.1.96), we get that

$$P\{S_{m-1,m-1}<x,\ S_m\geq x\}\leq C_8 m^{-1/2}$$

and hence

$$\max_{[n/2]+1\leq m\leq n-1} P\{S_{m-1,m-1}<x,\ S_m\geq x\}\}\leq C_8(n/2)^{-1/2}, \quad (3.1.111)$$

where C_8 depends on a, A, d and M only. From (3.1.98), (3.1.100), (3.1.105) and (3.1.111) we obtain that

$$0\leq P\{S_n<x\}-P\{S_{n,n}<x\}\leq C_9 n^{-1/2}, \quad (3.1.112)$$

where C_9 also depends on a, A, d and M only. To complete the proof of the theorem we must recall that

$$|P\{S_n-A_n<xB_n\}-\Phi(x)|\leq CL_{n,3}\leq CM/d^{3/2}n^{1/2}. \quad (3.1.113)$$

Finally (3.1.112) and (3.1.113) imply (3.1.97). The proof is completed.

Remark 3.1.13.

The analogous arguments can be used to show that if a weaker (than (3.1.96)) condition

$$\gamma_{p,k}=E|X_k-a_k|^p\leq M,\ k=1,2,\ldots.$$

is valid for some $2<p<3$, then one can get $n^{(p-2)/2}$ instead of $n^{-1/2}$ on the RHS of (3.1.97).

First passage times

We have investigated asymptotic distributions of maximal order statistics $S_{n,n}$ in some different situations. Very close to $S_{n,n}$ are the first passage times

$$T(r)=\min\{k\geq 0: S_k\geq r\}.$$

The integer valued random variable T(r) presents the time, when a level r is first reached. Of course, it is interesting to consider positive values of r only, because

$$P\{T(r)=0\}=1$$

for any $r \le 0$. In the sequel we will consider $r>0$. Distributions of random variables $S_{n,n}$ and $T(r)$ are dual:

$$P\{T(r) \le n\}=P\{S_{n,n} \ge r\}. \qquad (3.1.114)$$

Hence, we can apply results for $S_{n,n}$ given above to study the asymptotic behavior of the first passage time $T(r)$. We will prove two theorems for $T(r)$ obtained by Migai and Nevzorov (1976).

Let $X_1, X_2, \ldots$ are independent identically distributed random variables with a nonnegative expectation $a=EX_k$, a positive variance $\sigma^2=Var(X_k)$ and a finite moment

$$\gamma_3=E|X_k-a|^3, \ k=1,2,\ldots.$$

Theorem 3.1.14.

If the expectation $a=0$, then

$$\sup_x |P\{\sigma^2 T(r)<xr^2\}-H(x)| =O(1/r), \ r \to \infty, \qquad (3.1.115)$$

where the constant on the RHS of (3.1.115) depends on σ^2 and γ_3 only and

$$H(x)=\begin{cases} 0, & \textit{if } x \le 0 \\ 2(1-\Phi(x^{-1/2})), & \textit{if } x>0 \end{cases}. \qquad (3.1.116)$$

Theorem 3.1.15.

If the expectation $a>0$, then

$$\sup_x |P\{T(r)-r/a<x\sigma r^{1/2}a^{-3/2}\}-\Phi(x)| =O(r^{-1/2}), \ r \to \infty, \qquad (3.1.117)$$

with the constant on the RHS of (3.1.117) not depending on r.

Proof of Theorem 3.1.14.

We shall use non-uniform estimate (3.1.88) of Arak (1974):

$$\Delta_n(x)=|P\{S_{n,n}<y\sigma n^{1/2}\} - G(y)| \le C(\gamma_3/\sigma^3)/n^{1/2}(1+|y|^3),$$

where C is an absolute constant and function G is defined as follows:

$$G(y) =2\Phi(y)-1,$$

if $y>0$, and

$G(y)=0$,

if $y\leq 0$.

It is obvious that the statement of the theorem is valid for $x\leq 0$. Hence, we perform the proof only for positive values of x. Setting

$n=[xr^2/\sigma^2]$

in Arak's inequality , where [x] is the integer part of x, we have

$$|P\{S_{[xr^2],[xr^2]}\geq\sigma y[xr^2/\sigma^2]^{1/2}\}-2(1-\Phi(y))|\leq C(\gamma_3/\sigma^3)/(xr^2/\sigma^2)^{1/2}(1+|y|^3), \quad (3.1.118)$$

for $y\geq 0$. Let

$x\geq 2\sigma^2/r^2$

and let us substitute

$y=(r/\sigma)[xr^2/\sigma^2]^{-1/2}$

in (3.1.118). Taking into account equality (3.1.114) and the relation $x-1<[x]\leq x$, we get the following inequality:

$$|P\{\sigma^2\, T(r)<xr^2\}-H(x)|\leq M_1+M_2, \quad (3.1.119)$$

where

$$M_1=\frac{C\gamma_3}{r\sigma^2\sqrt{x-(\sigma^2/r^2)}(1+x^{-3/2})}$$

and

$$M_2=2|\Phi(\frac{r}{\sigma}\left[\frac{xr^2}{\sigma^2}\right]^{-1/2})-\Phi(x^{-1/2})|.$$

Evidently,

$$|\Phi(y)-\Phi(z)| \le \frac{1}{\sqrt{2\pi}}|y-z|\exp(-\min\{y^2/2, z^2/2\}) \quad (3.1.120)$$

if both values y and z simultaneously are positive or negative. It is easy to verify that

$$x_0 = \frac{r}{\sigma}\left[\frac{xr^2}{\sigma^2}\right]^{-1/2} \ge x^{-1/2} > 0$$

and

$$0 \le x_0 - x^{-1/2} \le \frac{r}{\sigma}\left(\frac{xr^2}{\sigma^2}-1\right)^{-1/2} - x^{-1/2} \le \frac{\sigma^2}{r^2}\left(x\sqrt{x-\frac{\sigma^2}{r^2}}\right)^{-1}. \quad (3.1.121)$$

Hence, applying (3.1.120) (with $y=x_0$, $z=x^{-1/2}$) and (3.1.121) we get

$$M_2 \le \sqrt{\frac{2}{\pi}\frac{\sigma^2}{r^2}}\exp\left\{-\frac{1}{2x}\right\}\left(x\sqrt{x-\frac{\sigma^2}{r^2}}\right)^{-1}.$$

Let us recall that we consider

$x \ge 2\sigma^2/r^2$.

Then,

$x-\sigma^2/r^2 \ge x/2$

and

$$M_2 \le \frac{2}{\sqrt{\pi}}\frac{\sigma^2}{r^2}x^{-3/2}\exp\left\{-\frac{1}{2x}\right\} \le \frac{2}{\sqrt{\pi}}\left(\frac{3}{e}\right)^{3/2}\frac{\sigma^2}{r^2}, \quad (3.1.122)$$

since for any p>0 (and in particular for p=3/2, as in our case),

$$\sup_{x\ge 0} x^p \exp(-x) = e^{-p}p^p.$$

To estimate M_1 we decompose the interval $[2\sigma^2/r^2, \infty)$ into $J_1=[1, \infty)$ and $J_2=[2\sigma^2/r^2, 1)$ (we assume that $2\sigma^2<r^2$ since otherwise (3.1.115) becomes a trivial relation).

Let $x \in J_1$. Then

$$M_1 \leq \frac{C\gamma_3}{r\sigma^2\sqrt{x-(\sigma^2/r^2)}} \leq \frac{C\sqrt{2}\gamma_3}{r\sigma^2}. \qquad (3.1.123)$$

Now let $x \in J_2$. In this situation,

$$M_1 \leq \frac{C\gamma_3\, x^{3/2}}{r\sigma^2\sqrt{x-(\sigma^2/r^2)}} = \frac{C\gamma_3\, x}{r\sigma^2\sqrt{1-(\sigma^2/xr^2)}} \leq \frac{C\sqrt{2}\gamma_3}{r\sigma^2}. \qquad (3.1.124)$$

Thus for $x \geq 2\sigma^2/r^2$, we get from (3.1.122)-(3.1.124) that

$$M_1+M_2=O(r^{-1})+O(r^{-2})$$

with the constant on the right side depending on σ^2 and γ_3 only. It remains to prove that the statement of the theorem is valid for $0 \leq x < 2\sigma^2/r^2$. In this case we have

$$P\{\sigma^2 T(r) < xr^2\} \leq P\{T(r)<2\} = P\{X_1 \geq r\} \leq \gamma_3/r^3$$

and

$$|P\{\sigma^2 T(r) < xr^2\} - H(x)| \leq P\{T(r)<2\} + H(2\sigma^2/r^2) =$$

$$P\{T(r)<2\} + 2\Phi(-r/\sigma\sqrt{2}) \leq \gamma_3/r^3 + 2\Phi(-r/\sigma\sqrt{2}) =$$

$$O(r^{-3}) = O(1/r),\ r \to \infty,$$

and that concludes the proof of theorem 3.1.14.

Proof of Theorem 3.1.15.

To prove this theorem we shall use Rogozin's estimate (3.1.91):

$$\sup_x |P\{S_{n,n}-a_n < x\sigma n^{1/2}\} - \Phi(x)| = O(n^{-1/2})$$

with the constant on the right side depending on a, σ, γ_3 and $E|X|$ only. For the sake of simplicity we assume that

$$n = x\sigma r^{1/2} a^{-3/2} + r/a$$

is an integer and substitute this value into (3.1.114). Then from (3.1.114) and (3.1.91) we get that

$$P\{T(r)-r/a \le x\sigma r^{1/2}a^{-3/2}\} - \Phi\left(\frac{x\sqrt{r}a^{-1/2}}{(x\sigma\sqrt{r}a^{-3/2}+r/a)^{1/2}}\right)=$$

$$P\left\{\frac{S_{n,n}-an}{\sigma\sqrt{n}} \ge -\frac{x\sqrt{r}a^{-1/2}}{(x\sigma\sqrt{r}a^{-3/2}+r/a)^{1/2}}\right\}-\left(1-\Phi\left(-\frac{x\sqrt{r}a^{-1/2}}{(x\sigma\sqrt{r}a^{-3/2}+r/a)^{1/2}}\right)\right)=$$

$$O((x\sigma r^{1/2}a^{-3/2}+r/a)^{-1/2}),\ r\to\infty, \qquad (3.1.125)$$

where the constant on the RHS of (3.1.125) does not depend on r. Now (3.1.125) implies that

$$P\{T(r)-r/a \le x\sigma r^{1/2}a^{-3/2}\} - \Phi(x)=O(M_1+M_2), \qquad (3.1.126)$$

where

$$M_1=(x\sigma r^{1/2}a^{-3/2}+r/a)^{-1/2}$$

and

$$M_2=|\Phi(x)-\Phi(x(1+x\sigma/\sqrt{ra})^{-1})|.$$

Let
$x \ge -\sqrt{ra}/2\sigma$.

Evidently, in this case

$$M_1 \le (2a/r)^{1/2}. \qquad (3.1.127)$$

Then, applying (3.1.120) , we get that

$$M_2 \le \frac{\sigma x^2}{\sqrt{2\pi a}(1+x\sigma/\sqrt{ra})}\exp\{-\frac{x^2}{2}\min[1,(1+\frac{x\sigma}{\sqrt{ra}})^{-1}]\}r^{-1/2}. \qquad (3.1.128)$$

It is easy to verify that (3.1.128) implies the estimate

$$M_2=O(r^{-1/2}),\ r\to\infty, \qquad (3.1.129)$$

with the constant on the RHS of (3.1.129) depending on a and σ^2. From (3.1.126), (3.1.127) and (3.1.129) it follows that the theorem is valid for $x \geq -\sqrt{ra}/2\sigma$.

It remains for us to consider $x < -\sqrt{ra}/2\sigma$. In this situation
$P\{T(r)-r/a \leq x\sigma r^{1/2}a^{-3/2}\} \leq P\{T(r)<r/2a\}=P\{S_{[r/2a],[r/2a]}>r\}$

and

$$|P\{T(r)-r/a \leq x\sigma r^{1/2}a^{-3/2}\}-\Phi(x)| \leq P\{S_{[r/2a],[r/2a]}>r\}+\Phi(-\sqrt{ra}/2\sigma)=$$

$$P\{\frac{S_{[r/2a],[r/2a]}-a[r/2a]}{\sigma\sqrt{[r/2a]}} > \frac{r-a[r/2a]}{\sigma\sqrt{[r/2a]}}\}+\Phi(-\sqrt{ra}/2\sigma). \tag{3.1.130}$$

Due to relation (3.1.91), we get from (3.1.130) that

$$|P\{T(r)-r/a \leq x\sigma r^{1/2}a^{-3/2}\}-\Phi(x)| \leq$$

$$1-\Phi(\frac{r-a[r/2a]}{\sigma\sqrt{[r/2a]}})+\Phi(-\sqrt{ra}/2\sigma)+O(\frac{r-a[r/2a]}{\sigma\sqrt{[r/2a]}})=$$

$$\Phi(\frac{a[r/2a]-r}{\sigma\sqrt{[r/2a]}})+\Phi(-\sqrt{ra}/2\sigma)+O(r^{-1/2}). \tag{3.1.131}$$

From (3.1.131), on estimating the tails of the standard normal distribution function Φ, we obtain that

$$|P\{T(r)-r/a \leq x\sigma r^{1/2}a^{-3/2}\}-\Phi(x)|=O(r^{-1/2}),\ r\to\infty.$$

The theorem is completely proved.

3.2. Distributions of Central Order Statistics Based on Sums

Above we discussed different results for maximal order statistics $S_{n,n}$. It turns out that methods , which have been used for $S_{n,n}$, can be also applied for other order statistics $S_{k,n}$.

Pollaczek- Spitzer- Wendel identities

In theorem 3.1.1 we have tied characteristic functions of maximal order statistics and random variables $S_n^+=\max\{0,S_n\}$:

$$\sum_{n=0}^{\infty} E\{\exp(itS_{n,n})\}z^n=\exp\{\sum_{n=1}^{\infty} E\exp(it S_n^+)z^n/n\}.$$

This result is a consequence of the following identity (Wendel (1960)), which is valid for $|w|<1$ and $|z|<1$:

$$\sum_{n=0}^{\infty} w^n \sum_{k=0}^{n} z^k E\exp\{itS_{n-k,n}^+ +iuS_n\}=$$

$$\{(1-z)(1-wzf(u))\}^{-1}\{\exp[\sum_{n=1}^{\infty} n^{-1}w^n(1-z^n)E(\exp\{it S_n^+ +iuS_n\})]-z\}, \qquad (3.2.1)$$

where

$$f(u)=E\exp\{iuX_1\}.$$

The first who has got a relation, which is similar to (132), was Pollaczek (1952). In his version of (3.2.1) u=0 (S_n is omitted) , the additional restriction on X's (the existence of a moment generating function) is supposed and the RHS has another (equivalent) form. Later, Pollaczek (1975) suggested a simplified proof of his formula, which did not require the existence of a moment generating function.

Spitzer (1956) has proved the partial case of (3.2.1) , when z=0:

$$\sum_{n=0}^{\infty} w^n E\exp\{itS_{n,n}+iuS_n\}=$$

$$\exp(\sum_{n=1}^{\infty} n^{-1}w^n E\exp\{it S_n^+ +iuS_n\}). \qquad (3.2.2)$$

Note that (3.1.1) immediately follows from (3.2.2). It is interesting that Pollaczek, Spitzer and Wendel have used different ways to prove their identities. The functional-theoretic treatment suggested by Pollaczek and the combinatorial approach of Spitzer were completed by the algebraic method of Wendel. A simple analytic proof of (3.2.1) based on using of a kind of the Wiener-Hopf decomposition has been suggested by de Smit (1973). Note that in Theorem 3.1.1 we applied a simple probabilistic approach of Heinrich to prove (3.1.1). Regarding all saying above it will be fair to call (3.1.1) as Pollaczek - Spitzer identity and (3.2.1) as Pollaczek - Spitzer- Wendel identity.

Wendel (1960) also gives the following identity , which is similar to (3.2.1):

$$\sum_{n=0}^{\infty} w^n \sum_{k=0}^{n} z^k E\exp\{itS_{n-k,n}+iuS_n\}=$$

$$\exp(\sum_{n=1}^{\infty} n^{-1}(w^n E\exp\{it S_n^+ +iuS_n\}+(wz)^n E\exp\{it S_n^- +iuS_n\})), \quad (3.2.3)$$

where

$$S_n^- =\min(0,S_n).$$

With the help of (3.2.3) Wendel (1960) proved the following result.

Theorem 3.2.1.

For any real t and u

$$E\exp\{itS_{k,n}+iuS_n\}=E\exp\{itS_{k,k}+iuS_k\}\ E\exp\{itS_{0,n-k}+iuS_{n-k}\}. \quad (3.2.4)$$

Note that (3.2.4) means that

$$(S_{k,n},\ S_k) \overset{d}{=} (S_{k,k},S_k)*(S_{0,n-k},\ S_{n-k}), \quad (3.2.5)$$

where * denote a convolution of random vectors $(S_{k,k},\ S_k)$ and $(S_{0,n-k},\ S_{n-k})$.

Remark 3.1.14.

In his paper Wendel (1960) has mentioned that a partial case of (135) (with u = 0) was proved combinatorially by Bohenblust, Spitzer and Welch and was brought to Wendel's attention by Spitzer.

The Proof of Theorem 3.2.1.

We will use relation (3.2.3). Taking z=0 we have from this equality that

$$\exp(\sum_{n=1}^{\infty} n^{-1}w^n E\exp\{it S_n^+ +iuS_n\})=\sum_{n=0}^{\infty} w^n E\exp\{itS_{n,n}+iuS_n\}. \quad (3.2.6)$$

On observing that

$$S_n^- =S_n- S_n^+,$$

we get from (3.2.6) that

$$\exp\Big(\sum_{n=1}^{\infty} n^{-1} (wz)^n \, E\exp\{it S_n^- + iuS_n\}\Big)=$$

$$\exp\Big(\sum_{n=1}^{\infty} n^{-1}(wz)^n \, E\exp\{-it S_n^+ + i(u+t)S_n\}\Big)=$$

$$\sum_{n=0}^{\infty} (wz)^n E\exp\{-itS_{n,n}+i(u+t)S_n\}=$$

$$\sum_{n=0}^{\infty} (wz)^n E\exp\{it(S_n-S_{n,n})+iuS_n\}. \tag{3.2.7}$$

Note that

$$S_n-S_{n,n}= -\max\{0,-X_n,-X_n-X_{n-1},\ldots,-X_n-\ldots-X_1\}=$$

$$\min\{0,X_n,X_n+X_{n-1},\ldots,S_n\}$$

and

$$(S_n-S_{n,n},\, S_n) \overset{d}{=} (S_{0,n},\, S_n). \tag{3.2.8}$$

From (3.2.3), (3.2.6), (3.2.7) and (3.2.8) it follows that

$$\sum_{n=0}^{\infty} w^n \sum_{k=0}^{n} z^k E\exp\{itS_{n-k,n}+ iuS_n\}=$$

$$\sum_{n=0}^{\infty} w^n E\exp\{itS_{n,n}+ iuS_n\} \sum_{k=0}^{\infty} (wz)^k E\exp\{itS_{0,k}+iuS_k\}. \tag{3.2.9}$$

Comparing coefficients of $w^n z^k$ in double power series on the left and right sides of (3.2.9) we get the following equality:

$$E\exp\{itS_{n-k,n}+ iuS_n\}=E\exp\{itS_{0,k}+iuS_k\}E\exp\{itS_{n-k,n-k}+ iuS_{n-k}\},$$

which is equivalent to (3.2.4). The theorem is proved.

To complete the list of the formulae for characteristic functions of order statistics $S_{k,n}$ given above we mention that Pollaczek (1975) has given without proof a rather complicated formula for the generating function

$$\sum_{n=0}^{\infty} z^n E\exp(itR_{n,n}+iuR_{n-1,n}),$$

where $R_{n,n}$ and $R_{n-1,n}$ are the first and second maximal order statistics based on sums $S_1,\ldots, S_n$ (without S_0) and $R_{0,0}=R_{-1,0}=R_{0,1}=0$.

Asymptotic distributions of $S_{k,n}$ in the i.i.d. case.

If $X_1, X_2,\ldots$ are i.i.d. random variables one can use the known asymptotic distributions of maximal and minimal order statistics $S_{n,n}$ and $S_{0,n}$ and the equality

$$S_{k,n} \overset{d}{=} S_{k,k} * S_{0,n-k}, \tag{3.2.10}$$

which is a partial case of (3.2.5), to get the limit distribution of the suitably normalized order statistics $S_{k,n}$.

Let $X, X_1, X_2,\ldots$ be i.i.d. random variables, $EX=a$ and $Var(X)=\sigma^2$, $0<\sigma<\infty$. We consider three cases ($a=0$, $a>0$ and $a<0$).

1). Let $a=0$. As we know , in this case

$$\sup_x |P\{S_{n,n} < x\sigma n^{1/2}\} - H_1(x)| \to 0,$$

as $n\to\infty$, where

$$H_1(x)=\max\{0, 2\Phi(x)-1\}.$$

Recalling that

$$S_{0,n} \overset{d}{=} - S_{n,n},$$

we have the analogous formula for minimal order statistics:

$$\sup_x |P\{S_{0,n} < x\sigma n^{1/2}\} - H_0(x)| \to 0, \ n\to\infty,$$

where

$H_0(x)=1-H_1(-x)=1-\max\{0,2\Phi(-x)-1\}=$

$1-\max\{0,1-2\Phi(x)\}=\min\{1,2\Phi(x)\}$.

Consider quantiles $S_{[\alpha n],n}$, where $0<\alpha<1$. For the sake of simplicity we suppose that αn is an integer. In this case (3.2.10) can be rewritten as

$$S_{\alpha n,n} \stackrel{d}{=} S_{\alpha n,\alpha n} * S_{0,(1-\alpha)n} \tag{3.2.11}$$

The random variables on the right side of (3.2.11) satisfy the following relations:

$$S_{\alpha n,\alpha n}/\ \sigma\,(\alpha n)^{1/2} \stackrel{d}{\rightarrow} \nu,$$

$$S_{0,(1-\alpha)n}/\sigma((1-\alpha)n)^{1/2} \stackrel{d}{\rightarrow} \mu,\ n\rightarrow\infty,$$

where random variables μ and ν have the distribution functions H_0 and H_1 correspondingly. Then (3.2.11) implies that

$$S_{\alpha n,n}/\sigma n^{1/2} \stackrel{d}{\rightarrow} \alpha^{1/2}\nu+(1-\alpha)^{1/2}\mu, \tag{3.2.12}$$

where ν and μ are independent. Note that

$$\nu \stackrel{d}{=} -\mu,$$

and hence $S_{\alpha n,n}/\sigma n^{1/2}$ and $-\ S_{\alpha n,n}/\sigma n^{1/2}$ have the same asymptotic distribution. Using (3.2.12) we will find this limit distribution.

Let $x\leq 0$.Then

$$H_\alpha(x)=\lim_{n\rightarrow\infty} P\{\ S_{\alpha n,n}/\sigma n^{1/2}<x\}=$$

$$\int_0^\infty P\{\mu<\frac{x-y\sqrt{\alpha}}{\sqrt{1-\alpha}}\}dP\{\nu<y\}=\int_0^\infty H_0(\frac{x-y\sqrt{\alpha}}{\sqrt{1-\alpha}})dH_1(y)=$$

$$4\int_0^\infty \Phi(\frac{x-y\sqrt{\alpha}}{\sqrt{1-\alpha}})\varphi(y)dy, \tag{3.2.13}$$

Φ and φ being the distribution function and density function of the standard normal law. Due to the fact that $S_{\alpha n,n}/\sigma n^{1/2}$ and $-S_{\alpha n,n}/\sigma n^{1/2}$ have the same limit distribution we get the following relation for positive values of x:

$$H_\alpha(x)=1-H_{1-\alpha}(-x)=1-4\int_0^\infty \Phi\left(\frac{-x-y\sqrt{1-\alpha}}{\sqrt{\alpha}}\right)\varphi(y)dy=$$

$$1-4\int_0^\infty \left(1-\Phi\left(\frac{x+y\sqrt{1-\alpha}}{\sqrt{\alpha}}\right)\right)\varphi(y)dy=$$

$$4\int_0^\infty \Phi\left(\frac{x+y\sqrt{1-\alpha}}{\sqrt{\alpha}}\right)\varphi(y)dy-1. \qquad (3.2.14).$$

On collecting (3.2.13) and (3.2.14) we have for any $0<\alpha<1$ the following expression:

$$H_\alpha(x)=\begin{cases} 4\int_0^\infty \Phi\left(\frac{x-y\sqrt{\alpha}}{\sqrt{1-\alpha}}\right)\varphi(y)dy, & if \quad x\le 0, \\ 4\int_0^\infty \Phi\left(\frac{x+y\sqrt{1-\alpha}}{\sqrt{\alpha}}\right)\varphi(y)dy-1, & if \ x>0 \end{cases} . \qquad (3.2.15)$$

Let us find the density function $h_\alpha(x)$ of the limit distribution. Note that

$$h_\alpha(x)=h_{1-\alpha}(-x)$$

Hence it suffices to consider $x\le 0$. In this case, differentiating of the RHS of (3.2.13) gives us the following equalities:

$$h_\alpha(x)=4(1-\alpha)^{-1/2}\int_0^\infty \varphi\left(\frac{x-y\sqrt{\alpha}}{\sqrt{1-\alpha}}\right)\varphi(y)dy=$$

$$\frac{2}{\pi\sqrt{1-\alpha}}\int_0^\infty \exp\left\{-\frac{y^2-2\sqrt{\alpha}\,xy+x^2}{2(1-\alpha)}\right\}dy=$$

$$\frac{2}{\pi\sqrt{1-\alpha}}\exp(-x^2/2)\int_0^\infty \exp\left\{-\frac{(y-\sqrt{\alpha}\ x)^2}{2(1-\alpha)}\right\}dy=$$

$$4\varphi(x)(1-\Phi(-\frac{\sqrt{\alpha}}{\sqrt{1-\alpha}}x))=4\varphi(x)\Phi(\frac{\sqrt{\alpha}}{\sqrt{1-\alpha}}x).$$

For x>0 we have

$$h_\alpha(x)=h_{1-\alpha}(-x)=4\varphi(-x)\Phi(-\frac{\sqrt{1-\alpha}}{\sqrt{\alpha}}x)=4\varphi(x)(1-\Phi(\frac{\sqrt{1-\alpha}}{\sqrt{\alpha}}x)).$$

Hence, for any $0<\alpha<1$,

$$h_\alpha(x)=\begin{cases}4\varphi(x)\Phi((\alpha/(1-\alpha))^{1/2}x), & if \quad x\leq 0,\\ 4\varphi(x)(1-\Phi((1/\alpha-1)^{1/2}x)), & if \quad x>0\end{cases}. \tag{3.2.16}$$

Note that (3.2.15) can be simplified for the case $\alpha=1/2$ (it corresponds to the median $S_{n/2,n}$).In this situation the limit distribution is symmetric,

$$h_{1/2}(-x)=h_{1/2}(x)=4\varphi(x)(1-\Phi(x)), x\geq 0,$$

and

$$H_{1/2}(x)=2\Phi^2(x),$$

if $x\leq 0$, and

$$H_{1/2}(x)=1-H_{1/2}(-x)=1-2\Phi^2(-x)=$$

$$1-2(1-\Phi(x))^2=4\Phi(x)-2\Phi^2(x)-1,$$

if $x>0$.

To characterize the limit distribution of the normalized quantile

$$S_{\alpha n,n}/\sigma n^{1/2}$$

we will find its expectation E_α and variance D_α. It follows from (3.2.12) that

$$E_\alpha=\alpha^{1/2}E\nu+(1-\alpha)^{1/2}E\mu$$

and

$D_\alpha=\alpha Var(\nu)+(1-\alpha)Var(\mu)$.

The equality

$$\nu \overset{d}{=} -\mu$$

implies that

$$E_\alpha=(\sqrt{\alpha}-\sqrt{1-\alpha})E\nu$$

and

$$D_\alpha= Var(\nu).$$

Simple calculations come to equalities

$$E\nu=(2/\pi)^{1/2},\ D\nu=1-2/\pi.$$

Finally, we get that

$$E_\alpha=(\sqrt{\alpha}-\sqrt{1-\alpha})(2/\pi)^{1/2}$$

and

$$D_\alpha=1-2/\pi$$

for any $0\le\alpha\le1$. It is interesting to see that D_α does not depend on α.

2). Consider now the case $a>0$. We again will use relation (3.2.11).
We discussed above that if $X, X_1, X_2,\ldots$ are independent identically distributed random variables and $EX<0$, then (see, for example, Feller (1971), ch.XII) sums S_n tend to $-\infty$, as $n\to\infty$, and

$$P\{\max(0,S_1,S_2,\ldots)<\infty\}=1.$$

Symmetrically, in our case sums S_n tend to ∞, as $n\to\infty$, and

$$P\{\min(0,S_1,S_2,\ldots)>-\infty\}=1.$$

Then

$0 \geq S_{0,(1-\alpha)n} \geq \theta$,

where a negative random variable $\theta=\min(0,S_1,S_2,\ldots)$ has a proper distribution, and

$$S_{0,(1-\alpha)n}/n^{1/2} \xrightarrow{d} 0, \tag{3.2.17}$$

as $n\to\infty$.

Recall that in our situation Theorem 3.1.6 is valid and it follows from (3.1.45) that

$$\sup_x |P\{S_{\alpha n,\alpha n}-a\alpha n<x\sigma(\alpha n)^{1/2}\}-\Phi(x)| \to 0, \tag{3.2.18}$$

as $n\to\infty$.

We can rewrite (3.2.11) as

$$(S_{\alpha n,n}- a\alpha n)/\sigma(\alpha n)^{1/2} \stackrel{d}{=} ((S_{\alpha n,\alpha n}-a\alpha n)/\sigma(\alpha n)^{1/2}) *(S_{0,(1-\alpha)n}/\sigma(\alpha n)^{1/2}). \tag{3.2.19}$$

We get from (3.2.17)-(3.2.19) that

$$\sup_x |P\{S_{\alpha n,n}-a\alpha n<x\sigma(\alpha n)^{1/2}\}-\Phi(x)| \to 0, \tag{3.2.20}$$

as $n\to\infty$.

3). Let $a<0$. In some sense this situation is symmetric to the case $a>0$. It is easy to see that now the asymptotic behavior of $S_{\alpha n,n}$ is determined by the second term on the right side of (3.2.11). Using the same arguments as in the previous situation we come to the following relation:

$$\sup_x |P\{S_{\alpha n,n}-a(1-\alpha)n<x\sigma((1-\alpha)n)^{1/2}\}-\Phi(x)| \to 0, \tag{3.2.21}$$

Relations (3.2.13) and (3.2.14), as well as (3.2.20) and (3.2.21), describe the asymptotic distributions of quantiles $S_{\alpha n,n}$ for the case, when i.i.d. random variables $X_1, X_2,\ldots$ have a finite variance. Equality (3.2.11) enables us to find the corresponding limit distributions of $S_{\alpha n,n}$ in all i.i.d. situations , for which we know the asymptotics of extremes $S_{0,n}$ and $S_{n,n}$.

Example 3.1.4

Nevzorov (1979) considered the Cauchy distribution with

$$F(x)=\frac{1}{2}+\frac{arctg(x-a)}{\pi}$$

and got that the limit density function for $S_{n,n}/n$ has the form (3.1.22):

$$g_{1,a}(x)=\frac{x^{\delta-1}}{\pi(1+(x-a)^2)^{(1+\delta)/2}}(1+a^2)^{h(x)}\exp\{-\frac{1}{\pi}\int_0^x\frac{\log v}{1+(v-a)^2}dv\},$$

where

$$\delta=\frac{1}{2}+\frac{arctga}{\pi}$$

and

$h(x)=(arctg(x-a)+arctga)/2\pi$.

The symmetry arguments show that the limit density function for $S_{0,n}/n$ coincides with

$$g_{0,a}(x)=\frac{x^{\lambda-1}}{\pi(1+(x+a)^2)^{(1+\lambda)/2}}(1+a^2)^{r(x)}\exp\{-\frac{1}{\pi}\int_0^x\frac{\log v}{1+(v+a)^2}dv\},$$

where

$$\lambda=\frac{1}{2}-\frac{arctga}{\pi}$$

and

$r(x)=(arctg(x+a)-arctga)/2\pi$.

From (3.2.11) we have the relation

$$\frac{S_{\alpha n,n}}{n} \stackrel{d}{=} \alpha\frac{S_{\alpha n,\alpha n}}{\alpha} * (1-\alpha)\frac{S_{0,(1-\alpha)n}}{1-\alpha}.$$

Hence the asymptotic density function $g_{\alpha,a}$ for $S_{\alpha n,n}/n$ coincides with the convolution of the densities

$g_{1,a}(x/\alpha)/\alpha$

and

$g_{0,a}(x/(1-\alpha))/(1-\alpha))$.

The Arcsine law and its generalization

Consider i.i.d. X's with zero expectations and unit variances.
Let $N_n(x)$ be a number of sums amongst $S_0,S_1,\ldots,S_n$, that exceed or equal x and denote $N_n=N_n(0+0)$ the number of positive sums. As it is known (see Feller (1971), ch. XII), the asymptotic distribution of N_n/n satisfies the following Arcsine law:

$$\lim_{n\to\infty} P\{N_n<n\alpha\}=(2/\pi)\arcsin\sqrt{\alpha}\ ,\ 0<\alpha<1. \quad (3.2.22)$$

Evidently, the distributions of order statistics $S_{k,n}$ and random variables N_n are dual, since (for the sake of simplicity we suppose that $n\alpha$ is an integer)

$$P\{N_n<m\}=P\{S_{n+2-m,n}<0\},\ m=1,\ldots,n+1. \quad (3.2.23)$$

From (3.2.13) and (3.2.23) we have

$$T(\alpha)=\lim_{n\to\infty} P\{N_n<n\alpha\}=\lim_{n\to\infty} P\{S_{(1-\alpha)n,n}<0\}=H_{1-\alpha}(0)=$$

$$4\int_0^\infty \Phi\left(-\frac{y\sqrt{1-\alpha}}{\sqrt{\alpha}}\right)\varphi(y)dy= 4\int_{-\infty}^0 \Phi\left(\frac{y\sqrt{1-\alpha}}{\sqrt{\alpha}}\right)\varphi(y)dy. \quad (3.2.24)$$

It is not difficult to prove that the RHS's of (3.2.22) and (3.2.24) coincide. It suffices to check that both the expressions are equal to 1/2 for $\alpha=1/2$ (that is evident) and to compare that their derivatives coincide for any $0<\alpha<1$. One can see that

$$T'(\alpha)=2\alpha^{-3/2}(1-\alpha)^{-1/2}\int_{-\infty}^0 y\varphi\left(\frac{y\sqrt{1-\alpha}}{\sqrt{\alpha}}\right)\varphi(y)dy$$

$$= \pi^{-1}\alpha^{-3/2}(1-\alpha)^{-1/2} \int_{-\infty}^{0} y\exp(-y^2/2\alpha)dy = 1/\pi\alpha^{1/2}(1-\alpha)^{1/2}$$

and the latter expression really coincides with the density function of the limit distribution (3.2.22).

To find the asymptotic distribution function of $N_n(x)$ we can use a more general than (3.2.23) relation

$$P\{N_n(x)<m\}=P\{ S_{n+2-m,n}<x\}, \tag{3.2.25}$$

from which one gets (recalling (3.2.15)) that

$$T(v, \alpha)=\lim_{n\to\infty} P\{N_n(v\sqrt{n})<n\alpha\}=\lim_{n\to\infty} P\{ S_{(1-\alpha)n,n}< v\sqrt{n}\}=H_{1-\alpha}(v)=$$

$$\begin{cases} 4\int_0^{\infty}\Phi(\frac{v-y\sqrt{1-\alpha}}{\sqrt{\alpha}})\varphi(y)dy, & if \quad v\le 0, \\ 4\int_0^{\infty}\Phi(\frac{v+y\sqrt{\alpha}}{\sqrt{1-\alpha}})\varphi(y)dy-1, & if \ v>0 \end{cases} . \tag{3.2.26}$$

This expression can be transformed to be closer to "arcsine" form. One can get that

$$\partial T(v,\alpha)/\partial\alpha= \pi^{-1}\alpha^{-1/2}(1-\alpha)^{-1/2}\exp\{-v^2/2\alpha\},\ v\le 0,$$

and

$$\partial T(v,\alpha)/\partial\alpha= \pi^{-1}\alpha^{-1/2}(1-\alpha)^{-1/2}\exp\{-v^2/2(1-\alpha)\},\ v>0.$$

Taking into account that

$$T(v,1)=2\Phi(v),\ v\le 0,$$

and

$$T(v,0)= 2\Phi(v)-1,\ v>0,$$

we finally have the following relations:

$$T(v, \alpha)=2\Phi(v)-\frac{1}{\pi}\int_{\alpha}^{1}\frac{1}{\sqrt{t(1-t)}}\exp\{-v^2/2t\}dt,$$

if $v \le 0$, and

$$T(v, \alpha) = 2\Phi(v) - 1 + \frac{1}{\pi}\int_0^{\alpha} \frac{1}{\sqrt{t(1-t)}} \exp\{-v^2/2(1-t)\}\, dt,$$

if $v>0$. Taking $v=0$ we get the classical Arcsine law:

$$T(0, \alpha) = \frac{1}{\pi}\int_0^{\alpha} \frac{1}{\sqrt{t(1-t)}}\, dt = (2/\pi) \arcsin \sqrt{\alpha}\ .$$

The invariance principle

Taking into account relations which are similar to (3.2.25) we can apply the invariance principle to find asymptotic distributions of quantiles $S_{\alpha n,n}$ for more general than i.i.d. sequences of random variables $X_1, X_2, \ldots$ .We will use another construction than one that was used above to apply the invariance principle for finding the limit distributions of extreme order statistics $S_{n,n}$ and $S_{0,n}$.. First, for the sake of simplicity, to describe the method we consider a sequence of i.i.d. random variables $X, X_1, X_2, \ldots$ with $EX=0$ and $Var(X) = \sigma^2$, $0 < \sigma < \infty$.

In section 3.1 we constructed broken lines

$$X_n(t) = (S_{[nt]} + (nt - [nt])X_{[nt]+1}) / \sigma n^{1/2},\ 0 \le t \le 1,$$

and applied the convergence of processes $X_n(t)$ to a Wiener process $W(t)$ on the set $C[0,1]$ of continuous paths. It turns out that one can use another form of the invariance principle that holds on the set $D=D[0,1]$ of functions $x(t)$ such that $x(t+0)=x(t)$ for any $0 \le t < 1$, and there exist limits $x(t-0) = \lim x(s)$, as $s \uparrow t$, for $0 < t \le 1$ (for details, see Billingsley (1968)). Due to the Donsker- Prohorov invariance principle,

$$\tilde{X}_n(t) \xrightarrow{d} W(t),\ 0 \le t \le 1, \tag{3.2.27}$$

where

$$\tilde{X}_n(t) = S_{k,n} / \sigma n^{1/2}, \tag{3.2.28}$$

if

$$(k-1) < t \le k/n,\ 0 \le k \le n,$$

and W(t) is a Wiener process. It means that for any continuous (everywhere , except may be a set of zero Wiener measure) functional $\varphi(x(t))$ defined on the set D[0,1] of functions x(t), $0\le t\le 1$,

$$P\{\varphi(X_n(t))<x\}\to P\{\varphi(W(t))<x\}, \qquad (3.2.29)$$

as $n\to\infty$. We know how to use (3.2.29) if we need to get the asymptotic distributions of

$$S_{0,n}=\min_{0\le k\le n} S_k$$

and

$$S_{n,n}=\max_{0\le k\le n} S_k\,.$$

In these cases we must find the distributions of

$$\inf_{0\le t\le 1} W(t)$$

and

$$\sup_{0\le t\le 1} W(t).$$

The question arises: what a functional φ can be taken to study the asymptotic distribution of quantiles $S_{\alpha n,n}$, $0<\alpha<1$?

Let us consider some occupation times for processes $X_n(t)$ and W(t), $0\le t\le 1$. We define

$$\delta_x(n)=\text{mes}\{t\in[0,1]:X_n(t)\ge x\}$$

and

$$D_x=\text{mes}\{\, t\in[0,1]:W(t)\ge x\}$$

for any level x, mes{A} being the Lebesque measure of a set A. Using the structure of $X_n(t)$ we can express the distributions of quantiles $S_{\alpha n,n}$ via the distributions of $\delta_x(n)$ as follows:

$$P\{\, S_{\alpha n,n}<x\sigma n^{1/2}\}=P\{\delta_x(n)<1-\alpha\}. \qquad (3.2.30)$$

For such functionals on D[0,1] as occupation times we can apply the invariance principle (see, for example, Billingsley (1968)). One gets the following limit relation as $n \to \infty$:

$$\lim P\{\delta_x(n)<1-\alpha\}=P\{D_x<1-\alpha\}. \tag{3.2.31}$$

It follows now from (3.2.30) and (3.2.31) that

$$\lim P\{ S_{\alpha n,n}<x\sigma n^{1/2}\}=P\{ D_x<1-\alpha\},\ n \to \infty. \tag{3.2.32}$$

Moreover, in the given situation we can get the joint asymptotic distributions of any quantiles

$$S_{\alpha(1)n,n},\ S_{\alpha(2)n,n},\dots,\ S_{\alpha(r)n,n}\ (0<\alpha(1)<\alpha(2)<\dots<\alpha(r)<1)$$

since (due to the invariance principle) we have that

$$\lim P\{ S_{\alpha(1)n,n}<x_1\sigma n^{1/2},\dots,\ S_{\alpha(r)n,n}<x_r\sigma n^{1/2}\}=$$
$$P\{ D_{x_1}<1-\alpha(1),\dots,D_{x_r}<1-\alpha(r)\},\ n \to \infty., \tag{3.2.33}$$

for any $x_1,\dots,x_r$.

Instead of (3.2.27) one can use other forms of the invariance principle for random processes

$$\tilde{X}_n(t)=S_{[tn],n}/b_n,$$

where b_n, n=1,2,..., are suitably chosen normalizing constants. The sequence of processes $\tilde{X}_n(t)$ in the limit relation

$$\tilde{X}_n(t) \xrightarrow{d} X(t),\ 0\le t\le 1 \tag{3.2.34}$$

may converge not only to a Wiener process W(t). As a limit process X(t) in (3.2.34) one can get , for instance, Brownian bridges, stable processes, processes of the type $W(\psi(t))$, where W(t) is a Wiener process and $\psi(t)$ is a continuous nondecreasing function, such that $\psi(0)=0$ and $\psi(1)=1$, etc. In any case when we have relation (3.2.34) with a certain limit process X(t), the knowing of the distribution of the functional

$$D_x= \text{mes}\{ t\in[0,1]:X(t)\ge x\}$$

suffices to find the asymptotic distribution of normalized quantiles $S_{\alpha n,n}/b_n$. Vice versa, if we know the limit distribution for $S_{\alpha n,n}/b_n$ we can get the corresponding expression for distributions of occupations times for X(t). For example, equalities (3.2.13), (3.2.14) and (3.2.32) imply that for a Wiener process W(t), the distribution of the functional

$$D_x = \text{mes}\{ t\in[0,1]: W(t)\geq x\}$$

is given as follows:

$$P\{ D_x < \alpha\} = H_{1-\alpha}(x),$$

where $H_\alpha(x)$ is determined in (3.2.15).

Note that relation (3.2.27) with

$$\widetilde{X}_n(t) = S_{[nt],n}/ b_n$$

and a Wiener process W(t) as limit may be true not only for i.i.d. X's. There are situations when (3.2.27) (possibly with other normalized constants b_n instead of $\sigma n^{1/2}$) is valid for some sequences of dependent or/and nonidentically distributed random variables $X_1, X_2, \ldots$. In these situations we can write immediately (since the limit process on the right side of (3.2.27) stays to be Wiener) that

$$\lim P\{ S_{\alpha n,n} < x b_n\} = P\{ D_x < 1-\alpha\} = H_\alpha(x),\ n \to\infty,$$

$H_\alpha(x)$ being the same limit distribution function as in the i.i.d. case.

As it is known, Prohorov (1956) extended the invariance principle of Donsker for nonidentically distributed X's.

Let $X_1, X_2, \ldots$ be independent random variables, $EX_k=0$ and $Var(X_k)= \sigma_k^2$, $k=1,2,\ldots$. Let also $X_1, X_2, \ldots$ satisfy the Lindeberg condition. Define

$$B_k^2 = \sum_{j=1}^{k} \sigma_j^2,$$

and let for any $n=1,2,\ldots$,

$$\overline{X}_n(t) = S_{kn,n} / B_n$$

for

$$t\in[B_k^2/B_n^2, B_{k+1}^2 / B_n^2),\ k=0,1,\ldots,n-1,$$

and

$\overline{X}_n(1) = S_{n,n}/B_n$.

The Donsker- Prohorov invariance principle states that under these circumstances

$$\overline{X}_n(t) \xrightarrow{d} W(t),\ 0 \le t \le 1. \tag{3.2.35}$$

Unfortunately the given form of statement (3.2.27) can be used to apply the previous arguments only if

$$B_k^2/B_n^2 = k/n.$$

Otherwise the relation

$$P\{S_{\alpha n,n} < xB_n\} = P\{\delta_x(n) < 1-\alpha\}, \tag{3.2.36}$$

which is substantial for our inferences, is not true. In some situations the appropriate substituting of t by $\tau(t)$, where $\tau(t)$ is such nondecreasing continuous function that

$$\tau(B_k^2/B_n^2) = k/n,\ k=0,1,\dots,n, \tag{3.2.37}$$

generates a new process $Y_n(t)$ defined as follows:

$$Y_n(t) = S_{kn,n}/B_n$$

for $t \in [k/n, (k+1/n))$. Then (3.2.35) can be rewritten as

$$Y_n(t) \xrightarrow{d} X(t),\ 0 \le t \le 1, \tag{3.2.38}$$

where the limit process on the right side has the form

$$X(t) = W(\tau(t)),$$

and we need to know the distribution of occupation times for the process $W(\tau(t))$ to get the asymptotic distribution of normalized quantiles. Since relation (3.2.38) is asymptotic, restrictions (3.2.37) can be weakened as follows:
for any $0 \le \beta \le 1$,

$$\tau(B_{[\beta n]}^2/B_n^2) \to \beta, \text{ as } n \to \infty.$$

For example, if

$$B_n^2 \sim \sigma^2 n, \; n \to \infty, \tag{3.2.39}$$

then we can take $\tau(x)=x$. Hence, if $X_1, X_2, \ldots$ are independent random variables with zero means and variances satisfying (3.2.39) and the Lindeberg condition holds then the asymptotic distribution for $S_{\alpha n,n}/\sigma n^{1/2}$ is the same as in the i.i.d. case.

One can see that it is very important for us to know how to find the distributions of occupation times for the appropriate limit processes. We will discuss below the case of Wiener processes.

Let

$$I(t,x)=\int_0^t f(W(s)+x)ds$$

and

$$\varphi(t,x,z)=E\exp\{izI(t,x)\}=E\exp\{iz\int_0^t f(W(s)+x)ds\}, \tag{3.2.40}$$

where f(x) is a piecewise continuous and bounded on any finite interval function and let

$$\tilde{\varphi}(\lambda,x,z)=\int_0^\infty e^{-\lambda t}\varphi(t,x,z)dt$$

be its Laplace transform. It turns out (see, for example, Skorohod and Slobodenjuk (1970), ch.3) that $\tilde{\varphi}(\lambda,x,z)$ satisfies the following differential equation

$$\frac{1}{2}\tilde{\varphi}''_{xx}(\lambda,x,z)+(izf(x)-\lambda)\tilde{\varphi}(\lambda,x,z)+1=0, \tag{3.2.41}$$

and a solution of the equation is bounded and have a continuous derivative .

Let us consider the case, when $f(x)=0$, if $x<0$, and $f(x)=1$, if $x\geq 0$. Then

$$I(t,-x)=\int_0^t f(W(s)-x)ds$$

coincides with the occupation time

$$D_{x,t}= \text{mes}\{ u\in[0,t]:W(u)\geq x\}.$$

First we must solve equation (3.2.41) separately for x<0 and x>0. In these two cases (3.2.41) transforms correspondingly to

$$\frac{1}{2}\tilde{\varphi}''_{xx}(\lambda,x,z)-\lambda\,\tilde{\varphi}(\lambda,x,z)+1=0,$$

for x<0, and

$$\frac{1}{2}\tilde{\varphi}''_{xx}(\lambda,x,z)+(iz-\lambda)\,\tilde{\varphi}(\lambda,x,z)+1=0,$$

for x>0.

The traditional methods give the following solution with arbitrary coefficients c_1, c_2, c_3 and c_4:

$$\tilde{\varphi}(\lambda,x,z)=c_3\exp(x\sqrt{2\lambda})+c_4\exp(-x\sqrt{2\lambda})+\frac{1}{\lambda}, \tag{3.2.42}$$

if x<0, and

$$\tilde{\varphi}(\lambda,x,z)=c_1\exp(x\sqrt{2\lambda-2iz})+c_2\exp(-x\sqrt{2\lambda-2iz})+\frac{1}{\lambda-iz}, \tag{3.2.43}$$

if x>0.

Since $\tilde{\varphi}(\lambda,x,z)$ must be bounded we get from (3.2.42) and (3.2.43) that $c_4=c_1=0$ and

$$\tilde{\varphi}(\lambda,x,z)=c_3\exp(x\sqrt{2\lambda})+\frac{1}{\lambda}, \tag{3.2.44}$$

if x<0, and

$$\tilde{\varphi}(\lambda,x,z)=c_2\exp(-x\sqrt{2\lambda-2iz})+\frac{1}{\lambda-iz}, \tag{3.2.45}$$

if x>0.

Now , recalling that $\tilde{\varphi}(\lambda,x,z)$ must be continuous , we substitute x=0 in (3.2.44) and (3.2.45), and equal the corresponding values:

$$c_3+\frac{1}{\lambda}=c_2+\frac{1}{\lambda-iz}.$$

Since the derivative $\tilde{\varphi}'_x$ also must be continuous we similarly find that

$$c_3\sqrt{2\lambda}=-\,c_2\sqrt{2\lambda-2iz}.$$

Hence,

$$c_2=(1-\sqrt{1-\frac{iz}{\lambda}})/(iz-\lambda)$$

and

$$c_3==(1/\lambda\sqrt{1-\frac{iz}{\lambda}})-1/\lambda.$$

We get that

$$\tilde{\varphi}(\lambda,x,z)=((1-\sqrt{1-\frac{iz}{\lambda}})/(iz-\lambda))\exp(-x\sqrt{2\lambda-2iz})+\frac{1}{\lambda-iz}, \tag{3.2.46}$$

if $x\geq 0$, and

$$\tilde{\varphi}(\lambda,x,z)=(\frac{1}{\lambda(1-\frac{iz}{\lambda})^{1/2}}-\frac{1}{\lambda})\exp(x\sqrt{2\lambda})+\frac{1}{\lambda}, \tag{3.2.47}$$

if $x<0$.

By that we have found the Laplace transform for function

$$\varphi(t,x,z)=E\exp\{iz\int_0^t f(W(s)+x)ds\},$$

which in our case coincides with characteristic function for the occupation time

$D_{-x,t}=\text{mes}\{\,u\in[0,t]:W(u)\geq -x\}.$

The further actions are to find $\varphi(t,x,z)$ using the tables of Laplace transforms or other methods. If we know $\varphi(t,x,z)$ or

$$\varphi(t,x,-iz)= E\exp\{-z \int_0^t f(W(s)+x)ds\},$$

then we can try to get the distribution function or the density distribution function for D_x, $-\infty<x<\infty$.

Consider the simplest case x=0. We have

$$\tilde{\varphi}(\lambda,0,z)= \frac{1}{\lambda\,(1-\frac{iz}{\lambda})^{1/2}} \quad . \tag{3.2.48}$$

The expansion in a series of the RHS of (3.2.48) and the relation

$$\int_0^\infty \exp(-\lambda t)t^k dt=k!/\lambda^{k+1},\ k=0,1,\ldots,$$

give us the following equality:

$$\int_0^\infty e^{-\lambda t}\varphi(t,0,z)dt = \int_0^\infty \exp(-\lambda t)\sum_{k=0}^\infty \frac{(2k-1)!!}{(2k)!!}\frac{(izt)^k}{k!}dt, \tag{3.2.49}$$

where

$$(2k-1)!!=\prod_{j=1}^k (2j-1)$$

and

$$(2k)!!= \prod_{j=1}^k 2j= 2^k k!\ .$$

Hence,

$$\varphi(t,0,z)=\sum_{k=0}^{\infty}\frac{(2k-1)!!}{(2k)!!}\frac{(izt)^k}{k!}. \qquad (3.2.50)$$

Since

$$\frac{(2k-1)!!}{(2k)!!}=\frac{2}{\pi}\int_0^{\pi/2}\sin^{2k}v\,dv,$$

we have that

$$\varphi(t,0,z)=\frac{2}{\pi}\int_0^{\pi/2}\sum_{k=0}^{\infty}\frac{(izt\sin^2 v)^k}{k!}dv=$$

$$\frac{2}{\pi}\int_0^{\pi/2}\exp\{izt\sin^2 v\}dv=\frac{2}{\pi}\int_0^{t}\exp\{izv\}d(\arcsin(v/t)^{1/2}). \qquad (3.2.51)$$

We see now that the RHS of (3.2.51) is the characteristic function, corresponding to the distribution function

$$T(x)=\frac{2}{\pi}\arcsin\sqrt{v/t},\ 0\le t\le v,$$

and hence

$$P\{I(1,0)<v\}=P\{D_{0,1}<v\}=\begin{cases}0, & if\ \ v<0,\\ \frac{2}{\pi}\arcsin\sqrt{v}, & if\ \ 0\le v\le 1,\\ 1, & if\ \ v>t\end{cases} \qquad (3.2.52)$$

where

$$D_{0,1}=\text{mes}\{u\in[0,1]:W(u)\ge 0\}.$$

Thus, (3.2.52) provides the classical Arcsine law. Converting analogously (3.2.46) and (3.2.47) one can get for $x\ge 0$ that

$$P\{D_{x,1}<1-\alpha\}=P\{D_{-x,1}<\alpha\}=1-\frac{4}{\pi}\arcsin\sqrt{\alpha}+\frac{2\sqrt{1-\alpha}}{\pi}\int_0^{\alpha}\Phi\left(\frac{x}{\sqrt{v}}\right)\frac{1}{1-v}\frac{1}{\sqrt{\alpha-v}}dv. \tag{3.2.53}$$

It means that we have one more formula for

$H_\alpha(x) = \lim P\{S_{\alpha n,n}<x\sigma n^{1/2}\}$, $n\to\infty$,

defined in (3.2.15):

$H_\alpha(x)= P\{D_{x,1}<1-\alpha\}$.

The RHS of (3.2.53) can be simplified if $\alpha=1/2$. Then, for any positive x we get that

$H_{1/2}(x) =1-H_{1/2}(-x)= P\{D_{x,1}<1/2\}=P\{D_{-x,1}<1/2\}$

$$=\frac{\sqrt{2}}{\pi}\int_0^{1/2}\Phi\left(\frac{x}{\sqrt{v}}\right)\frac{1}{1-v}\frac{1}{\sqrt{1/2-v}}dv$$

$$=\frac{4}{\pi}\int_0^{1}\Phi\left(\frac{x\sqrt{2}}{\sqrt{1-u^2}}\right)\frac{1}{1+u^2}du. \tag{3.2.54}$$

Earlier we have got that

$H_{1/2}(x) =1-H_{1/2}(-x)= 1-2\Phi^2(-x)$.

Hence the following identity holds for $x\geq 0$:

$$\frac{4}{\pi}\int_0^{1}\Phi\left(\frac{x\sqrt{2}}{\sqrt{1-u^2}}\right)\frac{1}{1+u^2}du=1-2\Phi^2(-x). \tag{3.2.55}$$

Considering in (3.2.40)

$f(x)=0$, if $|x|<v$, and $f(x)=1$, if $|x|>v$,

for some positive v, one can find the distribution of the occupation time

$D^*_t= \text{mes}\{u\in[0,t]:|W(u)|\geq v\}$

and hence the asymptotic distribution of order statistics

$S_{[0,n]} \le S_{[1,n]} \le \ldots \le S_{[n,n]}$

based on random variables

$|S_0|, |S_1|, \ldots, |S_n|$.

It turns out in the recent situation that

$$G_\alpha(x) = \lim_{n\to\infty} P\{S_{[\alpha n,n]} < x\sigma\sqrt{n}\} = 1 - \int_{1-\alpha}^{1} p_x(y)dy, \; x>0, \; 0<\alpha<1,$$

where

$$p_x(y) = \frac{2}{\pi}\sum_{k=1}^{\infty} \Big(\int_y^1 \cos(2k \arccos\sqrt{\frac{y}{\tau}})y^{-1/2}(\tau-y)^{-1/2}$$

$$\Big(\frac{(2k-1)x}{\sqrt{2\pi}(1-\tau)^{3/2}}\exp\{-\frac{x^2(2k-1)^2}{2(1-\tau)}\} + \frac{2k}{\tau}(1-\Phi(\frac{(2k-1)x}{\sqrt{1-\tau}})))d\tau\Big), \; x>0, \; y>0.$$

Since

$$\lim P\{ S_{\alpha(1)n,n} < x_1\sigma n^{1/2}, \ldots, S_{\alpha(r)n,n} < x_r\sigma n^{1/2} \} =$$

$$P\{ D_{x_1} < 1-\alpha(1), \ldots, D_{x_r} < 1-\alpha(r)\}, \; n \to\infty,$$

the method described above gives a way to find the joint distributions of quantiles $S_{\alpha(1)n,n}, \ldots, S_{\alpha(r)n,n}$. We will give below a scheme of the corresponding method for two quantiles.

To obtain the joint distribution of two occupation times, say D_{x_1} and D_{x_2}, $x_1<x_2$, instead of (3.2.40) we can introduce the function

$$\varphi(t,x,z_1,z_2) = E\exp\{i\int_0^t (z_1 f_1(W(s)+x) + z_2 f_2(W(s)+x))\,ds\}, \qquad (3.2.56)$$

where

$f_1(x)=0$, if $x<0$, and $f_1(x)=1$, if $x>0$,

$f_2(x)=0$, if $x<x_2-x_1$, and $f_2(x)=1$, if $x>x_2-x_1$.

Then, we denote the Laplace transform of (3.2.56):

$$\tilde{\varphi}(\lambda,x,z_1,z_2)=\int_0^\infty e^{-\lambda t}\varphi(t,x,z_1,z_2)dt \tag{3.2.57}$$

and find that $\tilde{\varphi}(\lambda,x,z_1,z_2)$ satisfies the following differential equation (compare with (3.2.41)):

$$\frac{1}{2}\tilde{\varphi}''_{xx}(\lambda,x,z_1,z_2)+(iz_1 f_1(x)+iz_2f_2(x)-\lambda)\tilde{\varphi}(\lambda,x,z_1,z_2)+1=0. \tag{3.2.58}$$

Denote

$x^*=x_2-x_1>0$.

Separately considering equation (3.2.58) for $x<0$, $0<x<x^*$ and $x>x^*$ we get that its solution is as follows:

$$\tilde{\varphi}(\lambda,x,z_1,z_2)=c_1\exp(x\sqrt{2\lambda})+c_2\exp(-x\sqrt{2\lambda})+1/\lambda,$$

if $x<0$,

$$\tilde{\varphi}(\lambda,x,z_1,z_2)=c_3\exp(x\sqrt{2\lambda-2iz_1})+c_4\exp(-x\sqrt{2\lambda-2iz_1})+1/(\lambda-iz_1),$$

if $0<x<x^*$, and

$$\tilde{\varphi}(\lambda,x,z_1,z_2)=c_5\exp(x\sqrt{2\lambda-2i(z_1+z_2)})+$$

$$c_6\exp(-x\sqrt{2\lambda-2i(z_1+z_2)})+1/(\lambda-i(z_1+z_2)), \text{ if } x>x^*,$$

where c_k, $k=1,\dots,6$, are arbitrary constants. Coefficients $c_1,\dots, c_6$ are determined uniquely if we recall that $\tilde{\varphi}(\lambda,x,z_1,z_2)$ must be bounded function (and hence $c_2=c_5=0$) and it and its derivative are continuous functions. Checking the second condition under $x=0$ and $x = x^*$, we get the following equations for c_1, c_3, c_4 and c_6:

$c_1+1/\lambda=c_3+c_4+1/(\lambda-iz_1)$,

$c_3\exp(x^*\sqrt{2\lambda-2iz_1})+c_4\exp(-x^*\sqrt{2\lambda-2iz_1})+1/(\lambda-iz_1)$

$= c_6\exp(-x^*\sqrt{2\lambda-2i(z_1+z_2)})+1/(\lambda-i(z_1+z_2))$,

$c_1\sqrt{2\lambda}=c_3\sqrt{2\lambda-2iz_1} - c_4\sqrt{2\lambda-2iz_1}$,

$c_3\sqrt{2\lambda-2iz_1}\exp(x^*\sqrt{2\lambda-2iz_1}) - c_4\sqrt{2\lambda-2iz_1}\exp(-x^*\sqrt{2\lambda-2iz_1})$

$= - c_6^*\sqrt{2\lambda-2i(z_1+z_2)}\exp(-x^*\sqrt{2\lambda-2i(z_1+z_2)})$.

Solving this system of equations one finds $\tilde{\varphi}(\lambda,x,z_1,z_2)$. Substituting $x=-x_1$ we get the function $\tilde{\varphi}(\lambda,-x_1,z_1,z_2)$ that coincides with the Laplace transform for the joint characteristic function

$$\mathrm{E}\exp\{i\int_0^t (z_1f_1(W(s)-x_1)+z_2 f_2(W(s)-x_1))\, ds\}=$$

$$\mathrm{E}\exp\{iz_1D_{x_1,t}+iz_2 D_{x_2,t}\}.$$

With the help of tables of Laplace transforms and characteristic functions one can find now the joint distribution of $D_{x_1,t}$ and $D_{x_2,t}$. The last stage in finding of the joint distribution of two quantiles is to use the equality

$$\lim_{n\to\infty} P\{ S_{\alpha(1)n,n}<x_1\sigma n^{1/2}, S_{\alpha(2)n,n}<x_2\sigma n^{1/2} \}=P\{ D_{x_1} <1-\alpha(1), D_{x_2} <1-\alpha(2)\}.$$

Indeed, this method is rather complicated but it enables us to get an explicit expression for joint distributions of at least two quantiles .

Asymptotic distributions of order statistics based on sums of shifted terms

Formulae (3.2.20) and (3.2.21) give the asymptotic distribution for order statistics $S_{\alpha n,n}$ in the case when the initial X's are independent identically distributed random variables and have positive (negative) expectations. As we know, it sufficies to consider only the case of positive means, because asymptotic distribution of $S_{\alpha n,n}$ for the case of negative means coincides with the distribution of $-S_{(1-\alpha)n,n}$ for X's with positive means.

Indeed, if $X_1, X_2, \ldots$ are positive random variables, then $S_{k,n} = S_k$. It turns out that if sums S_n tend to infinity (in probability) as $n \to \infty$, then the distribution of $S_{k,n}$ continues to stay close to the distribution of S_k. Consider this situation in details.

The following result is valid for arbitrary random variables $X_1, X_2, \ldots$.

Theorem 3.2.2.

For any sequence of random variables $X_1, X_2, \ldots$ the next inequalities hold:

$$_{k} + S'_{0,n-k} \leq S_{k,n} \leq S_{k,k} \tag{3.2.59}$$

and

$$S_{0,n-k} \leq S_{k,n} \leq S_{n-k} + S''_{k,k}, \tag{3.2.60}$$

Where

$$S'_{0,n-k} = \min\{0, X_{k+1}, X_{k+1} + X_{k+2}, \ldots, X_{k+1} + \ldots + X_n\}$$

and

$$S''_{k,k} = \max\{0, X_{n-k+1}, \ldots, X_n\}.$$

Proof.

It is easy to see that

$$S_{k,n} \geq \min\{S_k, S_{k+1}, \ldots, S_n\} = S_k + \min\{0, X_{k+1}, \ldots, X_{k+1} + \ldots + X_n\} = S_k + S'_{0,n-k}.$$

Then, $S_{k,n}$ presents the (k+1)-th minimum amongst sums $S_0, S_1, \ldots, S_n$ and indeed it is less or equal than the sum $S_{k,k}$ which is the (k+1) minimum amongst sums $S_0, S_1, \ldots, S_k$. These simple arguments prove (3.2.59). The similar way can be used to prove (3.2.60).

Remark 3.2.1.

If $X_1, X_2, \ldots$ are independent, then S_k and $S'_{0,n-k}$ also are independent as well as random variables S_{n-k} and $S''_{k,k}$.

We have mentioned above (see , for example, theorem 3.1.7) that in the situation, when sums S_n tend to infinity, distributions of $S_{k,k}$ and S_k are close one to another.

In this case it is convenient to use inequalities (3.2.59). A rather general result (see the statement including relation (3.1.46)) is connected with the case when there exists such a sequence $\varepsilon_n>0$, n=1,2,..., that

$$\max\{0,-X_1,\dots,-X_1-\dots-X_n\}/\varepsilon_n \xrightarrow{p} 0$$

and

$$Q(S_n, \varepsilon_n) \to 0,\ n\to\infty,$$

where

$$Q(Y, \varepsilon)=\sup_x P\{x\le Y<x+\varepsilon\}$$

denotes a concentration function of a random variable Y. It turns out that under the given conditions the following relation is true:

$$\sup_x |P\{S_n<x\}-P\{S_{n,n}<x\}|\to 0,\ n\to\infty.$$

It follows from (3.2.59) that

$$P\{S_{k,n}<x\}-P\{S_k<x\}=$$

$$P\{S_{k,n}<x,\ S_k\ge x+\varepsilon\}-P\{S_{k,n}\ge x,\ S_k<x+\varepsilon\}+P\{x\le S_k<x+\varepsilon\}\le$$

$$P\{S'_{0,n-k}<-\varepsilon\}+Q(S_k, \varepsilon) \qquad (3.2.61)$$

and

$$P\{S_{k,n}<x\}-P\{S_k<x\}\ge P\{S_{k,k}<x\}-P\{S_k<x\}. \qquad (3.2.62)$$

We will suppose that $0\le k\le n$ and $k(n)\to\infty$, as $n\to\infty$.

On combining (3.1.46), (3.1.61) and (3.1.62), one can see that the next statement is valid for an arbitrary sequence of random variables $X_1, X_2,\dots$:
if there exists such constants ε_n that

$$\min\{0, X_1,\dots, X_1+\dots+X_k\}/\varepsilon_n \xrightarrow{p} 0 \qquad (3.2.63)$$

$$\min\{0, X_{k+1},\dots, X_{k+1}+\dots+X_n\}/\varepsilon_n \xrightarrow{p} 0 \qquad (3.2.64)$$

and

$$Q(S_k, \varepsilon_n) \to 0, \ n \to \infty, \tag{3.2.65}$$

then

$$\sup_x |P\{S_k < x\} - P\{S_{k,n} < x\}| \to 0, \ n \to \infty. \tag{3.2.66}$$

Indeed, (3.2.66) holds for an arbitrary sequence of independent identically distributed X's with a positive mean EX=a, because in this case it suffices to choose any sequence ε_n, n=1,2,..., such that $\varepsilon_n = o(k^{1/2})$ and $\varepsilon_n \to \infty$, as $n \to \infty$.
In fact, (3.2.65) follows from the result of Petrov (1970) who has proved that

$$Q(S_n, \varepsilon) \le L(\varepsilon + 1) n^{-1/2}$$

for sums of independent identically distributed random variables, where L does not depend on ε and n. Then, we know that

$$M = \min\{0, X_1, X_1 + X_2, \ldots\}$$

is a proper random variable for i.i.d. random variables with positive means (see, for example, Feller (1971), ch.XII) and

$$P\{M < -\varepsilon_n\} \to 0,$$

if $\varepsilon_n \to \infty$. Hence both the conditions (3.2.63) and (3.2.64) are also valid.

Remark 3.2.2.

Let (3.2.63)-(3.2.65) hold for a sequence k=k(n), such that $0 \le k(n) \le n$ and $k(n) \to \infty$. If there exist such constants a_n and b_n and a limit distribution function T that

$$\sup_x | P\{S_n - a_n < x b_n\} - T(x)| \to 0,$$

then

$$\sup_x | P\{S_{k(n),n} - a_{k(n)} < x b_{k(n)}\} - T(x)| \to 0, \ n \to \infty.$$

The analogous conditions to provide the relation

$$\sup_x |P\{S_{n-k}<x\}-P\{ S_{k,n}<x\}|\to 0,\ n\to\infty,$$

can be also formulated if to use inequalities (3.2.60). It is evident that in the case when sums S_n tend to $-\infty$ the distribution of order statistics $S_{k,n}$ must be close to the distribution of the sum $S_{n-k.}$

REFERENCES

1. Akhundov, I.S. (1987). On the distribution of maximum normed sum. *Vestnik of Leningrad State Univ.* **8**, 99-102.

2. Aleshkevichene, A.K. (1977). Approximation of the distribution of the maximum of sums of random variables with a mean that is small in absolute value. *Dokl. Akad. Nauk. SSSR* **233**, 1017-1020.

3. Arak, T.V. (1973). On the rate of convergence of the distribution of the maximum of successive sums of independent random variables. *Dokl. Akad. Nauk SSSR* **208**, 11-13 (in Russian).

4. Arak, T.V. (1974). On the distribution of cumulative sums of independent random variables. *Theory Probab. Appl.* **19**, 245-266..

5. Arak, T.V. and Nevzorov, V.B. (1973). Some estimates for the maximum cumulative sum of independent random variables. *Theory Probab. Appl.* **18**, 384-387.

6. Baxter, G. (1958). An operator identity. *Pacific J.Math.* **8**, 649-663.

7. Baxter, G. and Donsker, M.D. (1957). On the distribution of the supremum functional for processes with stationary independent increments. *Trans. Amer. Math. Soc.* **85**, 73-87.

8. Baum, L.E. and Katz, M. (1965). Convergence rates in the law of large numbers. *Trans. Amer. Math. Soc.* **120**, 108-123.

9. Billingsley, P. (1956). The invariance principle for dependent random variables. *Trans. Amer. Math. Soc.* **83**, 250-268.

10. Borovkov, A.A. (1970). On factorization identities and properties of the distribution of the supremum of cumulative sums. *Theory Probab. Appl.* **15**, 377-418.

11. Bertoin, J. (1998). Darling-Erdos theorems for normalized sums of i.i.d. variables close to a stable law. *Ann. Probab.* **26**, 832-852.

12. Bikyalis, A. (1966). Estimates of the reminder term in the central limit theorem. *Lit. Matem. Sb.* **6**, 323-346.

13. Billingsley, P. (1956). The invariance principle for dependent random variables. *Trans. Amer. Math. Soc.* **83**, 250-268.

14. Billingsley, P. (1968). Convergence of probability measures. John Wiley, New York.

15. Borovkov, A.A. (1970). On factorization identities and properties of the distribution of the supremum of cumulative sums. *Theory Probab. Appl.* **15**, 377-418.

16. Chung, K.L. (1948). Asymptotic distribution of the maximum cumulative sum of independent random variables. *Bull. AMS* **54**, 1162-1170.

17. Chung, K.L. (1948). On the maximum partial sums of sequences of independent random variables. *Trans. Amer. Math. Soc.* **64**, 205-233.

18. Csaki, E. (1978). On the lower limits of maxima and minima of Wiener process and partial sums. *Z. Wahrscheinlichkeitsth.* **43**, 205-221.

19. Darling, D.A. (1956). The maximum of sums of stable random variables. *Trans. Amer. Math. Soc.* **83**, 164-169.

20. Darling, D.A. and Erdos, P. (1956). A limit theorem for the maximum of normalized sums of independent random variables. *Duke Math. J.* **23**, 143-156.

21. Davydov, Yu.A. (1970). Invariance principle for stationary processes. *Theory Probab. Appl.* **15**, 498-509.

22. Doney, R.A. (1982). On the exact asymptotic behavior of the distribution of ladder epochs. *Stoch. Proc. Appl.* **12**, 203-214.

23. Donsker, M. (1951). An invariance principle for certain probability theorems. *Mem. of AMS* **6**, 1-12.

24. Egorov, V.A. (1990). Functional law of the iterated logarithm for ordered sums. *Theory Probab. Appl.* **35**, 343-349.

25. Erdos, P. and Kac, M. (1946). On certain limit theorems in the theory of probability. *Bull. Amer. Math. Soc.* **52**, 292-302.

26. Heyde, C.C. (1966). Some renewal theorems with application to a first passage problem. *Ann. Math. Stat.* **37**, 699-710.

27. Heyde, C.C. (1967). A limit theorem for random walks with drift. *J.Appl. Probab.* **4**, 144-150.

28. Hirsh, W.M. (1965). A strong law for the maximum cumulative sum of independent random variables. *Comm. on Pure and Appl. Math.* **18**, 109-127.

29. Feller, W. (1971). An Introduction to Probability Theory and its Applications, **v.2**, 2nd ed., John Willey, New York.

30. Galambos, Ja. (1985). The asymptotic theory of extreme order statistics. 2nd ed., Krieger, Malabar, Florida.

31. Griffin, P.S. and Maller, R.A. (1998). On the rate of growth of the overshoot and the maximal partial sum. *Adv. Appl. Prob.* **30**, 181-196.

32. Heinrich, L. (1985). An elementary proof of Spitzer's identity. *Statistics* **16**, 249-252.

33. Heyde, C.C. (1969). On the maximum of sums of random variables and the supremum functional for stable processes. *J. Appl.Probab.* **6**, 419-429.

34. Jain, N.C. and Pruitt, W.E. (1973). Maxima of partial sums of independent random variables. . *Z. Wahrscheinlichkeitsth.* **27**, 141-151.

35. Kac, M. (1949). On distribution of certain Wiener functionals. *Trans. Amer. Math. Soc.* **65**, 1-13.

36. Kennedy, D.P. (1972). Estimates of the rate of convergence in limit theorems for the first passage times of random walks. *Ann. Math. Statist.* **43**, 2090-2094.

37. Lamperti, J. (1962). A new class of probability limit theorems. *J. Math. Mech.* **11**, 749-772.

38. Mark, A.M. (1949). Some probability limit theorems. *Bull. AMS* **55**, 885-900.

39. Migai, N.I. and Nevzorov, V.B. (1976). Limit theorems for the first passage time of a certain level. *Theory Probab. Appl.* **21**, 406-410.

40. Migai, N.I. and Nevzorov, V.B. (1977). On the first passage time of a certain parabolas. Vestnik of Leningrad State University **19**, 144-145.

41. Nagaev, S.V. (1969). An estimate of convergence rate for the distribution of the maximal sum of independent random variables. *Sibirian Math. J.* **10**, 614-633.

42. Nagaev, S.V. (1970). Asymptotic expansions for the distribution function of the maximum sum of independent identically distributed random variables. *Sibirian Math. J.* **11**, 381- 406.

43. Nagaev, S.V. (1970). On the speed of convergence of distribution of maximum sums of independent random variables. *Theory Probab. Appl.* **15**, 309-314.

44. Nevzorov, V.B. (1973). Joint distributions of random variables connected with fluctuations of a stable process. *Theory Probab. Appl.* **18**, 161-169.

45. Nevzorov, V.B. (1985). Order statistics based on sums of random variables. *Theory Probab. Appl.* **30**, 193-194.

46. Nevzorov, V.B. (1986). Limit theorems for order statistics based on the sums of random variables. *Proceedings of IV Vilnius conference on probability theory and mathematical statistics,* **v.2**. Utrecht, VNU Science Press, 365-376.

47. Nevzorov, V.B. (1987). On the first passage times of a certain level by a sequence S_n/n. *Vestnik of Leningrad State Univ.* **22**, 97-98.

48. Nevzorov, V.B. (1971a). On the speed of convergence of the distribution of the maximum of cumulative sums of independent random variables to the normal law. *Theory Probab. Appl.* **16**, 378-384.

49. Nevzorov, V.B. (1971b). On some estimates for the distribution of the maximum of cumulative sums. *Theory Probability and Mathematical Statistics* **5**, Kiev, 88-97.

50. Nevzorov, V.B. (1973). On the distribution of the maximal sum of independent terms. *Dokl. Akad. Nauk SSSR* **208**, 43-45.

51. Nevzorov, V.B. (1977). A class of limit distributions for maximum cumulative sums. *Zapiski Nauchn. Semin. LOMI* **72**, 92-97 (in Russian). Translated version in *J. Soviet. Math* **23** (1983), 2286-2290..

52. Nevzorov, V.B. (1979). Maximum of cumulative sums for the Cauchy distribution. *Zapiski Nauchn. Semin. LOMI* **85**, 169-174 (in Russian). Translated version in *J. Soviet. Math.* **20** (1982), 2221-2224.

53. Nevzorov, V.B. (1987). The distribution of the maximum term in a sequence of sample means. *Theory Probab. Appl.* **32**, 125-130

54. Nevzorov, V.B. and Petrov, V.V. (1969). On the distribution of the maximum cumulative sum of independent random variables. *Theory Probab. Appl.* **14**, 682-687.

55. Nevzorov, V.B. and Petrov, V.V. (1988). Inequalities for the maximum of partial sums and their applications. *Limit Theorems in Probability Theory and Related Fields.* Technische Universitat Dresden, 103-111.

56. Nevzorov, V.B. and Zhukova, E. (1996). Wiener process and order statistics. *J.Appl. Statist. Science* **3**, 317-323.

57. Pecherskii, A.A. and Rogozin, B.A. (1969). On joint distributions of random variables connected with fluctuations of a process with independent increments. *Theory Probab. Appl.* **14**, 410-423.

58. Petrov, V.V. (1970). On estimate of a concentration function of sums of independent random variables. *Theory Probab. Appl.* **15**, 718-721.

59. Petrov, V.V. (1975a). A generalization of a Levy inequality. *Theory Probab. Appl.* **20**, 140-144.

60. Petrov, V.V. (1975b). Sums of Independent Random Variables, Springer-Verlag, New York, Berlin, Heidelberg.

61. Pollaczek, F. (1952). Fonctions caracteristiques de certaines repartitions definies au moyen de la notion d'ordre. Application a la theorie des attentes. *C.R. Acad.Sci.Paris* **234**, 2334-2336.

62. Pollaczek, F. (1975). Order statistics of partial sums of mutually independent random variables. *J. Appl. Prob.* **12**, 390-395.

63. Prohorov, Yu.V. (1956). Convergence of stochastic processes and limit theorems in probability theory. *Theory Probab. Appl.***1**, 157-214.

64. Rogozin, B.A. (1966). Speed of convergence of the distribution of the maximum of sums of independent random variables to a limit law. *Theory Probab. Appl.* **11**, 438-441.

65. Rogozin , B.A. (1971). The distribution of the first ladder moment and height and fluctuations of a random walk. *Theory Probab. Appl.* **16**, 593-613.

66. Rosenkrantz, W. (1967). On rates of convergence for the invariance principle. *Trans. Amer. Math. Soc.* **129**, 542-552.

67. Sawyer, S. (1967). A uniform rate of convergence for the maximum absolute value of partial sums in probability. *Comm. on Pure and Appl. Math.* **20**, 647-658.

68. Sawyer, S. (1968). Uniform limit theorems for the maximum cumulative sum in probability. *Trans. Amer. Math. Soc.* **132**, 363-367.

69. Shorack, G.R. (1979). Extension of the Darling and Erdos theorem on the maximum of normalized sums. *Ann . Probab.* **7**, 1092-1096.

70. Shorack, G.R. (1979). Extension of the Darling and Erdos theorem on the maximum of normalized sums. *Ann . Probab.* **7**, 1092-1096.

71. Siegmund, D. (1969). Om moments of the maximum of normed partial sums. *Ann. Math. Stat.* **40**, 527-531.

72. Skorohod, A.V. (1956). Limit theorems for random processes. *Theory Probab. Appl.* **1**, 289-319.

73. Sreehari, M. (1968). A limit theorem for the maximum of cumulative sums. *Acta Math. Acad. Sci. Hung.* **19**, 117-120.

74. Sreehari, M. (1970). On a class of limit distributions for normalized sums of independent random variables. *Theory Probab. Appl.* **15**, 269-290.

75. Steiger, W.L. (1967). Some Kolmogoroff-type inequalities for bounded random variables. *Biometrica* **54**, 641-647.

76. Skorohod, A.V. and Slobodenjuk,N.P. (1970). Limit Theorems for Random Walks. Naukova Dumka, Kiev (in Russian).

77. De Smit, J.H.A. (1973). A simple analytic proof of the Pollaczek-Wendel identity for ordered partial sums. *Ann. Probab.* **1**, 348-351.

78. Spitzer, F. (1974). Principles of Random Walk, 2nd ed. Springer, New York.

79. Spitzer, F. (1956). A combinatorial lemma and its application to probability theory. *Trans. Amer. Math. Soc.* **82**, 323-339.

80. Teicher, H. (1973). A classical limit theorem without invariance or reflection . *Ann. Probab.* **1**, 702-704.

81. Wang Qiying (1996). On the maximal inequality. *Statistics & Probability Letters* **31**, 85-89.

82. Wendel, J.G. (1958). Spitzer's formula: a short proof. *Proc. Amer. Math. Soc.* **9**, 905- 908.

83. Wendel , J.G. (1960). Order statistics of partial sums. *Ann. Math. Stat.* **31**, 1034-1044

84. Zhukova, E.E. (1994). Monotone rearrangements of random processes. Zapiski Nauchn. Semin. LOMI **216**, 62-75 (in Russian). Translated version in *J. Math. Sci.*

Chapter 4

RECORD TIMES AND RECORD VALUES

4.1. EXTREMES AND RECORDS

Let $X_1,...,X_n$ be a random sample of a size n from an absolutely continuous population with a distribution function F. Wilks (1959) found the distribution of a number N(n) of additional observations that one must make to get the first value which exceeds

$$M_n=X_{n,n}=\max\{X_1,...,X_n\}.$$

It is easy to see that

$$P\{N(n)>m\}=P\{\max\{X_{n+1},...,X_{n+m}\}\le M_n\}$$

$$=\int_{-\infty}^{\infty} F^m(u)d(F^n(u))$$

$$=n\int_0^1 v^{m+n-1}dv=n/(n+m). \tag{4.1.1}$$

and then

$$EN(n)=\sum_{m=0}^{\infty} P\{N(n)>m\}=\sum_{m=0}^{\infty} n/(n+m)=\infty. \tag{4.1.2}$$

Evidently the distribution of a number of additional observations that one must make to get the first observation from the interval $(-\infty, m_n)$, where

$m_n=X_{1,n}=\min\{X_1,\dots,X_n\}$,
also coincides with (4.1.1).

Let now N(1,n) denote the minimal size of an additional sample which includes at least one observation out of the interval $[m_n, M_n]$. Then

$$P\{N(1,n)>m\}=P\{m_n\le X_{n+k}\le M_n,\ k=1,2,\dots,m\}$$

$$=\int_{-\infty}^{\infty}\int_{u}^{\infty}(F(v)-F(u))^m\, n(n-1)(F(v)-F(u))^{n-2}dF(u)dF(v)$$

$$=n(n-1)\int_{-\infty}^{\infty}\int_{u}^{\infty}(F(v)-F(u))^{m+n-2}\, dF(u)dF(v)$$

$$=(n(n-1)/(n+m)(n+m-1))\int_{-\infty}^{\infty}\int_{u}^{\infty}(n+m)(n+m-1)(F(v)-F(u))^{n+m-2}dF(u)dF(v)$$

$$=(n(n-1)/(n+m)(n+m-1))P\{-\infty<m_{n+m}<M_{n+m}<\infty\}$$

$$=n(n-1)/(n+m)(n+m-1) \tag{4.1.3}$$

and

$$EN(1,n)=\sum_{m=0}^{\infty}P\{N(n)>m\}=n(n-1)\sum_{m=0}^{\infty}1/(n+m-1)(n+m)$$

$$=n(n-1)\sum_{m=n}^{\infty}(1/(m-1)-1/m)=n. \tag{4.1.4}$$

Comparing (4.1.2) and (4.1.4) one can see that

$EN(n)=\infty$, $n=1,2,\dots$,

when

$EN(1,n)<\infty$

for any $n=1,2,\dots$.

Consider the problem of Wilks in the case when n=1. We fix a random variable X_1. Let its index be denoted as L(1) (L(1)=1). Then in a sequence $X_2, X_3, \ldots$ intercept the first random variable (denote it $X_{L(2)}$), which exceeds X_1 . It follows from (4.1.1) that $P\{L(2)>j\}=1/j$, j=1,2,...,

$$P\{L(2)=j\}=P\{L(2)>j-1\}-P\{L(2)>j\}=1/j(j-1),\ j=2,3,\ldots \tag{4.1.5}$$

and

$$EL(2)=\infty. \tag{4.1.6}$$

Together with the random index L(2) consider also the index

$$\tilde{L}(2)=\min\{j>1: X_j<X_1\}.$$

Evidently, (4.1.5) and (4.1.6) stay to be valid if L(2) is changed by $\tilde{L}(2)$. It is interesting to see that

$$P\{\min(L(2), \tilde{L}(2))=2\}=1$$

and

$$E(\min(L(2), \tilde{L}(2)))=2,$$

when

$$EL(2)= E\tilde{L}(2) = \infty.$$

Random variables L (2) and $\tilde{L}(2)$ present the upper record time and the lower record time respectively. The random variables

$$X(2)=X_{L(2)} \text{ and } \tilde{X}(2)= X_{\tilde{L}(2)}$$

are called the upper and the lower record values. Since

$$\min\{X_1,X_2,\ldots\}=-\max\{Y_1,Y_2,\ldots\},$$

where

$$Y_k\overset{d}{=}X_k,\ k=1,2,\ldots,$$

results for minima can be reformulated from the corresponding results for maxima. It is the reason why we restrict our consideration by upper record times and values, calling them simply as record times and record values. We have introduced record times L(1) and L(2). The next record times L(n) are defined as follows:

$$L(n+1)=\min\{j>L(n): X_j>X_{L(n)}\},\ n=2,3,\ldots, \quad (4.1.7)$$

that is, for every n=1,2,…, based on the sample $X_1,\ldots,X_{L(n)}$ of size L(n) and a maximum

$$M_{L(n)}=X_{L(n),L(n)}=\max\{X_1,\ldots,X_{L(n)}\},$$

we find the first random variable in a sequence

$$X_{L(n)+1},X_{L(n)+2},\ldots$$

(denote its index as L(n+1)), that exceeds $M_{L(n)}$. Random variables

$$X(n)=X_{L(n)}=X_{L(n),L(n)}=M_{L(n)},\ n=1,2,\ldots, \quad (4.1.8)$$

are called record values.

4.2. RECORD INDICATORS

Introduce record indicators

$$\xi_1=1$$

and

$$\xi_n=\mathbf{1}\{M_n>M_{n-1}\}=\mathbf{1}\{X_n \text{ is a record value}\},\ n=2,3,\ldots. \quad (4.2.1)$$

Hence, $\xi_n=1$, if n coincides with one of values L(1), L(2),... . Below we will give two important properties of record indicators.

Lemma 4.2.1.

(Renyi (1962)). Let $X_1,X_2,\ldots$ be independent random variables with a common continuous distribution function F. Then indicators ξ_1, $\xi_2,\ldots$ are independent and

$$P\{\xi_n=1\}=1/n,\ n=1,2,\ldots.$$

Lemma 4.2.2.

Under conditions of lemma 4.2.1, for any n=1,2,..., random variables ξ_1 , $\xi_2,\ldots,\xi_n$ and M_n are independent.

Proof.

Fix any $n\geq 2$. Due to the symmetry,

$$P\{\xi_n=1\}=P\{X_n=M_n\}=P\{X_k=M_n\},\ k=1,2,\ldots .$$

Since X's are independent and have a continuous distribution function, we get that

$$P\{X_i=X_j\}=0$$

for any $i\neq j$, and then

$$1=P\{M_n=X_1\}+\ldots+P\{M_n=X_n\}=nP\{M_n=X_n\}$$

and

$$P\{\xi_n=1\}=P\{X_n=M_n\}=1/n. \tag{4.2.2}$$

To prove both the statements the only we need is to show that for any x, r = 2,3,...
s =1,2,... and $1\leq\alpha(1)<\alpha(2)<\ldots<\alpha(s)\leq r$ the following equality is valid:

$$P\{\xi_{\alpha(1)}=1,\xi_{\alpha(2)}=1,\ldots,\xi_{\alpha(s)}=1,M_r<x\}=P\{\xi_{\alpha(1)}=1\}P\{\xi_{\alpha(2)}=1\}\cdots P\{\xi_{\alpha(s)}=1\}P\{M_r<x\}. \tag{4.2.3}$$

For the sake of simplicity we restrict our consideration by the cases s=1 and s=2. The proof of (4.2.3) for arbitrary s is analogous but more complicated.
Let s=1. Recalling (4.2.2) we see that

$$P\{\xi_{\alpha(1)}=1,M_r<x\}=$$

$$P\{\max(X_1,\ldots,X_{\alpha(1)-1})<X_{\alpha(1)}<x,\ \max(X_{\alpha(1)+1},\ldots,X_r)<x\}=$$

$$P\{\max(X_1,\ldots,X_{\alpha(1)-1})<X_{\alpha(1)}<x\}F^{r-\alpha(1)}(x)=$$

$$F^{r-\alpha(1)}(x)\int_{-\infty}^{x}F^{\alpha(1)-1}(u)dF(u)=F^r(x)/\alpha(1)=$$

$$P\{\xi_{\alpha(1)}=1\}P\{M_r<x\}.$$

Analogously, if s=2, we have

$P\{\xi_{\alpha(1)}=1, \xi_{\alpha(2)}=1, M_r<x\}=$

$P\{\max(X_1,...,X_{\alpha(1)-1})<X_{\alpha(1)}, \max(X_{\alpha(1)},\ldots,X_{\alpha(2)-1})<X_{\alpha(2)}<x,$

$\max(X_{\alpha(2)+1},\ldots,X_r)<x\}=$

$$F^{r-\alpha(2)}(x) \int_{-\infty}^{x} \int_{-\infty}^{u} F^{\alpha(1)-1}(v)\, F^{\alpha(2)-\alpha(1)-1}(u)dF(v)dF(u)=$$

$F^r(x)/\, \alpha(1)\,\alpha(2)=P\{\xi_{\alpha(1)}=1\}P\{\xi_{\alpha(2)}=1\}P\{M_r<x\}.$

Let us introduce now random variables

$$N(n)=\xi_1+...+\xi_n,\ n=1,2,\ldots.. \tag{4.2.4}$$

Note that N(n) is a number of records in the sequence $X_1,X_2,\ldots$.The statement of lemma 4.2.1 and (4.2.4) immediately imply that

$EN(n)=1+1/2+\ldots+1/n\sim\ln n$

and

$$Var(N(n))=\sum_{k=1}^{n} \left(\frac{1}{k}-\frac{1}{k^2}\right) \sim\ln n,\ n\to\infty.$$

The following equalities tie N(n) with record times L(n):

$N(L(n))=n,$

$$P\{L(n)>m\}=P\{N(m)<n\},\ n=1,2,\ldots,\ m=1,2,\ldots \tag{4.2.5}$$

and

$P\{L(n)=m\}=P\{N(m-1)=n-1,N(m)=n\}=$

$$P\{N(m-1)=n-1,\ \xi_m=1\}= P\{N(m-1)=n-1\}/m,\ 1\le m\le n. \tag{4.2.6}$$

Relations (4.2.4) - (4.2.6) enable us to express the distributions of random variables N(n) and L(n) via the distributions of independent record indicators. For instance, (4.2.4) implies that the generating function $P_n(s)$ of N(n) satisfies the following equalities:

$$P_n(s) = Es^{N(n)} = \prod_{j=1}^{n} Es^{\xi_j}$$

$$= \prod_{j=1}^{n} (1+(s-1)/j)=s(1+s)(2+s)...(n-1+s)/n! \quad (4.2.7)$$

and

$$P_n(-s)=(-1)^n s(s-1)...(s-n+1)/n! . \quad (4.2.8)$$

The equality (4.2.8) is given here because it enables to use Stirling numbers of the first kind, which are defined by equalities

$$x(x-1)...(x-n+1)= \sum_{k\geq 0} S_n^k x^k . \quad (4.2.9)$$

On comparing (4.2.8) and (4.2.9) one can see that

$$(-1)^k P\{N(n)=k\}= S_n^k /n!$$

and

$$P\{N(m)=k\}=(-1)^k S_m^k/m!=|S_m^k|/m!. \quad (4.2.10)$$

Now (4.2.6) and (4.2.10) imply that

$$P\{L(n)=m\}= P\{N(m-1)=n-1\}P\{\xi_m =1\}$$

$$= P\{N(m-1)=n-1\}/m=|S_{m-1}^{n-1}|/m!. \quad (4.2.11)$$

Relations (4.2.6) and (4.2.7) help us to find generating functions

$$Q_n(s)=Es^{L(n)}, n=1,2,\ldots,$$

of record times. Since $P\{L(1)=1\}=1$, it is clear that

$$Q_1(s)=s.$$

For $n=2,3,\ldots$, $|s|<1$ and $|z|<1$, due to (4.2.6) and (4.2.7), we have equalities

$$Q_n(s)=\sum_{m=1}^{\infty} P\{L(n)=m\}s^m$$

$$=\sum_{m=1}^{\infty} \frac{1}{m} P\{N(m-1)=n-1\}s^m$$

and

$$\sum_{n=1}^{\infty} Q_n(s)z^n=\sum_{m=1}^{\infty} \frac{s^m}{m}\sum_{n=1}^{\infty} P\{N(m-1)=n-1\}z^n$$

$$=z\sum_{m=1}^{\infty} \frac{s^m}{m} P_{m-1}(z)=z\sum_{m=1}^{\infty} \frac{s^m}{m!} z(1+z)\cdots(m-2+z).$$

$$=-\frac{z}{1-z}\sum_{m=1}^{\infty} \frac{(-s)^m}{m!}(1-z)(-z)(-1-z)\cdots(2-m-z). \tag{4.2.12}$$

Note that

$$\sum_{m=0}^{\infty} \frac{(-s)^m}{m!}(1-z)(-z)(-1-z)\cdots(2-m-z)=(1-s)^{1-z}=(1-s)\exp\{-z\ln(1-s)\} \tag{4.2.13}$$

and (4.2.12) and (4.2.13) imply that

$$(1-z)\sum_{n=1}^{\infty} Q_n(s)z^{n-1}=-(1-s)\exp\{-z\ln(1-s)\}+1. \tag{4.2.14}$$

Transforming the LHS of (4.2.14) in

$$\sum_{n=0}^{\infty} Q_{n+1}(s)z^n-\sum_{n=1}^{\infty} Q_n(s)z^n=s+\sum_{n=1}^{\infty} (Q_{n+1}(s)-Q_n(s))z^n,$$

we get the equality

$$1-s-\sum_{n=1}^{\infty} (Q_{n+1}(s)-Q_n(s))z^n=(1-s)\sum_{n=0}^{\infty} \frac{(-\ln(1-s))^n}{n!}z^n, \tag{4.2.15}$$

from which we obtain that

$$R_n=Q_n(s)-Q_{n+1}(s)=(1-s)\frac{(-\ln(1-s))^n}{n!},\ n=1,2,\ldots,$$

and

$$Q_n(s)=Q_1(s)-(R_1+\ldots+R_{n-1})=s-(1-s)\sum_{r=1}^{n-1}\frac{(-\ln(1-s))^r}{r!}=1-(1-s)\sum_{r=0}^{n-1}\frac{(-\ln(1-s))^r}{r!}. \tag{4.2.16}$$

Observing that values $(1-s)(-\ln(1-s))^r/r!$ coincide with Poisson probabilities $\lambda^r e^{-\lambda}/r!$ under $\lambda=-\ln(1-s)$ and using the well-known identity

$$\sum_{r=n}^{\infty}\lambda^r e^{-\lambda}/r!=\frac{1}{(n-1)!}\int_0^{\lambda} v^{n-1}e^{-v}dv,\ n=1,2,\ldots,\ \lambda>0,$$

one can transform (4.2.16) in another useful formula for $Q_n(s)$. In fact,

$$Q_n(s)=1-(1-s)\sum_{k=0}^{n-1}\frac{(-\ln(1-s))^k}{k!}$$

$$=(1-s)\sum_{k=n}^{\infty}\frac{(-\ln(1-s))^k}{k!}$$

$$=\frac{1}{(n-1)!}\int_0^{-\ln(1-s)} v^{n-1}\exp(-v)dv. \tag{4.2.17}$$

Lemma 4.2.2 states that for any n=1,2,... the vector of indicators $(\xi_1, \xi_2,\ldots, \xi_n)$ and maximal value M_n are independent. This argument enables us to connect the distribution of record values X(n) with the generating function of L(n). It is evident that

$$X(n)=M(L(n))=X_{L(n),L(n)}.$$

We get that

$$P\{X(n)<x\}=P\{M(L(n))<x\}=$$

$$\sum_{m=n}^{\infty}P\{M(L(n))<x|L(n)=m\}P\{L(n)=m\}=$$

$$\sum_{m=n}^{\infty} P\{M(m)<x|L(n)=m\}P\{L(n)=m\}. \tag{4.2.18}$$

Since

$$\{L(n)=m\}=\{\xi_1+\xi_2+...+\xi_{m-1}=n-1,\ \xi_m=1\},$$

lemma 4.2.2 implies that for any m=1,2,..., the event {L(n)=m} and the random variable M(m) are independent. Hence the RHS of (4.2.18) can be transformed as follows:

$$\sum_{m=n}^{\infty} P\{M(m)<x|L(n)=m\}P\{L(n)=m\}=$$

$$\sum_{m=n}^{\infty} P\{M(m)<x\}P\{L(n)=m\}=$$

$$\sum_{m=n}^{\infty} F^m(x)P\{L(n)=m\}=E(F(x))^{L(n)}. \tag{4.2.19}$$

Finally we get the relation

$$P\{X(n)<x\}= E(F(x))^{L(n)}= Q_n(F(x)), \tag{4.2.20}$$

and the corresponding expressions for the generating function $Q_n(s)$ are given in (4.2.16) and (4.2.17). Note that for the standard exponential distribution, when

$$F(x)=1\text{-}\exp(-x),\ x\geq 0,$$

we have the following simple equality

$$P\{X(n)<x\}=\frac{1}{(n-1)!}\int_0^x v^{n-1}\exp(-v)dv. \tag{4.2.21}$$

Thus, in the case of the exponential distribution, X(n) has the Gamma distribution with parameter n and

$$X(n)\overset{d}{=}X_1+\ldots+X_n.$$

4.3. DISTRIBUTIONS OF RECORD TIMES

In the previous section, using the properties of record indicators, we found that

$$P\{L(n)=m\}=|S_{m-1}^{n-1}|/m!,$$

where S_m^n are Stirling numbers defined in (5.2.9). Based on properties of Stirling numbers, Westcott (1977) have showed that

$$P\{L(n)=m\}\sim(\ln m)^{n-2}/m^2 (n-2)! \text{ as } m\to\infty.$$

In (4.2.17) there was given an expression for generating function of L(n):

$$Q_n(s)=\frac{1}{(n-1)!}\int_0^{-\ln(1-s)} v^{n-1}\exp(-v)dv.$$

The independence property of record indicators enables us to get joint and conditional distributions of record times L(n).

Theorem 4.3.1.

For any n=1,2,... and any integers $1=j(1)<j(2)<...<j(n)$ the following equality holds:

$$P\{L(1)=1,L(2)=j(2),...,L(n)=j(n)\}$$

$$= 1/(j(2)-1)(j(3)-1)...(j(n)-1)j(n). \tag{4.3.1}$$

Proof.

Evidently, the event on the left side of (4.3.1) coincides with the event

$$A=\{\xi_2=0,..., \xi_{j(2)-1}=0, \xi_{j(2)}=1, \xi_{j(2)+1}=0,...,$$

$$\xi_{j(3)-1}=0, \xi_{j(3)}=1,..., \xi_{j(n-1)-1}=0, \xi_{j(n-1)}=1, \xi_{j(n-1)+1}=0...,\xi_{j(n)-1}=0, \xi_{j(n)}=1\}.$$

Lemma 4.2.1 implies that

$$P\{A\}= P\{\xi_2=0\}... P\{\xi_{j(2)-1}=0\} P\{\xi_{j(2)}=1\} P\{\xi_{j(2)+1}=0\}... P\{\xi_{j(3)-1}=0\}$$

$$P\{\xi_{j(3)}=1\}... P\{\xi_{i(n-1)-1}=0\} P\{\xi_{j(n-1)}=1\} P\{\xi_{j(n-1)+1}=0\}...P\{\xi_{j(n)-1}=0\} P\{\xi_{i(n)}=1)$$

$$=\prod_{r=2}^{j(n)} P\{\xi_r=0\}\prod_{t=2}^{n} (P\{\xi_{j(t)}=1\}/P\{\xi_{j(t)}=0\})$$

$$=\prod_{r=2}^{j(n)} \frac{r-1}{r}\prod_{t=2}^{n} \frac{1}{j(t)-1}=\frac{1}{j(n)}\prod_{t=2}^{n} \frac{1}{j(t)-1},$$

and that coincides with the LHS of (4.3.1).

Remark 4.3.1.

In particular, we get from (4.3.1) that

$P\{L(2)=m\}=1/m(m-1)$, $m=2,3,\ldots$

(compare with (4.1.1) and (4.2.11). To obtain the distribution of L(n) for n>2 we also can use (4.3.1). It implies that

$P\{L(n)=k\}$

$$=\sum_{1<k(1)<\ldots<k(n-1)<k} (1/(k(2)-1)(k(3)-1)\ldots(k(n)-1)k(n)) \tag{4.3.2}$$

(compare with (4.2.11)).

We give also some useful and evident consequences of theorem 4.3.1.

Corollary 4.3.1.

For any $1=j(1)<j(2)<\ldots<j(n-1)<j<m$ the following equality is valid:

$$P\{L(n+1)=m|L(n)=j, L(n-1)=j_{n-1},\ldots, L(2)=j_2, L(1)=1\}=j/m(m-1). \tag{4.3.3}$$

Corollary 4.3.2.

For any $m>j\geq n$,

$$P\{L(n+1)>m|L(n)=j\}=j/m \tag{4.3.4}$$

and

$$P\{L(n+1)=m|L(n)=j\}=j/m(m-1). \tag{4.3.5}$$

Corollary 4.3.3.

On comparing (4.3.3) and (4.3.5) one can see that L(1), L(2),... is a Markov chain. Williams (1973) suggested the following representation for record times:

$$L(1)=1,\ L(n+1)=[L(n)\exp(\xi_n)]+1,\ n=1,2,..., \tag{4.3.6}$$

where ξ_1, ξ_2,... are independent random variables having the standard exponential distribution and [x] denotes the entire part of x. Since the standard exponential random variable Z and the uniform U([0,1]) random variable U are tied by the equalities

$$1-\exp(-Z)\overset{d}{=}U$$

and

$$Z\overset{d}{=}-\ln(1-U)\overset{d}{=}-\ln U,$$

the representation (4.3.6) can be rewritten in the following form:

$$L(1)=1,\ L(n+1)=[L(n)/U_n]+1,\ n=1,2,..., \tag{4.3.7}$$

where U_1, U_2,... are i.i.d. random variables , having the uniform U([0,1]) distribution. Let us prove (4.3.7). Since random variables U_1,U_2,... are independent and hence L(n), which is determined by the first (n-1) elements of this sequence only, does not depend on U_n, it is evident that the sequence of random variables given in (4.3.7) is a Markov chain. Let us find the transition probabilities for this Markov chain , keeping in mind, that L(n) and U_n are independent random variables:

$$P\{L(n+1)=m|L(n)=j\}=P\{[L(n)/U_n]+1=m|L(n)=j\}=$$

$$P\{[j/U_n]+1=m|L(n)=m\}=$$

$$P\{[j/U_n]+1=m\}=P\{m-1\leq j/U_n<m\}=j/(m-1)m. \tag{4.3.8}$$

Comparing transition probabilities (4.3.5) and (4.3.7) and noting that L(1)=1 in both the definitions (4.1.7) and (4.3.7), we see that the record times given in (4.1.7) and random variables defined in (4.3.7) (as well as in (4.3.6)) have the same distribution.

From (4.3.7) one can see that quotients

$$T_n=L(n)/L(n+1),\ n=1,2,...,$$

are close to uniformly distributed random variables. To stress this closeness we will prove the following result.

Theorem 4.3.2.

a) The relation

$$\lim P\{T_n<x\}=x,\ n\to\infty, \tag{4.3.9}$$

holds for any $0<x<1$.

If $x=1/m$, where $m=1,2,\ldots$, then

$$P\{T_n<x\}=x. \tag{4.3.10}$$

Proof.

Using (4.3.5) we obtain that

$$P\{T_n<x\}=P\{L(n)<xL(n+1)\}$$

$$=\sum_{i=N}^{\infty} P\{L(n)<xL(n+1)|L(n)=i\}P\{L(n)=i\}$$

$$=\sum_{i=N}^{\infty} P\{L(n+1)>[i/x]|L(n)=i\}P\{L(n)=i\}$$

$$=\sum_{i=N}^{\infty} (i/[i/x])\, P\{L(n)=i\}. \tag{4.3.11}$$

If $1/x$ is an integer then

$$\sum_{i=N}^{\infty} (i/[i/x])\, P\{L(n)=i\}=\sum_{i=N}^{\infty} x\, P\{L(n)=i\}$$

$$=x\sum_{i=N}^{\infty} P\{L(n)=i\}=x,$$

that proves the second statement of the theorem.

For arbitrary $0<x<1$ the following inequalities are evidently hold:

$$x \le i/[i/x] < x + x/[i/x],$$

and then

$$x \le \sum_{i=N}^{\infty} (i/[i/x])\, P\{L(n)=i\} < x + x \sum_{i=N}^{\infty} (1/[i/x])\, P\{L(n)=i\} < x + x/[n/x]. \tag{4.3.12}$$

It immediately follows from (4.3.12) that

$$P\{T_n<x\} \to x$$

for any fixed $0<x<1$, as $n\to\infty$. The theorem is proved.

Remark 4.3.2.

We got in theorem 4.3.2 that the asymptotic distribution of the random variables T_n is uniform. Moreover, Shorrock (1972) proved that for any fixed $r=2,3,\ldots$ random variables

$$T_{k,n}=L(n+k-1)/L(n+k),\ k=1,2,\ldots,r,$$

are asymptotically independent as $n\to\infty$. Galambos and Seneta (1975) considered quotients

$$L(n)/L(n-1)$$

and defined integer valued random variables R_n as follows:

$$R_n-1< L(n)/L(n-1) \le R_n,\ n=2,3,\ldots .$$

They have proved that $R_2,R_3,\ldots$ are independent identically distributed random variables and

$$P\{R_n=j\}=P\{L(2)=j\}=1/j(j-1),\ j\ge 2,$$

for any $n=2,3,\ldots$.

4.4. Distributions of Record Values (Continuous Case)

In section 4.2 (formula (4.2.20)) we found the distributions of record values X(n). Let us get the joint distribution function and joint density function for X(1),X(2),…,X(n). To avoid technical problems we consider the case n=2. Let random variables X_1, X_2,… have a continuous d.f. F. The following equalities hold:

$$P\{X(1)<x_1,\ X(2)<x_2\}$$

$$=\sum_{m=2}^{\infty} P\{X_1<x_1,\ \max\{X_2,\dots,X_{m-1}\}\le X_1,\ X_1<X_m<x_2\}$$

$$=\sum_{m=2}^{\infty}\int_{-\infty}^{\min(x_1,x_2)} F^{m-2}(u)\,(F(x_2)-F(u))\,dF(u)$$

$$=\int_{-\infty}^{\min(x_1,x_2)} (F(x_2)-F(u))/(1-F(u))dF(u)$$

$$=\int_{0}^{F(\min(x_1,x_2))} (F(x_2)-u)/(1-u)du$$

$$=(1-F(x_2))\ln(1-F(\min(x_1,x_2))+F(\min(x_1,x_2)). \tag{4.4.1}$$

Consider the case $x_1<x_2$ and suppose that F is an absolutely continuous d.f. with a density function f. Then

$$P\{X(1)<x_1,\ X(2)<x_2\}=(1-F(x_2))\ln(1-F(x_1))+F(x_1) \tag{4.4.2}$$

and the joint density function $f_{1,2}$ of random variables X(1) and X(2) is defined as follows:

$$f_{1,2}(x_1,x_2)=f(x_1)f(x_2)/(1-F(x_1)). \tag{4.4.3}$$

Since X(1)<X(2), then evidently

$$f_{1,2}(x_1,x_2)=0 \text{ for } x_1\ge x_2.$$

Let us denote

$R(x)=f(x)/(1-F(x))$.

The analogous arguments show that the joint density function of record values $X(1),X(2),\ldots,X(n)$ is given as

$$f_{1,2,\ldots,n}(x_1,x_2,\ldots,x_n)=R(x_1)R(x_2)\ldots R(x_{n-1})f(x_n), \quad (4.4.4)$$

if $x_1<x_2<\ldots<x_n$, and

$$f_{1,2,\ldots,n}(x_1,x_2,\ldots,x_n) = 0, \text{otherwise}.$$

Let $Z(1)<Z(2)<\ldots$ denote record values corresponding to the standard exponential distribution with d.f.

$$F(x)=\max\{0, 1-\exp(-x)\}$$

and the density function

$$f(x)= e^{-x}, x\geq 0.$$

It follows from (4.4.4) that the joint density function

$$g_{1,2,\ldots,n}(x_1,x_2,\ldots,x_n)$$

of the exponential record values has the form

$$g_{1,2,\ldots,n}(x_1,x_2,\ldots,x_n) = \exp(-x_n), \quad (4.4.5)$$

if $0\leq x_1<x_2<\ldots<x_n$,

and

$$g_{1,2,\ldots,n}(x_1,x_2,\ldots,x_n)=0,$$

otherwise.

The standard transformation of (4.4.5) show that the joint density function

$$h_{1,2,\ldots,n}(v_1,v_2,\ldots,v_n)$$

of the exponential inter-record values

$$V(1)=Z(1),\ V(2)=Z(2)-Z(1),\ldots,V(n)=Z(n)-Z(n-1)$$

is given by the expression

$$h_{1,2,\ldots,n}(v_1,v_2,\ldots,v_n)=\exp\{-(v_1+\ldots+v_n)\},\ v_1\geq 0,\ldots,\ v_n\geq 0. \tag{4.4.6}$$

The latest equality implies that random variables V(1), V(2),... are independent and have the same standard exponential distribution. It really means that we have got the following results.

Theorem 4.4.1.

For any n =1,2,...,

$$(Z(1),Z(2)-Z(1),\ldots,Z(n)-Z(n-1)) \stackrel{d}{=} (\omega_1,\ \omega_2,\ldots,\ \omega_n), \tag{4.4.7}$$

where ω_k, k=1,2,..., are i.i.d. random variables , having the standard exponential distribution.

Corollary 4.4.1.

For any n=1,2,...,

$$(Z(1),Z(2),\ldots,Z(n)) \stackrel{d}{=} (\omega_1,\ \omega_1+\omega_2,\ldots,\ \omega_1+\omega_2+\ldots+\omega_n). \tag{4.4.8}$$

Remark 4.4.1

Note, that we can also get the formula (4.2.21) as a consequence of (4.4.8).

Equalities (4.4.7) and (4.4.8) determine the structure of the exponential record values. At the same time one can use (4.4.7) and (4.4.8) to present the probability structure of record values for any continuous distribution. The way to apply these equalities is based on the following result.

Representation 4.4.1.

Let $X(1)<X(2)<\ldots$ be record values based on a sequence of independent random variables with a continuous d.f. F and $U(1)<U(2)<\ldots$ denote record values, corresponding to the uniform U([0,1]) distribution. Then for any n=1,2,... ,

$$(F(X(1)),\ldots,F(X(n))) \stackrel{d}{=} (U(1),\ldots,U(n)) . \tag{4.4.9}$$

Proof.

Evidently, for any random variable X with a continuous d.f. F, the probability integral transformation U=F(X) produces a random variable U , having the standard U([0,1]) uniform distribution. Since this transform does not change the order of X's, the values $U(1)<U(2)<\dots$ of records in a sequence $U_i=F(X_i)$ coincide with values $F(X(1))<F(X(2))<\dots$.

A more fundamental way to prove (4.4.9) is to compare expressions for joint distribution functions of the uniform records U(1),...,U(n) and arbitrary record values X(1),...,X(n). For example, let n=2. We know from (4.4.1) that

$$P\{X(1)<x_1,\ X(2)<x_2\}=(1-F(x_2))\ln(1-F(\min(x_1,x_2))+ F(\min(x_1,x_2)) .$$

For the standard uniform distribution it can be rewritten as

$$P\{U(1)<x_1,\ U(2)<x_2\}=(1-x_2)\ln(1-\min(x_1,x_2))+ \min(x_1,x_2), \tag{4.4.10}$$

if $0<x_1<1$ and $0<x_2<1$. Let G(x) be the inverse function of d.f. F. Then (4.4.1) implies that

$$P\{F(X(1))<x_1,\ F(X(2))<x_2\}=P\{X(1)<G(x_1),X(2)<G(x_2)\}$$

$$= (1-F(G(x_2)))\ln(1-F(\min(G(x_1),G(x_2)))+ F(\min(G(x_1),G(x_2)))$$

$$= (1-x_2)\ln(1-\min(x_1,x_2))+ \min(x_1,x_2). \tag{4.4.11}$$

The statement (4.4.9) for n=2 immediately follows from (4.4.10) and (4.4.11). For arbitrary n one needs to use analogously the general formula for joint d.f. for record values, which can be obtained if to develop as in (4.4.1) the following equality:

$$P\{X(1)<x_1,\dots,\ X(n)<x_n\}$$

$$= \sum_{m(1)=1}^{\infty}\ \sum_{m(2)=1}^{\infty} \cdots \sum_{m(n-1)=1}^{\infty} P\{X_1<x_1,\ \max\{X_2,\dots,X_{m(1)}\}\le X_1<X_{1+m(1)}<x_2,$$

$$\max\{ X_{m(1)+2},\dots,X_{m(1)+m(2)}\} \le X_{1+m(1)}<X_{1+m(1)+m(2)}<x_3,\dots,$$

$$\max\{ X_{m(1)+\dots+m(n-2)+2},\dots,X_{m(1)+m(n-1)}\} \le X_{1+m(1)+\dots+m(n-2)}<X_{1+m(1)+m(2)+\dots+m(n-1)}<x_n\}.$$

The following results are simple consequences of (4.4.9).

Representation 4.4.2.

If F is a continuous d.f. and G is the inverse of F, then

$$(X(1),\ldots,X(n)) \underset{=}{d} (G(U(1)),\ldots,G(U(n))). \tag{4.4.12}$$

Representation 4.4.3.

Let record values $X(1)<X(2)<\ldots$ and $Y(1)<Y(2)<\ldots$ correspond to continuous distribution functions F and H respectively. Then

$$(X(1),\ldots,X(n)) \underset{=}{d} (G(H(Y(1))),\ldots,G(H(Y(n)))), \tag{4.4.13}$$

where G is the inverse function of F .

Representation 4.4.4.

Let $X(1)<X(2)<\ldots$ be record values based on a sequence of independent random variables with a continuous d.f. F and let $Z(1)<Z(2)<\ldots$ be the exponential record values given in theorem 4.4.3. Then for any $n=1,2,\ldots$,

$$(X(1),\ldots,X(n)) \overset{d}{=} (H(Z(1)),\ldots,H(Z(n))), \tag{4.4.14}$$

where $H(x)=G(1-e^{-x})$ and G is the inverse of F.

Remark 4.4.2.

Combining (4.4.8) and (4.4.14) we come to the next equality:

$$(X(1),X(2),\ldots,X(n)) \overset{d}{=} (H(\omega_1), H(\omega_1+\omega_2),\ldots, H(\omega_1+\omega_2+\ldots+\omega_n)), \tag{4.4.15}$$

that expresses distributions of arbitrary record values via distributions of sums of i.i.d. exponential random variables $\omega_1, \omega_2,\ldots$.

Recalling of the formula (4.4.4) comes to the following expression for the joint density function of record values $X(1),X(2),\ldots,X(n)$:

$$f_{1,2,\ldots,n}(x_1,x_2,\ldots,x_n)=R(x_1)R(x_2)\ldots R(x_{n-1})f(x_n),$$

if $x_1<x_2<\ldots<x_n$,

and

$f_{1,2,\dots,n}(x_1,x_2,\dots,x_n) = 0$, otherwise,

where

$R(x)=f(x)/(1-F(x))$.

This equality implies that

$$P\{X(1)<x_1,X(2)<x_2,\dots,X(n)<x_n\}=\int\dots\int\prod_{j=1}^{n-1} R(u_j)f(u_n)du_1\dots du_n, \tag{4.4.16}$$

with integration on the RHS of (4.4.16) proceeded over the set

$$B=\{u_j<x_j,\ j=1,2,\dots,n\ ,\ -\infty<u_1<\dots<u_n<\infty\}. \tag{4.4.17}$$

As we can see, the latest equality holds for any absolutely continuous distribution. Combining (4.4.16) for the case of the standard uniform distribution and (4.4.12) one can show that for record values corresponding to any continuous distribution function F the following equality holds:

$$P\{X(1)<x_1,X(2)<x_2,\dots,X(n)<x_n\}=\int\dots\int\prod_{j=1}^{n-1}\frac{dF(u_j)}{1-F(u_j)}dF(u_n), \tag{4.4.18}$$

where integration on the RHS of (4.4.18) also holds over B, defined in (4.4.17).

Let us again consider the initial random variables $X_1,X_2,\dots$ having an absolutely continuous d.f. F and a density function f. With the help of (4.4.4) we can find the conditional density function $f(x_n|x_1,x_2,\dots,x_{n-1})$ of X(n), given that

$X(1)=x_1,\ X(2)=x_2,\dots,\ X(n-1)=x_{n-1}$.

It is easy to see that

$$f(x_n|x_1,x_2,\dots,x_{n-1})= f_{1,2,\dots,n}(x_1,x_2,\dots,x_n)/f_{1,2,\dots,n-1}(x_1,x_2,\dots,x_{n-1})$$

$$= R(x_1)R(x_2)\dots R(x_{n-1})f(x_n)/R(x_1)R(x_2)\dots R(x_{n-2})f(x_{n-1})$$

$$= f(x_n)/(1-F(x_{n-1})),\ x_n>x_{n-1}, \tag{4.4.19}$$

and then

$P\{X(n)>x|\ X(1)=x_1,\ X(2)=x_2,\dots,\ X(n-1)=x_{n-1}\}$

$$= (1-F(x))/(1-F(x_{n-1})) \qquad (4.4.20)$$

for $x>x_{n-1}$.

Noting that for any n=2,3,... the conditional density function

$f(x_n|x_1,x_2,\ldots,x_{n-1})$

depends on x_{n-1} and does not depend on $x_1,x_2,\ldots,x_{n-1}$, one can state that X(1), X(2),... is a Markov chain, at least for absolutely continuous distribution function F.

Applying (4.4.13) with arbitrary continuous d.f.'s F and H, one can prove that this Markov property remains true for any continuous d.f. F. The analogous arguments show that (4.4.20) is also valid for any continuous d.f. F.

4.5. Distributions of Record Values (Discrete Case)

Let now X, X_1, X_2,... be i.i.d. discrete random variables. Without loss of generality we will suppose that X's take values 0,1,.... Let

$p_k=P\{X=k\}$

and

$$q_k=P\{X \geq k\}=\sum_{j=k}^{\infty} p_j\,,\ k=0,1,\ldots.$$

To provide the existence (with probability 1) of any record value X(n) in the sequence X_1, X_2,... we need the following restriction:

$$q_k>0\,,\ k=0,1,2,\ldots. \qquad (4.5.1).$$

Unlike the continuous case where independent record indicators ξ_1, ξ_2,... play a rather essential role, for discrete random variables we will use the independence property of another indicators. Let random indicators η_n, n=0,1,..., be defined as follows:

$\eta_n=1$, if n is a record value in the sequence X_1, X_2,... , i.e. there exists such an integer m , that X(m)=n, and $\eta_n=0$, otherwise.

The next result belongs to Shorrock (1972).

Theorem 4.5.1.

The indicators η_0, η_1,... are independent and

$$P\{\eta_n=1\}=P\{X=n|X\geq n\}=p_n/q_n,\ n=0,1,\dots . \tag{4.5.2}$$

Proof.

Equality (4.5.2) is evident, since

$$P\{\eta_n=1\}=P\{X_1=n\}+P\{X_1<n,X_2=n\}+P\{X_1<n,X_2<n,X_3=n\}+\dots$$

$$=P\{X=n\}(1+P\{X<n\}+P^2\{X<n\}+\dots)=P\{X=n\}/(1-P\{X<n\})=p_n/q_n .$$

To prove the independence of η_0, $\eta_1,\dots$ we can apply the following arguments. Since the indicators take two values, 0 and 1, the only we need is to show that for any r=2,3,... and any $0\leq\alpha(1)<\alpha(2)<\dots<\alpha(r)$,

$$P\{\eta_{\alpha(1)}=1,\eta_{\alpha(2)}=1,\dots,\eta_{\alpha(r)}=1\}=\prod_{k=1}^{r} p_{\alpha(k)}/q_{\alpha(k)}. \tag{4.5.3}$$

Let M(r-1) denote the time of the appearance of a record value which equals $\alpha(r-1)$.

1) Then

$$P\{\eta_{\alpha(1)}=1,\eta_{\alpha(2)}=1,\dots,\eta_{\alpha(r)}=1\}$$

$$=\sum_{m=r-2}^{\infty} P\{\eta_{\alpha(1)}=1,\eta_{\alpha(2)}=1,\dots,\eta_{\alpha(r)}=1,M(r-1)=m\}$$

$$=\sum_{m=r-2}^{\infty}\sum_{s=1}^{\infty} P\{\eta_{\alpha(1)}=1,\ \eta_{\alpha(2)}=1,\ \dots,\ \eta_{\alpha(r-1)}=1,\ M(r-1)=m,$$

$$X_{m+1}<\alpha(r-1),\dots,X_{m+s-1}<\alpha(r-1),\ X_{m+s}=\alpha(r)\}$$

$$=\sum_{m=r-2}^{\infty}\sum_{s=1}^{\infty} P\{\eta_{\alpha(1)}=1,\eta_{\alpha(2)}=1,\dots,\eta_{\alpha(r-1)}=1,M(r-1)=m\}$$

$$P\{X_{m+1}<\alpha(r-1)\}\dots P\{X_{m+s-1}<\alpha(r-1)\}P\{X_{m+s}=\alpha(r)\}$$

$$=\sum_{m=r-2}^{\infty} P\{\eta_{\alpha(1)}=1,\eta_{\alpha(2)}=1,\dots,\eta_{\alpha(r-1)}=1,M(r-1)=m\}$$

$$\sum_{s=1}^{\infty} P\{X_{m+1}<\alpha(r-1)\}...P\{X_{m+s-1}<\alpha(r-1)\}P\{X_{m+s}=\alpha(r)\}$$

$$=\sum_{m=r-2}^{\infty} P\{\eta_{\alpha(1)}=1,\eta_{\alpha(2)}=1,...,\eta_{\alpha(r-1)}=1,M(r-1)=m\}P\{X=\alpha(r)\}/P\{X\geq\alpha(r-1)\}$$

$$=P\{\eta_{\alpha(1)}=1,\eta_{\alpha(2)}=1,...,\eta_{\alpha(r-1)}=1\}(p_{\alpha(r)}/q_{\alpha(r)})\,. \quad (4.5.4)$$

Consequently applying (4.5.4), one comes to (4.5.3) and that completes the proof of the theorem.

The statement of theorem 4.5.1 implies the following useful representation for record values in a sequence of discrete random variables $X_1, X_2, \ldots$.

Representation 4.5.1.

For any m=0,1,... and n=1,2,... ,

$$P\{X(n)>m\}=P\{\eta_0+\eta_1+...+\eta_m<n\}. \quad (4.5.5)$$

and

$$P\{X(n)=m\}=P\{\eta_0+\eta_1+...+\eta_{m-1}=n-1,\eta_m=1\}$$

$$=P\{\eta_0+\eta_1+...+\eta_{m-1}=n-1\}p_m/q_m\,. \quad (4.5.6)$$

The following result is also based on the independence property of indicators $\eta_0,\eta_1,\ldots$.

Theorem 4.5.2

The joint distribution of record values X(1)<X(2)<... is given by the equalities:

$$P\{X(1)=i(1),X(2)=i(2),...,X(n)=i(n)\}$$

$$=p_{i(n)}\cdot\prod_{r=1}^{n-1}\frac{p_{i(r)}}{q_{i(r)+1}},\ 0\leq i(1)<i(2)<...<i(n). \quad (4.5.7)$$

Proof.

Any event

$A=\{X(1)=i(1),X(2)=i(2),...,X(n)=i(n)\}$

can be expressed in terms of independent indicators $\eta_0,\eta_1,\ldots$ in such a manner:
$A=\{\eta_0=0, \eta_1=0,\ldots, \eta_{i(1)-1}=0,\eta_{i(1)}=1, \eta_{i(1)+1}=0,\ldots, \eta_{i(n)-1}=0, \eta_{i(n)}=1\}$.
Then from (4.5.2), observing that

$P\{\eta_r=0\}=q_{r+1}/q_r$ and $q_0=1$,

we find finally that

$$P\{A\}=\prod_{r=0}^{i(n)} P\{\eta_r=0\}\cdot\prod_{r=1}^{n} (P\{\eta_{i(r)}=1\}/P\{\eta_{i(r)}=0\}$$

$$=\prod_{r=0}^{i(n)}\frac{q_{r+1}}{q_r}\cdot\prod_{r=1}^{n}\frac{p_{i(r)}}{q_{i(r)+1}}=q_{i(n)+1}\cdot\prod_{r=1}^{n}\frac{p_{i(r)}}{q_{i(r)+1}}=p_{i(n)}\cdot\prod_{r=1}^{n-1}\frac{p_{i(r)}}{q_{i(r)+1}}.$$

From theorem 4.5.2 we have the following consequence.

Corollary 4.5.1.

For any $j>k>k_{n-1}>\ldots>k_1$

$P\{X(n+1)=j|X(n)=k, X(n-1)= k_{n-1},\ldots, X(1)=k_1\}$

$$= P\{X(n+1)=j|X(n)=k\}= P\{\eta_{k+1}=0, \eta_{k+2}=0,..., \eta_{j-1}=0, \eta_j=1\}=p_j/q_{k+1} \qquad (4.5.8)$$

and

$$P\{X(n+1)>j|X(n)=k\}= q_{j+1}/q_{k+1}. \qquad (4.5.9)$$

Proof.

From the definition of conditional probabilities we get that

$P\{X(n+1)=j|X(n)=k, X(n-1)= k_{n-1},\ldots, X(1)=k_1\}=$

$P\{X(n+1)=j,X(n)=k, X(n-1)= k_{n-1},\ldots, X(1)=k_1\}/ P\{X(n)=k, \ldots, X(1)=k_1\}$.

Equalities (4.5.7) immediately imply that

$$P\{X(n+1)=j|X(n)=k, X(n-1)= k_{n-1},\ldots, X(1)=k_1\}= p_j/q_{k+1}. \tag{4.5.10}$$

On the other hand, the RHS of (4.5.10) evidently coincides with the probability

$$P\{\eta_{k+1}=0, \eta_{k+2}=0,\ldots, \eta_{j-1}=0, \eta_j=1\}=P\{\eta_j=1\}\cdot \prod_{r=k+1}^{j-1} P\{\eta_r=0\}.$$

It comes also from (4.5.10) that the conditional probability there does not depend on values of random variables X(1),..., X(n-1) and hence

$$P\{X(n+1)=j|X(n)=k\}= P\{X(n+1)=j|X(n)=k, X(n-1)= k_{n-1},\ldots, X(1)=k_1\}= p_j/q_{k+1}$$

and then

$$P\{X(n+1)>j|X(n)=k\}= \sum_{r=j+1}^{\infty} p_r/ q_{k+1}=q_{j+1}/ q_{k+1}.$$

By that all the statements of the corollary are proved.

Remark 4.5.1.

In fact we have proved that X(1), X(2),... is a Markov chain with transition probabilities

$$P\{X(n+1)=j|X(n)=k\}= p_j/ q_{k+1}.$$

Geometric distribution

Consider a sequence of independent geometrically $G_1(p)$ distributed random variables, i.e. random variables, taking values 1,2,... with probabilities

$$p_k=P\{X=k\}=(1-p)p^{k-1},\ k=1,2,\ldots,\ 0<p<1.$$

In this case

$$q_k=P\{X\geq k\}=p^{k-1},\ k\geq 1,$$

and we obtain from (4.5.7) that

$$P\{X(1)=i(1),X(2)=i(2),\dots,X(n)=i(n)\}=(1-p)^n p^{i_n-n} \qquad (4.5.11)$$

for any $1 \le i(1) < i(2) < \dots < i(n)$.
The simple transformation of (4.5.11) gives us the next relation:

$$P\{X(1)=i_1,\ X(2)-X(1)=i_2,\dots,X(n)-X(n-1)=i_n\}=(1-p)^n\, p^{\,i_1+\dots+i_n-n}$$

$$=P\{X=i_1\}P\{X=i_2\}\dots P\{X=i_n\}. \qquad (4.5.12)$$

By that, the following result is valid.

Theorem 4.5.3.

Let $X_1, X_2,\dots$ be independent random variables, taking values 1,2,... with probabilities

$$p_k=P\{X_j=k\}=(1-p)p^{k-1},\ j=1,2,\dots,\ k=1,2,\dots .$$

Then for any n=1,2,…,

$$\{X(1),\ X(2)-X(1),\dots,\ X(n)-X(n-1)\} \overset{d}{=} \{X_1, X_2, \dots, X_n\}.$$

Corollary 4.5.2.

The statement of theorem 4.5.3 means that inter-record differences X(1), X(2)-X(1), X(3)-X(2),… for the geometric distribution $G_1(p)$, are independent and have the same geometric distribution as the initial X's. Hence, for any n=1,2,…,

$$X(n) \overset{d}{=} X_1+X_2+\dots+X_n.$$

Note that the sum

$$S_n=Y_1+\dots+Y_n$$

of geometrically $G_0(p)$ (taking values 0,1,... with probabilities $(1-p)p^k$, k=0,1,...) distributed random variables $Y_1,Y_2,\dots$ has the negative binomial NB(n, q) distribution with parameters n and q=1-p. Since

$$X(n) \overset{d}{=} S_n+n,$$

we get that X(n)-n also has the negative binomial NB(n, q) distribution and

$$P\{X(n)=n+m\}=\binom{m+n-1}{m}(1-p)^n p^m,\ m=0,1,\ldots. \qquad (4.5.13)$$

Another approach to obtain (4.5.13) is to use the general formula (4.5.6). In our situation

$$P\{X(n)=m\}=p_m P\{\eta_1+\ldots+\eta_{m-1}=n-1\}/q_m,$$

where $\eta_1,\ldots,\eta_{m-1}$ are independent random indicators such that

$$P\{\eta_k=1\}=1-P\{\eta_k=0\}=(1-p),$$

and

$$p_m/q_m=(1-p).$$

Then the sum $\eta_1+\ldots+\eta_{m-1}$ has the binomial B(m-1,1-p) distribution,

$$P\{\eta_1+\ldots+\eta_{m-1}=n-1\}=\binom{m-1}{n-1}(1-p)^{n-1}p^{m-n},$$

if $1\le n\le m$, and

$$P\{X(n)=m\}=\binom{m-1}{n-1}(1-p)^n p^{m-n},\ m\ge n. \qquad (4.5.14)$$

Indeed , the relations (4.5.13) and (4.5.14) coincide.

Weak records

For sequences of discrete random variables, the so-called weak records have self-dependent meaning.

Definition.

Weak record times $L_w(n)$ and weak record values $X_w(n)$ are given by the following recurrent relations:

$$L_w(1)=1,\ L_w(n+1)=\min\{j>L_w(n): X_j \ge \max\{X_1,X_2,\ldots,X_{j-1}\}\},$$

$$X_w(n)=X_{L_w(n)},\ n=1,2,\ldots. \qquad (4.5.15)$$

Unlike the classical (strong) records the definitions (4.5.15) register any repetition of the recent maximal value as a new record. Indeed, for continuous distributions, sequences of strong and weak records coincide with probability one. There is an essential reason why weak records gain an advantage over strong records in sequences of discrete random variables. Unlike the strong records, which are defined with probability 1 only if the population distribution has not the last point of growth, weak records are defined for any distribution.

Without loss of generality we again will consider independent identically distributed random variables $X, X_1, X_2, \ldots$ taking values $0,1, \ldots$ with probabilities

$$p_k = P\{X=k\}.$$

Let us introduce random indicators

$$\eta^w_n,\ n=0,1, \ldots ,$$

as follows:

$\eta^w_n=1$, if n is a weak record value, i.e. $X_w(m)=n$ for any $m=1,2,\ldots$, and $\eta^w_n=0$, otherwise. For any $n = 0,1,2, \ldots$, let μ_n denote a number of weak records in a sequence $X_1,X_2,\ldots$, taking a value n .

The following results have been proved for random variables η^w_n and μ_n (the proofs of these results and their generalizations for the so-called k-th weak records see in Vervaat (1973) and Stepanov (1992)).

Theorem 4.5.4.

The indicators $\eta^w_0, \eta^w_1,\ldots$ are independent and

$$P\{\eta^w_n=1\}= P\{X=n|X\geq n\},$$

if $p_n>0$,

and

$$P\{\eta^w_n=1\}=0,$$

if $p_n=0$.

Remark 4.5.2.

The statements of theorems 4.5.1 and 4.5.4 practically coincide and that is evident since

$$\{\eta^{w}_{n}=1\}=\{\eta_n=1\}.$$

Theorem 4.5.5.

Suppose additionally that restriction (4.5.1) holds. Then random variables μ_0, μ_1, μ_2,... are independent and

$$P\{\mu_n=m\}=(1-r_n)r_n^{m},\ n=0,1,...,m=0,1,..., \tag{4.5.16}$$

Where

$r_n=P\{X=n\}/P\{X\geq n\}$,

if $p_n=P\{X=n\}>0$,

and

$P\{\mu_n=0\}=1$, if $p_n=0$.

Remark 4.5.3.

One can see that if (4.5.1) holds, then random variables μ_n have geometric distributions $G_0(r_n)$ with parameters

$r_n=P\{X=n\}/P\{X\geq n\}$, $n=0,1,\ldots$,

and

$$P\{\eta^{w}_{n}=1\}=P\{\mu_n>0\}. \tag{4.5.17}$$

If condition (4.5.1) fails then there exists such a point n* that

$P\{X<n^*\}<P\{X\leq n^*\}=1$.

In this situation $r_{n^*}=1$ and $P\{\mu_{n^*}=\infty\}=1$.

Representation 4.5.2.

For any m=0,1,... and n =1,2,... , the following equality holds:

$$P\{X_w(n)>m\}=P\{\mu_0+\mu_1+...+\mu_m<n\}. \tag{4.5.18}$$

Example 4.5.1. (geometric distribution).

Let $X_1, X_2, \ldots$ be independent and have the geometric $G_0(p)$ distribution. Then we find from (4.5.16) that for any n=0,1,2, …, the r.v. μ_n has the geometric $G_0(1-p)$ distribution. Hence the sum $\mu_0+\mu_1+\ldots+\mu_m$ has the negative binomial NB(m+1, 1-p) distribution and

$$P\{X(n)>m\}=P\{\mu_0+\mu_1+\ldots+\mu_m<n\}=\sum_{r=0}^{n-1}\binom{m+r}{m}(1-p)^r p^{m+1},\ m=0,1,\ldots.$$

4.6. Joint Distributions of Record Times and Record Values

There are some useful relations including record times L(n), record values X(n) and inter-record times

$\Delta(1)=L(1)=1,\ \Delta(n)=L(n)-L(n-1),\ n=2,3,\ldots$.

Let us consider some of them.
It is evident, that for any $x_1,\ldots,x_n$ and $1=k(1)<k(2)<\ldots<k(n)$ the following equalities hold:

$$P\{X(1)<x_1, X(2)<x_2,\ldots,X(n)<x_n,\ L(1)=1, L(2)=k(2),\ldots,L(n)=k(n)\}=$$

$$P\{X_1<x_1, \max(X_2,\ldots,X_{k(2)-1})\le X_1<X_{k(2)}<x_2,\ldots,$$

$$\max(X_{k(n-1)+1},\ldots,X_{k(n)-1})\le X_{k(n-1)}<X_{k(n)}<x_n\}=$$

$$\int_{-\infty}^{y_1}\ldots\int_{-\infty}^{y_n} h(u_1,\ldots,u_n)dF(u_1)\ldots dF(u_n), \tag{4.6.1}$$

where

$$y_k=\min\{x_k, x_{k+1},\ldots,x_n\},\ k=1,2,\ldots,n,$$

and

$$h(u_1,\ldots,u_n)=\prod_{r=1}^{n-1} F^{k(r+1)-k(r)-1}(u_r),$$

if $-\infty<u_1<\ldots<u_n<\infty$,

and

$h(u_1,...,u_n) = 0$,otherwise.

In the sequel we will consider that X's have an absolutely continuous distribution with a density function f. Introduce a function $f_{(n)}(k(1),...,k(n),x_1,...,x_n)$, which is a joint density function with respect to record values $X(1),X(2),...,X(n)$ and a probability distribution with respect to discrete record times $L(1),L(2),...,L(n)$. On differenciting (4.6.1) with respect to $x_1,x_2,\ldots,x_n$, we obtain that

$$f_{(n)}(k(1),...,k(n),x_1,...,x_n) = (F(x_1))^{k(2)-k(1)-1}(F(x_2))^{k(3)-k(2)-1}...(F(x_{n-1}))^{k(n)-k(n-1)-1}f(x_1)f(x_2)...f(x_n), \quad (4.6.2)$$

if $-\infty<x_1<x_2<...<x_n<\infty$, $1=k_1<k_2<...<k_n$,

and

$f_{(n)}(k(1),...,k(n),x_1,...,x_n) = 0$, otherwise.

Now we can use (4.6.2) to prove the following result.

Theorem 4.6.1.

Let independent random variables $X_1, X_2,\ldots$ have a continuous distribution function F. Then inter-record times $\Delta(1)$, $\Delta(2)$,... are conditionally independent under condition that record values $X(1),X(2),...$ are fixed and

$$P\{\Delta(n)=m|\ X(k)=x_k,\ k=1,2,\ldots\}=(1-F(x_{n-1}))(F(x_{n-1}))^{m-1}, \quad (4.6.3)$$

$m=1,2,...,\ n=2,3,...$.

Proof.

For the sake of simplicity we will prove the theorem under the additional restriction that X's have an absolutely continuous distribution with a density function f. Together with the function $f_{(n)}(k(1),...,k(n),x_1,...,x_n)$ let us consider one more density-probability function $h_n(m(1),...,m(n),x_1,...,x_n)$, which is a joint density function for record values $X(1)$, $X(2),\ldots,X(n)$ and a probability function for inter-record times $\Delta(1),\Delta(2),...,\Delta(n)$. Comparing with (4.6.2), one gets the formula

$$h_n(m(1),...,m(n),x_1,...,x_n)= (F(x_1))^{m(2)-1}(F(x_2))^{m(3)-1}...(F(x_{n-1}))^{m(n)-1}f(x_1)f(x_2)...f(x_n), \quad (4.6.4)$$

if

$-\infty<x_1<x_2<...<x_n<\infty$, $m(1)=1$, $m(r)>0$, $r = 2,...,n$.

Recalling that the joint density function of record values $X(1),X(2),\ldots,X(n)$ is given as

$$f_{1,2,\ldots,n}(x_1,x_2,\ldots,x_n)=R(x_1)R(x_2)\ldots R(x_{n-1})f(x_n),$$

where

$$R(x)=f(x)/(1-F(x)).$$

We obtain from (4.6.4) that

$$P\{\Delta(2)=m(2),..., \Delta(n)=m(n)|X(1)=x_1,...,X(n)=x_n\}=$$

$$h_n(m(1),...,m(n),x_1,...,x_n)/ f_{1,2,\ldots,n}(x_1,...,x_n) =$$

$$(F(x_1))^{m(2)-1}...(F(x_{n-1}))^{m(n)-1}f(x_1)...f(x_n)/R(x_1)...R(x_n)(1-F(x_n))$$

$$= (F(x_1))^{m(2)-1}(1-F(x_1))(F(x_2))^{m(3)-1}(1-F(x_2))...(F(x_{n-1}))^{m(n)-1}(1-F(x_{n-1})). \quad (4.6.5)$$

It is evident now that both the statements of the theorem are valid.

Corollary 4.6.1.

The probabilities on the RHS of (4.6.3) corresponds to the geometric $G_1(F(x_{n-1}))$ distribution. It follows from (4.6.3) that

$$P\{\Delta(n)>m|\ X(1),X(2),...\}=(F(X(n-1)))^m,\ m=1,2,\ldots,\ n=2,3,\ldots,$$

and

$$P\{\Delta(n)(1-F(X(n)))>y|X(1),X(2),...\}=(F(X(n-1)))^{N(y,n)}, \quad (4.6.6)$$

where

$$N(y,n)=[y/(1-F(X(n)))].$$

Since for any y and fixed values X(1),X(2),... the RHS of (4.6.6) converges asymptotically to

$$\exp\{-y(1-F(X(n-1)))/(1-F(X(n)))\},\ n\to\infty,$$

the simple arguments show that

$\lim P\{\Delta(n)(1-F(X(n)))>y\}= E\exp\{-y(1-F(X(n-1)))/(1-F(X(n)))\}$, as $n\to\infty$.

Using properties of the probability integral transformation we obtain that the quotient

$(1-F(X(n-1)))/(1-F(X(n)))$

has the same distribution as

$\exp\{Z(n)-Z(n-1)\}$,

$Z(1), Z(2),\ldots$ being the exponential record values. We know from 4.4.7 that the inter-record value $Z(n)-Z(n-1)$ has the standard exponential distribution and hence

$U = \exp(-(Z(n)-Z(n-1))$

is an uniformly $U(0,1)$ distributed random variable. Finally we obtain that

$G(y)= \lim P\{\Delta(n)(1-F(X(n)))<y\}=1-E\exp\{-y/U\}=$

$$1-\int_0^1 \exp(-y/x)\,dx = 1-\int_1^\infty z^{-2}\exp(-zy)\,dz,\ y>0.$$

4.7. Moments of Record Times and Record Values

We know from chapter 4.2 that the number of records $N(n)$ can be presented as a sum $\xi_1+\ldots+\xi_n$ of independent record indicators ξ_k, taking values 0 and 1 with probabilities $1-1/k$ and $1/k$, correspondingly. It immediately implies that

$$EN(n)=\sum_{k=1}^{n}\frac{1}{k},$$

$$EN^2(n)=\sum_{k=1}^{n}\left(\frac{1}{k}-\frac{1}{k^2}\right)+\left(\sum_{k=1}^{n}\frac{1}{k}\right)^2$$

and

$$Var(N(n))=\sum_{k=1}^{n}\left(\frac{1}{k}-\frac{1}{k^2}\right).$$

Since

$$\sum_{k=1}^{n} \frac{1}{k} = \ln n + C + 1/2n + O(1/n^2)$$

and

$$\sum_{k=1}^{n} \frac{1}{k^2} = \pi^2/6 + O(1/n),$$

as $n \to \infty$, where $C=0.5772\ldots$ is Euler's constant, the following useful relations for moments of $N(n)$ are valid:

$$EN(n) = \ln n + C + O(1/n),$$

$$EN^2(n) = \ln^2 n + (2C+1)\ln n + C^2 + C - \pi^2/6 + \ln n/n + O(1/n),$$

$$\mathrm{Var}(N(n)) = \ln n + C - \pi^2/6 + O(1/n), \qquad n \to \infty.$$

Note also that

$$EN(m)N(n) = EN^2(m) + EN(m)(EN(n) - EN(m))$$

and

$$\mathrm{Cov}(N(m), N(n)) = \mathrm{Var}(N(m)), \text{ if } m \le n.$$

Now we will discuss ways to find moment characteristics of record times $L(n)$. In chapter 4.1 we obtained (formula 4.1.6) that $EL(2) = \infty$. Evidently,

$$L^{\alpha}(n) \ge L(2) > 1,$$

if $\alpha \ge 1$ and $n=2,3,\ldots$. Hence,

$$EL^{\alpha}(n) = \infty$$

for any $\alpha \ge 1$ and $n=2,3,\ldots$. Below we will give some expressions for moments $EL^{\alpha}(n)$, $\alpha < 1$, and logarithmic moments $E\ln L(n)$.

For any $\beta < 1$ let us introduce random variables

$$T_n(\beta) = (1-\beta)^n \Gamma(L(n)+1) / \Gamma(L(n)-\beta+1), \quad n \ge 1, \tag{4.7.1}$$

where

$$\Gamma(s)=\int_0^\infty x^{s-1}e^{-x}dx$$

is the gamma-function. Also let $\mathfrak{I}_n$ denote a σ-algebra of events generated by random variables L(1), ..., L(n), n=1,2,.... The next result is a partial case of a more general theorem (see theorem 4.9.3 below), which was proved by Nevzorov (1990).

Theorem 4.7.1.

For any fixed $\beta<1$ a sequence of random variables

$T_n(\beta)$, n=1,2,...,

is a martingale with respect to a sequence of σ-algebras $\mathfrak{I}_n$, i.e.

$$E(T_{n+1}(\beta)|\mathfrak{I}_n)=T_n(\beta),\ n=1,2,.... \tag{4.7.2}$$

As a consequence of (4.7.2), due to the Markov property of the sequence of record times L(1),L(2),..., the following result have been obtained.

Corollary 4.7.1.

For any $n\geq m\geq 1$ and $\beta<1$ it holds that

$$E(T_n(\beta)\mid T_m(\beta))=T_m(\beta) \tag{4.7.3}$$

and

$$E(T_n(\beta))=E(T_1(\beta))=(1-\beta)\Gamma(L(1)+1)/\Gamma(L(1)-\beta+1)=$$

$$(1-\beta)\Gamma(2)/\Gamma(2-\beta)=1/\Gamma(1-\beta),\ n=1,2,.... \tag{4.7.4}$$

From (4.7.4) we have the relation

$$E(\Gamma(L(n)+1)/\Gamma(L(n)-\beta+1))=(1-\beta)^{-n}/\Gamma(1-\beta), \tag{4.7.5}$$

which is true for any $\beta<1$ and n=1,2,....

As it is known, the factorial moments of a positive order r=1,2,... of a random variable X are defined as

$$\mu_{(r)}(X) = EX(X-1)\ldots(X-r+1). \tag{4.7.6}$$

Note that (4.7.6) can be rewritten as

$$\mu_{(r)}(X) = E(\Gamma(X+1)/\Gamma(X-r+1)). \tag{4.7.7}$$

The latest definition lets us to introduce the factorial moments of a negative order as follows:

$$\mu_{(-r)}(X) = E(\Gamma(X+1)/\Gamma(X+r+1)) = E(1/(X+1)(X+2)\ldots(X+r)),\ r=1,2,\ldots. \tag{4.7.8}$$

Now consider factorial moments of negative orders for record times. Let

$$m(r,n) = E\mu_{(-r)}(L(n)) = E(\Gamma(L(n)+1)/\ \Gamma(L(n)+r+1)) = E(1/(L(n)+1)\ldots(L(n)+r)),\ r=1,2,\ldots.$$

It follows immediately from (4.7.5) that

$$m(r,n) = (1+r)^{-n}/\Gamma(1+r) = 1/(r!(1+r)^{n}),\ r=1,2,\ldots,\ n=1,2,\ldots, \tag{4.7.9}$$

and , in particular, for any $n=1,2,\ldots$, the next equalities hold:

$$m(1,n) = E(1/(L(n)+1)) = 2^{-n},$$

$$m(2,n) = E(1/(L(n)+1)(L(n)+2)) = (3^{-n})/2,$$

$$m(3,n) = E(1/(L(n)+1)(L(n)+2)(L(n)+3)) = (4^{-n})/6. \tag{4.7.10}$$

On combining (4.7.5) and Stirling's approximation

$$\Gamma(z) = \exp(-z)z^{z-1/2}(2\pi)^{1/2}(1+1/(12z)+1/(288z^{2})+O(z^{-3})),\ z\to\infty,$$

Nevzorov and Stepanov (1988) showed that for any $\beta>0$,

$$E(L(n))^{1-\beta} = \frac{1}{\Gamma(\beta)}\ \{\beta^{-n} + \frac{\beta-1}{2}\ (\beta+1)^{-n} + O((\beta+2)^{-n})\},\ n\to\infty. \tag{4.7.11}$$

Now we will find the logarithmic moments $E\ln L(n)$.

The next result is a particular case of a more general statement, which has been obtained by Nevzorov (1987).

Theorem 4.7.2.

Let

$f(n)=1+1/2+...+1/n$, $n\geq 1$.

Then for any n=1,2,..., the following equality is valid:

$$Ef(L(n))=n. \qquad (4.7.12)$$

Proof.

It is known from chapter 4.2 that

$P\{L(n)=j|L(n-1)=i\}= i/j(j-1)$.

Then

$$Ef(L(n))=\sum_{i=n-1}^{\infty} E\{f(L(n))|L(n-1)=i\}P\{L(n-1)=i\}. \qquad (4.7.13)$$

The conditional expectation $E\{f(L(n))|L(n-1)=i\}$ is given as follows:

$$E\{f(L(n))|L(n-1)=i\}=\sum_{j=i+1}^{\infty} f(j)P\{L(n)=j|L(n-1)=i\}$$

$$=i\sum_{j=i+1}^{\infty} f(j)/j(j-1)$$

$$=i\sum_{j=i+1}^{\infty} f(j)/(j-1)-i\sum_{j=i+1}^{\infty} f(j)/j$$

$$=i\sum_{j=i+1}^{\infty} f(j)/(j-1)-i\sum_{j=i+1}^{\infty} (f(j+1)-1/(j+1))/j$$

$$=i\sum_{j=i+1}^{\infty} f(j)/(j-1)-i\sum_{j=i+2}^{\infty} f(j)/(j-1)+i\sum_{j=i+1}^{\infty} (1/j-1/(j+1))$$

$$=f(i+1)+i/(i+1)=f(i)+1. \qquad (4.7.14)$$

In (4.7.14) the fact, that $L(n-1)\geq n-1$, was used.

Really, we have got that

$E(f(L(n))|L(n-1)) = f(L(n-1))+1.$

Equalities (4.7.13) and (4.7.14) imply that

$$Ef(L(n))=\sum_{i=n-1}^{\infty} (f(i)+1)\,P\{L(n-1)=i\}=Ef(L(n-1))+1. \tag{4.7.15}$$

Hence,

$Ef(L(n))= Ef(L(n-1))+1=\ldots=Ef(L(1))+(n-1)=f(1)+n-1=n,$

and the theorem is proved.
The next argument, which we will use, is the following well-known relation:

$$\ln n=f(n)-C-1/2n+O(1/n^2),\ n\to\infty, \tag{4.7.16}$$

$C=0.5772\ldots$ being Euler's constant.

As it will be seen below the next form of (4.7.16) is more convenient for our purposes:

$$\ln n=f(n)-C-1/2(n+1)+O(1/(n+1)(n+2)),\ n\to\infty. \tag{4.7.17}$$

After substituting n by $L(n)$ in (4.7.17) we immediately come to the following relation for the logarithmic moment:

$$E\ln L(n)=Ef(L(n))-C-1/2\ m(1,n)+O(m(2,n)),\ n\to\infty, \tag{4.7.18}$$

where $m(1,n)$ and $m(2,n)$ are given in (4.7.10) and the expression for $Ef(L(n))$ was obtained in (4.7.12). Finally we get that

$$E\ln L(n)=n-C-2^{-(n+1)}+O(3^{-n}),\ n\to\infty, \tag{4.7.19}$$

where $C=0.5772\ldots$ is Euler's constant.

In the next chapter we will prove that $\ln L(n)$ has an asymptotically normal distribution rather than $L(n)$. Hence the relations for the logarithmic moment of record times , given above, are important for us.

Next we will give some remarks on different moment characteristics of record values. We know from chapter 4.2 that if independent random variables $X_1, X_2, \ldots$ have a continuous distribution function F, then

$$P\{X(n)<x\}=\frac{1}{(n-1)!}\int_0^{R(x)} v^{n-1}\exp(-v)dv,$$

where

$R(x)=-\ln(1-F(x))$.

This formula gives the following expression for moments of X(n):

$$E(X(n))^{\alpha}=\frac{1}{(n-1)!}\int_{-\infty}^{\infty} x^{\alpha}(R(x))^{n-1}\exp(-R(x))dR(x). \quad (4.7.20)$$

Transforming the RHS of (4.7.20) we obtain that

$$E(X(n))^{\alpha}=\frac{1}{(n-1)!}\int_0^1 (G(u))^{\alpha}(-\ln(1-u))^{n-1}du. \quad (4.7.21)$$

One more useful expression, being equivalent to (4.7.20) and (4.7.21), is the following:

$$E(X(n))^{\alpha}=\frac{1}{(n-1)!}\int_{-\infty}^{\infty} u^{\alpha}(-\ln(1-F(u)))^{n-1}dF(u). \quad (4.7.22)$$

Since

$(-\ln(1-F(u)))\to\infty$ if $u\to\infty$,

and

$(-\ln(1-F(u)))\to 0$ if $u\to-\infty$, it is evident that there exist such distribution function F that

$E|X|^{\alpha}=E|X(1)|^{\alpha}<\infty$,

when

$E|X(n)|^{\alpha}=\infty$, $n=2,3,\ldots$,

as well as there are distributions, for which

$E|X(n)|^{\alpha}<\infty$

for any n=2,3,…, when

$E|X|^{\alpha}= E|X(1)|^{\alpha} = \infty.$

The next result gives a sufficient condition of existence of moments for record values.

Theorem 4.7.3.

If $\int_{-\infty}^{\infty} |x|^{\alpha+\delta} dF(x)<\infty,$

for some $\alpha>0$ and $\delta>0$, then

$E|X(n)|^{\alpha}< \infty$

for any n=1,2,….

Proof.

From (4.7.21) , by Holder's inequality, we get that

$(n-1)!\ E|X(n)|^{\alpha}=$

$$\int_{0}^{t} (|G(u)|^{\alpha+\delta})^{\alpha/(\alpha+\delta)} ((-\ln(1-u))^{(n-1)(\alpha+\delta)/\delta})^{\delta/(\alpha+\delta)} du \leq$$

$$\left(\int_{0}^{t} |G(u)|^{\alpha+\delta} du\right)^{\alpha/(\alpha+\delta)} \left(\int_{0}^{t} (-\ln(1-u))^{(n-1)(\alpha+\delta)/\delta} du\right)^{\delta/(\alpha+\delta)}. \quad (4.7.23)$$

Note that

$$\int_{0}^{t} |G(u)|^{\alpha+\delta} du = E|X|^{\alpha+\delta}<\infty$$

and

$$\int_{0}^{t} (-\ln(1-u))^{(n-1)(\alpha+\delta)/\delta} du= \int_{0}^{\infty} v^{(n-1)(\alpha+\delta)/\delta} e^{-v} dv=$$

$\Gamma(\frac{(n-1)(\alpha+\delta)}{\delta}+1)$.

Then

$E|X(n)|^{\alpha} \le C(\alpha,\delta,n)E|X|^{\alpha+\delta} < \infty$,

where

$$C(\alpha,\delta,n)=\Gamma(\frac{(n-1)(\alpha+\delta)}{\delta}+1)/(n-1)!.$$

By that we completed the proof of the theorem.
To find a formula for product moments of record values, say

$E(X(m))^{\alpha}(X(n))^{\beta}$,

it is convenient to use the relation (4.4.15) :

$$(X(1),X(2),...,X(n)) \stackrel{d}{=} (H(\omega_1), H(\omega_1+\omega_2),..., H(\omega_1+\omega_2+...+\omega_n)),$$

where

$H(x)=G(1-e^{-x})$

is the inverse of the function $R(x)=-\ln(1-F(x))$ and $\omega_1, \omega_2,\ldots$ are i.i.d. exponential $E(1)$ random variables. Since

$$(X(m),X(n)) \stackrel{d}{=} (H(\omega_1+...+\omega_m), H(\omega_1+\omega_2+...+\omega_n)),\ m<n,$$

and sums $(\omega_1+...+\omega_m)$ and $(\omega_{m+1}+...+\omega_n)$ are independent and have gamma distributions with parameters m and n-m correspondingly, we obtain for $1 \le m<n$ and any α and β that

$$E(X(m))^{\alpha}(X(n))^{\beta}=\int_0^{\infty}\int_0^{\infty} (H(x))^{\alpha}(H(x+y))^{\beta}x^{m-1}y^{n-m-1}\exp(-(x+y))dxdy/(m-1)!(n-m-1)!. \quad (4.7.24)$$

Below we will list (without proofs) some interesting results for covariances and correlation coefficients of record values in the continuous case (see Nagaraja and Nevzorov (1996) and Nevzorov (1992)):

a) If $0<E(X(m))^2<\infty$ and $0<E(X(n))^2<\infty$, $m<n$, then the correlation coefficient

$$\rho = \rho(X(m),X(n))$$

is positive and does not exceed $(m/n)^{1/2}$. Moreover, the equality

$$\rho =(m/n)^{1/2}$$

holds if and only if $X_1,X_2,\dots$ have an exponential distribution;

b) Let g be an arbitrary function, such that

$$E(g(X(n)))^2<\infty$$

and

$$E(g(X(n+1)))^2<\infty$$

for some $n=1,2,\dots$. Then

$$Cov(g(X(n)),g(X(n+1)))\geq 0;$$

c) For any $n>0,m>0$, $|n-m|\geq 2$, there exists a function g such that

$$Cov(g(X(m)),g(X(n)))<0.$$

Remark 4.7.1.

It is interesting to note (compare with the statement b) above) that there are examples of discrete distributions for which

$$Cov(g(X(n)),g(X(n+1)))<0.$$

4.8. ASYMPTOTIC BEHAVIOR OF RECORD TIMES AND RECORD VALUES

We know from chapter 4.2 that numbers of records N(n) can be expressed as sums of independent record indicators:

$$N(n)=\xi_1+\dots+\xi_n,\ n=1,2,\dots.$$

This representation has helped us to find generating functions and some moments of N(n). For example, we have found that

$A(n)=EN(n)=1+1/2+\ldots+1/n$

and

$$B_n^2=Var(N(n))=\sum_{k=1}^{n}\left(\frac{1}{k}-\frac{1}{k^2}\right).$$

Hence,

$A(n)\sim\ln n$

and

$B_n^2\sim\ln n,$

as $n\to\infty$. More sharp asymptotic relations for expectations and variances of N(n) are given in chapter 4.7. Having a representation of N(n) via sums of independent summands one can use the very deeply developed theory of sums of independent random variables to study the asymptotic behavior of N(n) . Note also that in the given situation record indicators $\xi_1, \xi_2,\ldots$ are bounded ($0\le\xi_n\le1$, n=1,2,...) random variables , and this fact essentially simplifies the inspection of different moment conditions, which provide the validity of the classical limit theorems. Hence the important asymptotic relations for sums of independent random variables are true automatically for N(n). As a rule, the only thing we must to do is to check that our relation stays true when expectations A(n) and variances B_n^2 are replaced by ln n. We list some of these statements, which hold under the assumption that $n\to\infty$.

a) Central limit theorem (CLT):

$$\sup_x|P\{N(n)-\ln n<x\sqrt{\ln n}\}-\Phi(x)|\to0, \tag{4.8.1}$$

Φ being the distribution function of the standard normal law;

b)Uniform estimate in CLT:

$$\sup_x|P\{N(n)-\ln n<x\sqrt{\ln n}\}-\Phi(x)|\le C/\sqrt{\ln n}; \tag{4.8.2}$$

C being an absolute constant (Shiganov's (1982) constant 0.7915 can be taken here as C);

c) Nonuniform estimate in CLT:

$|P\{N(n)-\ln n<x\sqrt{\ln n}\}-\Phi(x)|\leq C/((1+|x|^3)\sqrt{\ln n})$, (4.8.3)

where C is an absolute constant;

d) Strong Law of Large Numbers (SLLN):

$P\{\lim (N(n)/\ln n)=1\}=1$; (4.8.4)

e) Law of Iterative Logarithm (LIL):

$$P\{\limsup\frac{N(n)-\ln n}{(2\ln n\ \ln\ln\ln n)^{1/2}}=1\}=1 \quad (4.8.5)$$

and

$$P\{\liminf\frac{N(n)-\ln n}{(2\ln n\ \ln\ln\ln n)^{1/2}}=-1\}=1. \quad (4.8.6)$$

In chapter 4.2 we mentioned that the relation

$P\{L(n)>m\}=P\{N(m)<n\}$, n=1,2,..., m=1,2,...

ties distributions of N(n) and L(n). Due to this equality the most part of the limit theorems given above, were worked over (Renyi (1962)) into the corresponding theorems for record times as follows:

a) CLT:

$$\sup_x|P\{\ln L(n)-n<x\sqrt{n}\}-\Phi(x)|\to 0; \quad (4.8.7)$$

b) SLLN:

$P\{\lim \ln L(n)/n=1\}=1$; (4.8.8)

c) LIL:

$$P\{\limsup\frac{\ln L(n)-n}{(2n\ \ln\ln n)^{1/2}}=1\}=1 \quad (4.8.9)$$

and

$$P\{\liminf \frac{\ln L(n) - n}{(2n \ \ln\ln n)^{1/2}} = -1\} = 1. \tag{4.8.10}$$

It is interesting to see that relations (4.8.7)-(4.8.10) stay true if to replace L(n) in all of them by inter- record times

$$\Delta(n) = L(n) - L(n-1),\ n = 2,3,\ldots, \tag{4.8.11}$$

(see, for example, Neuts (1967), Holmes and Strawderman (1969), Strawderman and Holmes (1970))

Since

$$X(n) = M_{L(n)},$$

where

$$M_n = \max\{X_1,\ldots,X_n\},$$

the limit distributions of record values must be close to the analogous distributions of maximal order statistics. As it is known, there are three types of asymptotic distributions for suitable centering and normalizing maxima

$$(M(n) - b(n))/a(n).$$

They are:

$$\Lambda(x) = \exp(-\exp(-x)),$$

$$\Phi_\alpha(x) = \begin{cases} 0, & if \quad x < 0, \\ \exp(-x^{-\alpha}), & if \quad x > 0 \end{cases};$$

and

$$\Psi_\alpha(x) = \begin{cases} \exp(-(-x)^\alpha), & if \quad x < 0, \\ 1, & if \quad x > 0. \end{cases}$$

There is no doubt that under the corresponding random centering and normalizing we will obtain the same limit distributions for random variables

$$(X(n) - b(L(n)))/a(L(n)).$$

The more interesting question arises: what types of asymptotic distributions can we get for normalized record values?

The answer becomes simple if we consider the exponential records Z(1), Z(2),... , which correspond to the distribution function

$$F(x)=\max\{0,1-e^{-x}\}.$$

From chapter 4.4 we know that

$$Z(n)\overset{d}{=} \omega_1+\omega_2+\ldots+\omega_n,$$

$\omega_1,\omega_2,\ldots$ being independent exponentially E(1) distributed random variables.
Note also that

$$E\omega_n=Var(\omega_n)=1,\ n=1,2,\ldots.$$

Recalling the central limit theorem for sums of i.i.d. random variables , one gets immediately that

$$P\{Z(n)-n<xn^{1/2}\} \to \Phi(x), \qquad (4.8.12)$$

where $\Phi(x)$ is the distribution of the standard normal law.

It is interesting to compare (4.8.12) with the corresponding asymptotic relation for the exponential maximal order statistics M_n:

$$P\{ M_n-\ln n<x\}\to\Lambda(x)=\exp(-\exp(-x)), \qquad (4.8.13)$$

which implies that

$$P\{Z(n)-\ln L(n)<x\}\to\Lambda(x). \qquad (4.8.14)$$

From (4.18.12) and (4.18.14) one can see that asymptotic distributions of record values do not need to coincide under random and nonrandom normalizing.

As it is known, record values, corresponding to arbitrary continuous distribution function F, are expressed via the exponential record values. Hence, we have the following equalities:

$$P\{(X(n)-B(n))/A(n)<x\}$$

$$=P\{Z(n)<R(xA(n)+B(n))\}$$

$$=P\{(Z(n)-n)/n^{1/2}<(R(xA(n)+B(n))-n)/n^{1/2}\}, \qquad (4.8.15)$$

where

$R(x)=-\ln(1-F(x))$.

It follows from (4.8.15) that existence (for some constant A(n) and B(n)>0) of a nondegenerate limit distribution function

$$G(x)=\lim P\{X(n)-B(n)<xA(n)\},\ n\to\infty, \tag{4.8.16}$$

is equivalent to existence of a limit function

$$g(x)=\lim (R(xA(n)+B(n))-n)/n^{1/2}, \tag{4.8.17}$$

having at least two points of growth. Moreover, the limits G(x) and g(x) must satisfy the following relation:

$$G(x)=\Phi(g(x)), \tag{4.8.18}$$

where

$$\Phi(x)=(2\pi)^{-1/2}\int_{-\infty}^{x} \exp(-t^2/2)dt.$$

Tata (1969) (see also Resnick (1973)) found that there are three possibilities (up to linear transformations) for the limit functions g(x). They are:

$$\begin{aligned} &1)\ g_1(x)=x;\\ &2)\ g_2(x)=\gamma\ln x,\ \gamma>0,\ \text{if } x>0, \text{ and } g_2(x)=-\infty,\ \text{if } x<0;\\ &3)\ g_3(x)=-\gamma\ln(-x),\ \gamma>0,\ \text{if } x<0, \text{ and } g_3(x)=\infty,\ \text{if } x>0. \end{aligned} \tag{4.8.19}$$

It is not a problem to prove that each of functions $\Phi(g_i(x))$, i=1,2,3, where $g_i(x)$ are defined in (4.8.19), can be a limit distribution function for record values under a suitable choice of the initial distribution function F and centering and normalizing constants A(n) and B(n). Finally, we obtain that for records, as well as for maximal order statistics, there are only three types of limit distribution functions.

Remark 4.8.1.

It is interesting to compare that the set of all possible nondegenerate distribution functions for records consists of three types of functions $\Phi(g_i(x))$, when the corresponding set of limits for maxima also consists of three types of distribution functions and these functions have the form

$\exp(-\exp(-g_i(x)))$, $i=1,2,3$,

where $g_i(x)$ are defined in (4.8.19).

Returning to relation (4.8.14) one can see that the exponential record values Z(n) are close to ln L(n). Note that (4.8.14) stays true if to replace Z(n) by the random variable

$$\tau_n=-\ln(1-F(X(n))),$$

where X(n) are records, corresponding to some continuous d.f. F. Moreover, Nevzorov (1995) obtained the following estimate.

$$P\{Z(n)-\ln L(n)<x\}\rightarrow\Lambda(x). \qquad (4.8.20)$$

Theorem 4.8.1.

Let

$$\Lambda(x)=\exp(-\exp(-x)).$$

Then for any q, $0<q<1$, the following equality is valid:

$$|P\{\tau_n-\ln L(n)<x\}-\Lambda(x)|\le r_n(x), \qquad (4.8.21)$$

where

$$r_n(x)=\Lambda(x)(e^{-2x}2^{2-n}+2e^{-4x}3^{1-n}/(1-q)),$$

if $x>x_n$, and

$$r_n(x)=\exp\{-(3qn/2)^{1/2}\}(1+3qn2^{1-n}+q^2n^23^{3-n}/2(1-q)),$$

if $x<x_n$, with

$$x_n=-\ln(3nq/2)/2.$$

4.9. Generalizations of Record Times and Record Values

For i.i.d. random variables $X_1,X_2,\ldots$ and any $k=1,2,\ldots$ Dziubdziela and Kopocinsky (1976) have introduced the so-called k-th record times L(n,k) and k-th record values X(n,k) as follows:

$$L(1,k)=k,\ L(n+1,k)=\min\{j>L(n,k):\ X_j>X_{j-k,j-1}\},\ n\ge1, \qquad (4.9.1)$$

and

$$X(n,k)=X_{L(n,k)-k+1,L(n,k)},\ n\geq 1. \qquad (4.9.2)$$

A more natural way to determine the k-th records is to consider a nondecreasing sequence of k-th maximal order statistics

$$-\infty<X_{1,k}\leq X_{2,k+1}\leq\ldots\leq X_{n-k,n-1}\leq X_{n-k+1,n}\leq\ldots\ .$$

and to separate the terms of this sequence which are preceded by the sign of the strong inequality. The subsequence

$$X_{1,k}<X_{n(2)-k+1,n(2)}<X_{n(3)-k+1,n(3)}<\ldots,$$

which is a result of such operations , generates a sequence of k-th record values when the sequence of random indices n(1)=k , n(2),... generates k-th record times

$$L(1,k)<L(2,k)<\ldots\ .$$

Indeed, random variables L(n,1) and X(n,1) coincide with classical record times L(n) and record values X(n) correspondingly.

We will use N(n,k) to denote the numbers of k-th records in a sequence X_1, X_2, ...Note that N(n,1)=N(n). We will also need the so-called sequential ranks R_n, which are defined for arbitrary X_n in a sequence X_1, X_2,... as follows:

$$R_n=\sum_{k=1}^{n} \mathbf{1}\{X_n\geq X_k\},\ n=1,2,\ldots. \qquad (4.9.3)$$

Let $\xi_n^{(k)}$, n=k, k+1,..., denote indicators of k-th records, i.e. $\xi_n^{(k)}=1$, if X_n is a k-th record value, and $\xi_n^{(k)}=0$ otherwise. Note that events $\{\xi_n^{(k)}=1\}$ and $\{R_n\geq n-k+1\}$ coincide. Hence, the indicators of k-th records can be defined also as follows:

$$\xi_n^{(k)}=\mathbf{1}\{R_n\geq n-k+1\},\ n=k,k+1,\ldots\ .$$

In chapter 4.2 we formulated and proved the lemma of Renyi about the independence of record indicators $\xi_n=\xi_n^{(1)}$. In fact, Renyi (1962) (see also Barndorf-Nielsen (1963)) obtained a more general result, which we give without proof.

Theorem 4.9.1.

Let $X_1, X_2, \ldots$ be a sequence of independent random variables with a common continuous distribution function. Then sequential ranks $R_1, R_2, \ldots$ are independent and for any $n=1,2,\ldots$,

$$P\{R_n=k\}=1/n, \; k=1,2,\ldots,n. \quad (4.9.4)$$

Remark 4.9.1.

Indeed, for any $k=1,2, \ldots$, the independence of events

$$\{\xi_n^{(k)}=1\}=\{R_n \geq n-k+1\}, \; n=k,k+1,\ldots,$$

and thereby the independence of the k-th record indicators $\xi_n^{(k)}$ immediately follow from theorem 4.9.1. Additionally, one can obtain from this theorem that

$$P\{\xi_n^{(k)}=1\}=P\{R_n \geq n-k+1\}=k/n, \; k=1,2,\ldots n, \; n=k,k+1,\ldots. \quad (4.9.5)$$

Analogously to the case $k=1$, we can use the following representation of random variables $N(n,k)$ via sums of independent indicators:

$$N(n,k)= \xi_k^{(k)}+\ldots+\xi_n^{(k)}, \quad (4.9.6)$$

The distributions of the k-th record times also are expressed in terms of k-th record indicators:

$$P\{L(n,k)>m\}=P\{N(m,k)<n\}=$$

$$P\{\xi_k^{(k)}+\ldots+\xi_m^{(k)}<n\}, \quad (4.9.7)$$

$$P\{L(n)=m\}=P\{N(m-1,k)=n-1,N(m,k)=n\}=$$

$$P\{N(m-1,k)=n-1, \; \xi_m^{(k)}=1\}=$$

$$kP\{N(m-1,k)=n-1\}/m, \; n=1,2,\ldots, \; m=k, \; k+1,\ldots . \quad (4.9.8)$$

The independence of indicators $\xi_n(k)$ helps us to find joint distributions of the k-th record times as follows:

$$P\{L(1,k)=k, \; L(2,k)=m(2), \ldots, L(n,k)=m(n)\}$$

$$= P\{\xi_k^{(k)}=1, \xi_{k+1}^{(k)}=0,\ldots, \xi_{m(2)-1}^{(k)}=0, \xi_{m(2)}^{(k)}=1,\ldots, \xi_{m(n)-1}^{(k)}=0, \xi_{m(n)}^{(k)}=1\}$$

$$=P\{\xi_k^{(k)}=1\}P\{\xi_{k+1}^{(k)}=0\}\cdots P\{\xi_{m(2)-1}^{(k)}=0\}P\{\xi_{m(2)}^{(k)}=1\}\cdots P\{\xi_{m(n)-1}^{(k)}=0\}P\{\xi_{m(n)}^{(k)}=1\}$$

$$= \frac{k!(m(n)-k)!}{m(n)!}\cdot\frac{k^{n-1}}{(m(2)-k)(m(3)-k)\cdots(m(n)-k)} \tag{4.9.9}$$

if $k<m(2)<\ldots<m(n)$.

Analogously to the case k=1 (see chapter 4.3), we can use (4.9.9) to prove that a sequence L(1,k),L(2,k),... forms a Markov with transition probabilities

$P\{L(n+1,k)>r|L(n,k)=m\}$

$= P\{L(n+1,k)>r|L(n,k)=m,L(n-1,k)=m_{n-1},\ldots,L(2,k)=m_2,L(1,k)=k\}$

$= (m+1-k)\ldots(r-k-1)(r-k)/(m+1)\ldots(r-1)r,\ r\geq m\geq n+k-1,$(4.9.10)

and

$$P\{L(n+1,k)= r|L(n,k)=m\}= (m+1-k)\ldots(r-k-1)k/(m+1)\ldots(r-1)r,\ r>m\geq n+k-1. \tag{4.9.11}$$

In section 4.2 for the generating function

$Q_n(s)=Es^{L(n)}$

the following expressions were given:

$$Q_n(s) =1-(1-s)\sum_{k=0}^{n-1}\frac{(-\ln(1-s))^k}{k!}$$

$$= \frac{1}{(n-1)!}\int_0^{-\log(1-s)} v^{n-1}\exp(-v)dv.$$

Nevzorov (1990) suggested the martingale approach to find generating functions of the k-th record times and he has obtained the following result.

Theorem 4.9.2.

For any k=1,2,...,

$Q_{n,k}(s) = Es^{L(n,k)} = r(-\ln(1-s))$,

where

$$r(v)=\frac{k^n}{(n-1)!}\int_0^v x^{n-1}e^{-kx}(1-e^{-(v-x)})^{k-1}dx. \tag{4.9.12}$$

Let us consider moment characteristics of random variables N(n,k) and L(n,k). Essentially, there are no new effects when we find moments of numbers of records N(n,k). In fact, recalling that record indicators $\xi_n^{(k)}$ are independent and noting that

$E\xi_n(k)=k/n$

And

$Var(\xi_n(k))=k/n - k^2/n^2$, n=k,k+1,... ,

one gets that

$$EN(n,k)=E(\xi_k(k)+...+\xi_n(k))= k\sum_{m=k}^{N} 1/m \tag{4.9.13}$$

and

$$Var(N(n,k))= EN(n,k) - k^2\sum_{m=k}^{n} 1/m^2, \quad m=k,k+1,\dots . \tag{4.9.14}$$

In the asymptotic situation, when $n\to\infty$, it is more convenient to use the following relations instead of exact equalities (4.9.13) and (4.9.14):

$$EN(n,k)= k(\ln n-\sum_{m=1}^{k-1} 1/m+C)+O(1/n), \tag{4.9.15}$$

C=0.5772... being Euler's constant, and

$$Var(N(n,k))= k(\ln n-\sum_{m=1}^{k-1} 1/m+\gamma)-k^2(\pi^2/6-\sum_{m=1}^{k-1} 1/m^2)+O(1/n) . \quad (4.9.16)$$

It is more interesting to compare moment characteristics of the classical (k=1) record times L(n) and the k-th record times L(n,k). As it has been shown above,

$$EL^{\alpha}(n)= \infty$$

for any $\alpha\geq 1$ and n=2,3,... .In particular,

$$EL(n)= \infty, n=2,3,\ldots .$$

Probably, the latter fact is one of the reasons, why record statistics are not very popular amongst statisticians as yet. The situation changes for k-th record times, when $k\geq 2$. It turns out that EL(n,k) exists for any n=1,2,..., if $k\geq 2$. Moreover, any moment $EL^{\alpha}(n,k)$ exists if $k>\alpha$.

The following result is very important for finding of moment characteristics of k-th record times.

Let $\Im_n^{(k)}$ denote a σ-algebra of events, generated by the k-th record times L(1,k),...,L(n,k) and let

$$T_n(\beta)=(k-\beta)^n\Gamma(L(n,k)+1)/ k^n \Gamma(L(n,k)-\beta+1), n\geq 1.$$

Theorem 4.9.3. (Nevzorov (1990)).

For any fixed k=1,2,... and $\beta < k$ the sequence

$$T_n(\beta), n=1,2,\ldots$$

is a martingale with respect to the sequence of σ-algebras $\Im_n^{(k)}$, and

$$ET_n(\beta) = ET_1(\beta) = (k-\beta)\,\Gamma(k+1)/k\Gamma(k-\beta+1). \quad (4.9.17)$$

Remark 4.9.3.

Taking r=1,2,..., k-1 instead of β, one can see that

$$T_n(r)= (k-r)^n\Gamma(L(n,k)+1)/ k^n \Gamma(L(n,k)-r+1)=$$

$$(1-r/k)^n L(n,k)(L(n,k)-1)\cdots(L(n.k)-r+1), n\geq 1. \quad (4.9.18)$$

Hence, it follows from (4.9.17) that

$$\mu_{(r)}(L(n,k)) = k^{n-1}k!/(k-r)^{n-1}(k-r)!, \qquad (4.9.19)$$

where

$$\mu_{(r)}(L(n,k))= EL(n,k)(L(n,k)-1)\cdots(L(n.k)-r+1)$$

denote the factorial moment of the k-th record time L(n,k) of order r. Interesting to see that (4.9.19) stays true for negative r as well. For r=-1,-2,... the factorial moments are defined as follows:

$$\mu_{(r)}(L(n,k))=E\Gamma(L(n,k)+1)/\ \Gamma(L(n,k)-r+1)=E(1/(L(n,k)+1)(L(n,k)+2)\cdots(L(n,k)-r+1)).$$

Mention here the most important partial cases of (4.9.17).

Corollary 4.9.1.

For any n=1,2,... the following relations are true:

$$\mu_{(1)}(L(n,k))=EL(n,k)=k^n/(k-1)^{n-1},\ k=2,3,\ldots;$$

$$\mu_{(2)}(L(n,k))=EL(n,k)(L(n,k)-1)=k^n(k-1)/(k-2)^{n-1},\ k=3,4,\ldots;$$

$$\mu_{(-1)}(L(n,k))=E(1/(L(n,k)+1))=k^{n-1}/(k+1)^n.$$

Note also that

$$E(L(n,k))^2=\mu_{(1)}(L(n,k))+\ \mu_{(2)}(L(n,k)),\ k=3,4,\ldots,$$

and

$$Var(L(n,k))=\ \mu_{(2)}(L(n,k))+\mu_{(1)}(L(n,k))-(\mu_{(1)}(L(n,k)))^2.$$

Some other moment characteristics, which were found for L(n) in chapter 4.7, also have been obtained for the k-th record times. For example the following relation holds (see Nevzorov and Stepanov (1988)) for any k=1,2, ...:

$$E\ln L(n,k)=f(k)+(n-1)/k-C-k^{n-1}/2(k+1)^n+O((k/(k+2))^n),\ n\to\infty,$$

where C=0.5772... is Euler's constant and

$$f(k)=1+1/2+\ldots+1/k.$$

Since the distributions of N(n,k) and L(n,k) are expressed via distributions of sums of independent indicators $\xi_n^{(k)}$ in the same manner, as N(n) and L(n) were expressed in terms of record indicators ξ_n, one can easily obtain the corresponding generalizations of limit relations (4.8.1)-(4.8.10) with some natural corrections of normalizing constants. For example, it appears that for any k=1,2,... the random variables

$$(N(n,k)-k\ln n)/(k\ln n)^{1/2}$$

and

$$(k\ln L(n,k)-n)/n^{1/2}$$

asymptotically have the standard normal distribution.

The situation with the k-th record values is very close to the classical (k=1) record values. As we have set above, the classical record values Z(n), corresponding to the standard exponential distribution, satisfy the following relation:

$$(Z(1),Z(2),\ldots,Z(n)) \stackrel{d}{=} (\omega_1, \omega_1+\omega_2,\ldots, \omega_1+\omega_2+\ldots+\omega_n),$$

where $\omega_1,\omega_2,\ldots$ are independent random variables having the common distribution function

$$F(x)=\max\{0,1-\exp(-x)\}.$$

The results of Dziubdziela and Kopocinsky (1976) and Deheuvels (1984) imply that

$$(kZ(1,k),kZ(2,k),\ldots,kZ(n,k)) \stackrel{d}{=} (\omega_1, \omega_1+\omega_2,\ldots, \omega_1+\omega_2+\ldots+\omega_n), \qquad (4.9.20)$$

where Z(n,k) denote the k-th record values for the standard exponential distribution. Hence, for any k=1,2,... and n=1,2,...,

$$Z(n) \stackrel{d}{=} kZ(n,k) \stackrel{d}{=} \omega_1+\omega_2+\ldots+\omega_n. \qquad (4.9.21)$$

To find the probability structure of k-th record values X(n,k) for arbitrary continuous distribution function F we need to use the probability integral transformation , which helps to obtain that

$$(X(1,k),\ldots,X(n,k)) \stackrel{d}{=} (H(Z(1,n)),\ldots,H(Z(n,k))), \qquad (4.9.22)$$

where $H(x)=G(1-e^{-x})$ and G is the inverse of F. Since the sum on the RHS of (4.9.21) has a gamma distribution, one can get from (4.9.21) and (4.9.22) that

$$P\{X(n,k)<x\}=\frac{1}{(n-1)!}\int_0^{-k\log(1-F(x))} v^{n-1}\exp(-v)dv. \quad (4.9.23)$$

Note that we can get the expression on the RHS of (4.9.23) considering the classical (k=1) record values Y(n) for a sequence of i.i.d. random variables $Y_1,Y_2,\ldots$ with d.f.

$G(x)= 1-(1-F(x))^k$.

It is easy to see that G(x) is the cdf. of the random variable

$Y=\min\{X_1,X_2,\ldots,X_k\}$.

The following result belongs to Deheuvels (1984).

Theorem 4.9.4.

Let independent random variables X_1, X_2,... have a common continuous d.f. F and let Y(n) denote the classical record values in a sequence

$Y_1=\min\{X_1,\ldots,X_k\}$, $Y_2=\min\{X_{k+1},\ldots, X_{2k}\},\ldots$.

Then for any k=1,2,... the following relation holds

$$\{X(n,k)\}_{n=1}^{\infty}\stackrel{d}{=}\{Y(n)\}_{n=1}^{\infty}. \quad (4.9.24)$$

The latest theorem enables us to reformulate for the k-th record values many results , which are known for the usual record values. Let us list some of such sequences.

Corollary 4.9.2.

Recalling the situation for the classical record values and applying theorem 4.9.4 one can see that that the set of all possible nondegenerate distribution functions for the k-th record values also consists of functions $\Phi(g_i(x))$, i=1,2,3, where $g_i(x)$ are defined in (4.8.19).

Corollary 4.9.3.

For any continuous distribution function F and arbitrary k=1,2,..., a sequence X(1,k),X(2,k),... forms a Markov chain and

$$P\{X(n+1,k)>x|X(n,k)=y\}=((1-F(x))/(1-F(y)))^k,\ x>y.$$

Moreover, comparing the Markov properties of order statistics and k-th records we obtain that for any n=1,2,... and m=1,2,... the following equality holds almost surely for x>y:

$$P\{X(n+1,k)>x|X(n,k)=y\}=P\{X_{m+1,m+k}>x|X_{m,m+k}=y\}.$$

Further generalizations of the classical records and k-th record values connected with the so-called k_n-records, which have been introduced by Nevzorov (1986). We give here only definitions of k_n-record times and k_n record values. A more detailed information about k_n-records can be found in Deheuvels and Nevzorov (1994).

Definition 4.9.1.

Let k_1, k_2,... be a sequence of integers such that $0\le k_n\le n$. Let also $X_1,X_2,\ldots$ be independent random variables with a common continuous d.f. F and let R_n denote the sequential rank of X_n amongst random variables $X_1,X_2,\ldots,X_n$, n=1,2,... Then k_n-record times $L_k(n)$ and k_n-record values $X_k(n)$ are defined as follows:

$$L_k(0)=0,\ L_k(n)=\min\{m>L_k(n-1):\ R_m\ge m-k_m+1\}$$

and

$$X_k(n)=X_{L_k(n)-k_n+1,L_k(n)},\ n=1,2,\ldots.$$

Remark 4.9.4.

The main properties of k_n-records are also based on the independence of the corresponding record indicators.

4.10. NONSTATIONARY RECORD MODELS

Unlike the records in the previous chapters here we will study record times and record values in sequences of X's , which may have different distributions. The first two of the considering schemes include elements of stationary as well as nonstationary models.

Pfeifer's scheme

Let $\{X_{nk}, n\geq 1, k\geq 1\}$ be a sequence of series of independent random variables and let random variables $X_{n1}, X_{n2},\ldots$ in the n-th series have a common distribution function F_n, $n\geq 1$. Firstly, we recursively introduce a sequence of inter-record times $\Delta(n)=L(n)-L(n-1)$ as follows:

$$\Delta(1)=1,\ \Delta(n+1)= \min\{k: X_{n+1,k}>X_{n,\Delta(n)}\},\ n=1,2,\ldots .$$

Then , following Pfeifer (1982, 1984), record times $L(n)$ and record values $X(n)$ are defined as

$$L(n)= \Delta(1)+\ldots+ \Delta(n)$$

and

$$X(n)=X_{n,\Delta(n)},\ n=1,2,\ldots .$$

This record model reflects situations where conditions of an experiment (say, sports competitions) change after an occurrence of a new record event. Different results for this scheme have been obtained by Pfeifer (1982, 1984), Deheuvels (1984) and Kamps (1994,1995).

We mention here one of these results. It turns out that if

$$F_n(x)=1-\exp\{-x/\lambda_n\},\ x>0,$$

in Pfeifer's scheme, where $\lambda_n>0$, $n=1,2,\ldots$, then inter-record values

$$T(1)=X(1),\ T(2)=X(2)-X(1),\ldots$$

are independent and

$$P\{T(n)<x\}= F_n(x),\ n=1,2,\ldots .$$

Balabekyan- Nevzorov's record model

One more record scheme has been suggested by Balabekyan and Nevzorov (1986). It can be illustrated as follows. Let m athletes have in succession n starts each. Then the distribution functions, which correspond to their results

$X_1,\ldots, X_m,\ldots,X_{(n-1)m+1},\ldots, X_{nm},$

form a sequence

$F_1,\ldots, F_m,\ldots, F_{(n-1)m+1},\ldots, F_{nm},$

such that

$F_{rm+k} = F_k$, $k=1,\ldots,m$, $r=0,\ldots,n-1$,

that is, a group of m different distribution functions $F_1,\ldots,F_m$ (an element of nonstationarity) is repeated n times (an element of stationarity) in this sequence. Hence we can see that Balabekyan-Nevzorov's scheme, as well as Pfeifer's one, also contains elements of stationarity and nonstationarity at the same time.

It turns out that the number of records N(nm) in a sequence $X_1,\ldots, X_{nm}$ has the same asymptotic distribution as the number of records N(n) in the classical record scheme.

Theorem 4.10.1.

Let $F_1, F_2,\ldots, F_m$ be continuous distribution functions. Then the following relation holds as $n\to\infty$:

$$\sup_x | P\{N(nm) - \ln n < x\sqrt{\ln n}\} - \Phi(x) | \to 0, \tag{4.10.1}$$

where $\Phi(x)$ is the distribution function of the standard normal law.

Proof.

The main idea of the method is to compare N(nm) with the number of records $N_1(n)$ in the sequence

$Y_1=\max\{X_1,\ldots,X_m\}$, $Y_2=\max\{X_{m+1},\ldots,X_{2m}\},\ldots,Y_n=\max\{X_{m(n-1)+1},\ldots,X_{nm}\}$.

Since $Y_1, Y_2,\ldots$ is a sequence of independent random variables with a common continuous distribution function

$$G=\prod_{r=1}^{m} F_r, \tag{4.10.2}$$

it follows from (4.8.1) that

$$\sup_x |P\{N_1(n) - \ln n < x\sqrt{\ln n}\} - \Phi(x)| \to 0. \quad (4.10.3)$$

Note that

$N_1(n)=\xi_1+...+\xi_n$,

where ξ_k is an indicator of the occurrence at least one record in the k-th group of results , which includes random variables

$$X_{m(k-1)+1},...,X_{mk},\ k=1,2,... .n. \quad (4.10.4)$$

Let ν_k denote the real number of records in group (4.10.4) and let

$\eta_k=\nu_k-\xi_k=\max\{0,\nu_k-1\}$.

Introduce sums

$N_2(n)=\eta_1+...+\eta_n$, n=1,2,...

and note that

$$N(nm)=(\xi_1+...+\xi_n)+(\eta_1+...+\eta_n)=N_1(n)+N_2(n),\ n=1,2,.... \quad (4.10.5)$$

The next step is to show that the second term ($N_2(n)$) on the RHS of (4.10.5) is negligible with respect to $N_1(n)$ and then the asymptotic distributions of N(nm) and $N_1(n)$ coincide. It turns out that the following estimate is valid.

Lemma 4.10.1.

For any m≥1 and n≥1,

$$0 \le E_n = E(\eta_1 + ... + \eta_n) \le \frac{2n(m-1)}{n+1} \quad (4.10.6)$$

Proof.

Evidently,

$E_n= E\eta_1+...+E\eta_n$.

Let us estimate $E\eta_r$ for any r=1,2,.... It is easy to see that

$$E\eta_r \le (m-1)P\{\eta_r > 0\} \tag{4.10.7}$$

and, in particular,

$$E\eta_1 \le (m-1). \tag{4.10.8}$$

Indeed, (4.10.8) stays true for any expectation $E\eta_r$, but we need more exact estimates for $r>1$. The event $\{\eta_r > 0\}$ implies that we have at least two record values amongst random variables

$$X_{m(r-1)+1}, \ldots, X_{mr}.$$

It means that there exists an integer j, $1 \le j \le m-1$, such that

$$\max\{X_{m(r-1)+1}, \ldots, X_{mr+j-1}\} \le \max\{X_1, \ldots, X_{m(r-1)}\} <$$

$$X_{mr+j} < \max\{X_{m(r-1)+j+1}, \ldots, X_{m(r+1)}\}. \tag{4.10.9}$$

Note that distribution functions of four random variables involved in (4.10.9) are

$$F_1(x)F_2(x)\cdots F_{j-1}(x),\ G^{r-1}(x),\ F_j(x)$$

and

$$F_{j+1}(x)F_{j+2}(x)\cdots F_m(x)$$

correspondingly, where G is defined in (4.10.2). We will propose that

$$F_1(x)F_2(x)\cdots F_{j-1}(x) = 1 \tag{4.10.10}$$

if $j = 1$. Note also that

$$1 - F_{i+1}(x)F_{j+2}(x)\cdots F_j(x) \le 1 - G(x), \tag{4.10.11}$$

for any $0 \le i < j \le m$. Now we can write for any $r=2, \ldots, n-1$, that

$$P\{\eta_r > 0\} = \sum_{j=1}^{m-2} P\{\max(X_{m(r-1)+1}, \ldots, X_{mr+j-1})$$

$$\le \max(X_1, \ldots, X_{m(r-1)}) < X_{mr+j} < \max(X_{m(r-1)+j+1}, \ldots, X_{m(r+1)})\}$$

$$= \sum_{j=1}^{m-1} \int_{-\infty}^{\infty} d(G^{r-1}(x))\, F_1(x)F_2(x)\cdots F_{j-1}(x)$$

$$\cdot \int_{x}^{\infty} d(F_j(u))(1- F_{j+1}(u)F_{j+2}(u)\cdots F_m(u)). \qquad (4.10.12)$$

Due to inequality (4.10.11), the RHS of (4.10.12) can be estimated as follows:

$$\sum_{j=1}^{m-1} \int_{-\infty}^{\infty} d(G^{r-1}(x))\, F_1(x)F_2(x)\cdots F_{j-1}(x)$$

$$\cdot \int_{x}^{\infty} d(F_j(u))(1- F_{j+1}(u)F_{j+2}(u)\cdots F_m(u))$$

$$\leq \int_{-\infty}^{\infty} d(G^{r-1}(x)) \sum_{j=1}^{m-1} F_1(x)F_2(x)\cdots F_{j-1}(x) \int_{x}^{\infty} d(F_j(u))(1-G(u))$$

$$\leq \int_{-\infty}^{\infty} \sum_{j=1}^{m-1} F_1(x)F_2(x)\cdots F_{j-1}(x)(1-F_j(x))\,(1-G(x))d(G^{r-1}(x)). \qquad (4.10.13)$$

Taking into account (4.10.10) and (4.10.11), we get that

$$\sum_{j=1}^{m-1} F_1(x)F_2(x)\cdots F_{j-1}(x)(1-F_j(x)) = F_1(x)F_2(x)\cdots F_{j-1}(x)F_j(x) \leq 1-G(x). \qquad (4.10.14)$$

It follows directly from (4.10.12)-(4.10.14) that

$$P\{\eta_r > 0\} \leq \int_{-\infty}^{\infty} (1-G(x))^2 d(G^{r-1}(x)) = \int_0^1 (1-v)^2 d(v^{r-1}) = 2/r(r+1)$$

and hence

$$E_n \leq (m-1)\Big(1+\sum_{r=2}^{n} \frac{2}{r(r+1)}\Big) = 2n(m-1)/(n+1).$$

That proves the statement of the lemma.

Finally, to prove (4.10.1) we need to consider the following identity:

$$\frac{N(nm)-\ln n}{\sqrt{\ln n}}=\frac{N_1(n)-\ln n}{\sqrt{\ln n}}+\frac{N_2(n)}{\sqrt{\ln n}}. \qquad (4.10.15)$$

In (4.10.15) the first term on the RHS has asymptotically normal distribution (due to (4.10.3)) and the second term converges to zero in probability. The latest fact follows immediately from the statement of lemma 4.10.1, since for any $\varepsilon>0$,

$$P\{\frac{N_2(n)}{\sqrt{\ln n}}\geq\varepsilon\}\leq\frac{EN_2(n)}{\varepsilon\sqrt{\ln n}}=\frac{2n(m-1)}{\varepsilon(n+1)\sqrt{\ln n}}\to 0, \qquad (4.10.16)$$

as $n\to\infty$

Remark 4.10.1.

Analyzing the proof of theorem 4.10.1 one can see that the value m need not to be fixed. It may be permitted to increase to a certain degree with n, like

$m=m(n)=o(\sqrt{\ln n})$, $n\to\infty$.

Note also that the analogous methods enable us to obtain estimates of the convergence rate in the limit relation (4.10.1) (see Balabekyan and Nevzorov (1986) and Rannen (1991)).

Record models with trend

One of the simplest way to get a nonstationary record scheme is to take a sequence of independent random variables $X_1, X_2,\ldots$ with a common distribution function F and then consider random variables

$Y_n=X_n+c(n)$, $n=1,2,\ldots$,

where constants c_n provide a nonrandom trend. Indeed, the most natural is the situation with the linear trend $c(n)=cn$. It is clear that if the coefficient c is negative, then the number of records in the sequence $Y_1, Y_2,\ldots$ is finite with probability 1. The zero value of c gives us the well-known classical scheme. Therefore, the most interesting is the case $c>0$. Different results for record schemes with a trend have been obtained by Foster and Stuart (1954), Foster and Teichroew (1955), Ballerini and Resnick (1985, 1987), Smith and Miller (1986), de Haan and Verkade (1987), Smith (1988), etc. Unfortunately the record indicators ξ_n in such schemes (when $c\neq 0$) lose the convenient for a statistician the independence property. Moreover, in this situation, unlike the classical i.i.d. model, the probabilities

$p_n=P\{\xi_n=1\}=P\{Y_n \text{ is a record}\}, n=2,3,...$ (4.10.17)

depend on the distribution function F. Therefore, some new problems arise in the given schemes. For example , for schemes with a positive linear trend c(n)=cn,
Ballerini and Resnick (1985) introduced a new characteristic, the so-called asymptotic record rate p, which is defined as a limit of the probabilities p_n from (4.10.17), and found it for some distributions. They got that

$p=0.72506...$

for the standard normal distribution and c=1.

Yang's model

Three nonstationary models given above, to some extend, have elements of stationarity. The first actual nonstationary scheme is the model suggested by Yang (1975), who reasonably supposed that the dynamics of sports records is due in some degree to the increase in the population of the world. He has proposed that the population increases geometrically and its size m(n) after n four - year periods between two consecutive Olympic games is given as $m\lambda^n$, where m=m(0) is the initial population size . The coefficient λ was chosen as $2^{1/9}=1.08$, since the population of the world doubled for 9 periods between Olympic games of 1900 and 1936. Therefore, Yang considered independent random variables X_n having distribution functions $F_n=F^{m(n)}$, n=1,2,..., where $m(n)=m\lambda^n$. Note that F_n corresponds to the random variable

$Y_n=\max\{X_1,...,X_{m(n)}\}$,

where $X_1,X_2,...$ are independent random variables with a common continuous distribution function F.

For inter-record times $\Delta(n)=L(n)-L(n-1)$, n=2,3,..., Yang has got the following results for his record scheme.

Theorem 4.10.2.

Let

$S(k)=m(1)+m(2)+...+m(k)$, k=1,2,... ,

where $m(n)=m\lambda^n$, m>0 and $\lambda>1$ are arbitrary constants. Then

$P\{\Delta(1)>j\}=m(1)/S(j+1)$

and

$$P\{\Delta(n)>j\}=\sum_{k_1=2}^{\infty}\sum_{k_2=k_1+1}^{\infty}\cdots\sum_{k_{n-1}=k_{n-2}+1}^{\infty}\frac{m(1)m(k_1)\ldots m(k_{n-1})}{S(k_1-1)\ldots S(k_{n-1}-1)S(k_{n-1}+j)},$$

j=0,1,..., n=2,3,... .

Theorem 4.10.3.

For any j=1,2,... the following equality holds:

$$p_j=\lim P\{\Delta(n)=j\}=(\lambda-1)\,\lambda^{-j}, \tag{4.10.18}$$

as $n\to\infty$.

These results were applied to analyze records of Olympic games of 1900-1936 years. Indeed, the hypothesis that the increasing population is the main reason for the rapid breaking of Olympic records has been rejected. Analyzing the paper of Yang, Nevzorov (1981,1985) has set that like the classical i.i.d. scheme, the model of Yang saves the independence property of record indicators. Moreover, record indicators stay independent for arbitrary positive coefficients m(1),m(2),... in Yang's scheme. This argument has simplified the proofs of theorems 5.10.2 and 5.10.3 and led to a more general record schemes, which are called F^{α}-scheme.

F^{α}-schemes

Definition 4.10.1.

A sequence of independent random variables $X_1,X_2,\ldots$ with distribution functions $F_1,F_2,\ldots$ forms the F^{α}-scheme if

$$F_k=F^{\alpha(k)},\ k=1,2,\ldots,$$

where F is any continuous distribution function and $\alpha(1)$, $\alpha(2)$,... are arbitrary positive constants.

As F, we can take F_1. It is a reason why we can take $\alpha(1)=1$ without loss of generality. If $\alpha(n)=\alpha$, n=1,2,..., then the F^{α}-scheme coincides with the classical i.i.d. record model. We have mentioned above that record indicators in the F^{α}-scheme are independent. The following lemma is a generalization of the corresponding result of

Renyi for sequences of independent random variables with a common continuous diastribution function.

Let $\xi_1,\xi_2,\ldots$ be record indicators, i.e.

$$\xi_n = 1\{X_n \text{ is a record value}\},\ n=1,2,\ldots,$$

and

$$S_n=\alpha(1)+\ldots+\alpha(n),\ n=1,2,\ldots.$$

Lemma 4.10.2.

Record indicators $\xi_1,\xi_2,\ldots$ in the F^{α}-scheme are independent and

$$P\{\xi_n=1\} = 1-P\{\xi_n=0\}=\alpha(n)/(\alpha(1)+\ldots+\alpha(n))= (S(n)-S(n-1))/S(n),\ n=1,2,\ldots. \qquad (4.10.19)$$

The next lemma (Ballerini and Resnick (1987)) also generalizes the corresponding result for indicators and maximal values in sequences of i.i.d. random variables.

Lemma 4.10.3.

Under conditions of lemma 5.10.2, random variables ξ_1 , $\xi_2,\ldots$, ξ_n and M_n = $\max\{X_1,X_2,\ldots,X_n\}$ are independent for any $n=1,2,\ldots$.

Note that the proof of the lemmas practically repeats the proof of the corresponding results from chapter 4.2.

It turns out that properties of record indicators given above in some sense characterize F^{α}-schemes. The following two results are due to Nevzorov (1986, 1990, 1993).

Theorem 4.10.4.

Let X_1, X_2, ..., X_n be independent random variables with continuous distribution functions $F_1,F_2,\ldots,F_n$ and

$$0<F_j(a)<F_j(b)<1,\ 1\le j\le n-1, \qquad (4.10.20)$$

for some a and b, $-\infty<a<b<\infty$. If the record indicator ξ_n and the vector $(\xi_1,\xi_2,\ldots,\xi_{n-1})$ are independent for any distribution function F_n, then there exist positive constants $\alpha(2),\ldots,\alpha(n-1)$ such that

$$F_j=(F_1)^{\alpha(j)},\ 2\le j\le n-1. \qquad (4.10.21)$$

Moreover, indicators $\xi_1,\xi_2,\ldots,\xi_{n-1}$ are also independent.

Remark 4.10.2.

Restriction (4.10.20) in theorem 4.10.4 is essential. In fact, we can take any sequence of random variables $X_1,\ldots,X_{n-1}$ with non-overlapping supports (a_j, a_{j+1}), $a_1<a_2<\ldots<a_n$. Then (4.10.20) fails. Evidently, in this situation

$$P\{\xi_k=1\}=1,\ k=1,2,\ldots,n-1,$$

i.e. the indicators $\xi_1,\xi_2,\ldots,\xi_{n-1}$ have degenerate distributions, and the vector $(\xi_1,\xi_2,\ldots,\xi_{n-1})$ is independent with any random variable ξ_n, when (4.10.21) does not hold.

Theorem 4.10.5.

Let $X_1, X_2, \ldots, X_n$ be independent random variables with continuous distribution functions $F_1,F_2,\ldots,F_n$ and

$$0<F_j(a)<F_j(b)<1,\ 1\le j\le n-1,$$

for some a and b, $-\infty<a<b<\infty$. If $M_r = \max\{X_1,X_2,\ldots,X_r\}$ and the vector $(\xi_1,\xi_2,\ldots,\xi_r)$ are independent for any $r=2,3,\ldots,n$, then there exist positive constants $\alpha(2),\ldots,\alpha(n-1)$ such that

$$F_j=(F_1)^{\alpha(j)},\ 2\le j\le n-1,$$

and the record indicators $\xi_1,\xi_2,\ldots,\xi_n$ are also independent.

The independence property of the indicators enables us to reformulate for the F^α-scheme (probably, with some slight changing) the most part of results, which have been obtained for the classical record model.

Indeed, in the F^α-scheme, as well as in i.i.d. sequences, the number of records

$$N(n)=\xi_1+\xi_2+\ldots+\xi_n,\ n=1,2,\ldots.$$

can be expressed as a sum of independent record indicators. It follows immediately that

$$A(n)=EN(n)=\sum_{j=1}^{N} p_j, \tag{4.10.22}$$

and

$$B(n)=Var(N(n))=A(n)-\sum_{j=1}^{n} p_j^2,$$

where

$$p_n = P\{\xi_n=1\} = \alpha(n)/(\alpha(1)+\ldots+\alpha(n)). \qquad (4.10.23)$$

Evidently,

$$P\{N(n)\to\infty\}=1, \; n\to\infty,$$

if and only if $A(n)\to\infty$. It follows from Dini's test that both the sequences $S(n)$ and $A(n)$ simultaneously have a finite limit or they tend to infinity, as $n\to\infty$. Hence,

$$P\{N(n)\to\infty\}=1$$

(and then all record times and record values exist with probability one) only when

$$S(n)\to\infty, \; n\to\infty. \qquad (4.10.24)$$

Further, we will study the F^α-schemes, coefficients of which satisfy (4.10.24).
The standard operations with the record indicators give us the following expression for joint distributions of record times:

$$P\{L(1)=1, L(2)=m(2), \ldots, L(n)=m(n)\} = \left(\prod_{r=2}^{n} \frac{\alpha(m(r))}{S(m(r)-1)}\right) / S(m(n)), \qquad (4.10.25)$$

for

$$1=m(1)<m(2)<\ldots<m(n).$$

Moreover, as in the classical record schemes, it follows from (4.10.25) that the sequence of record times $L(1), L(2), \ldots$ forms a Markov chain with transition probabilities

$$P\{L(n)=j \mid L(n-1)=i\} = S(i)(1/S(j-1) - 1/S(j)) \qquad (4.10.26)$$

and

$$P\{L(n)>j \mid L(n-1)=i\} = S(i)/S(j), \; j>i. \qquad (4.10.27)$$

Deheuvels and Nevzorov (1993) used the Markov property of record times to prove the following result. Let $\Im_n$ denote a σ-algebra of events generated by random variables $L(1), \ldots, L(n)$, $n=1,2,\ldots$.

Theorem 4.10.6.

Sequences of random variables

$V_1(n)=A(L(n))-n$

and

$V_2(n)=\{A(L(n))-n\}^2-B(L(n)),\ n=1,2,\ldots$

form martingales with respect to σ-algebras $\Im_n$, n=1,2,..., and

$EV_1(n)=EV_2(n)=0$

for any n=1,2,....

As simple consequences of the statements of theorem 4.10.6 one obtains that

$EA(L(n))=n,\ n=1,2,\ldots.$

and

$$Var(A(L(n)))=n-E\sum_{j=1}^{L(n)} p_j^2,$$

where p_j are defined in (4.10.23). Moreover, if

$$D=\sum_{j=1}^{\infty} p_j^2<\infty,$$

then

$$Var(A(L(n)))=n-D+o(1),$$

as $n\to\infty$.

Without any problems one can obtain limit theorems for N(n) and then for L(n). In particular, using the standard arguments of limit theorems of the probability theory we immediately obtain that

$(N(n)-A(n))/B(n)$

has asymptotically normal distribution if $B(n)\to\infty$, as $n\to\infty$. If we suppose additionally that $p_n\to 0$ and

$$\sum_{j=1}^{n} p_j^2 = 0(\sum_{j=1}^{n} p_j), \ n\to\infty,$$

then

$$\ln S(n) = -\sum_{j=2}^{n} \ln(S_{j-1}/S_j) = -\sum_{j=2}^{n} \ln(1-p_j)$$

$$= \sum_{j=2}^{n} p_j + O(\sum_{j=1}^{n} p_j^2)$$

and

$$\ln S(n) \sim \sum_{j=1}^{n} p_j = A(n), \ n\to\infty.$$

Such relations help to prove (under additional restrictions on coefficients $\alpha(n)$) that

$$(N(n)-\ln S(n))/(\ln S(n))^{1/2}$$

has asymptotically normal distribution. The latest fact, due to equality

$$P\{L(m)<n\} = P\{N(n)>m\},$$

implies the asymptotic normality of

$$(\ln S(L(n))-n)/n^{1/2},$$

as $n\to\infty$. These and related results concerning central limit theorem for the number of records $N(n)$ and the record times $L(n)$ have been proved by Nevzorov (1986, 1995). Some limit theorems for record values have been obtained in Nevzorov (1995). Their proof has been based on the equality

$$P\{X(n)<x\} = E\{F(x))^{L(n)}\}$$

and it has used limit theorems for $L(n)$. In particular, it was proved (under some rather mild restrictions on coefficients $\alpha(n)$) that the set of all possible asymptotic distributions of the suitably normalized record values X(n) consists of three functions $\Phi(g_i(x))$, i=1,2,3, where $g_i(x)$ are given in (4.8.19).

4.11. RECORDS IN SEQUENCES OF DEPENDENT RANDOM VARIABLES

As we have seen, there is a good progress in studying records in sequences of independent random variables X_1, X_2,... having in general different distributions, when there are only a few papers which deal with dependent X's. Below we will list some of "dependent" record schemes.

Sequences of exchangeable random variables

The simplest situation, which practically does not require additional theory, is the case of exchangeable X's. The following result shows that the Renyi's lemma from chapter 4.2 stays true for exchangeable random variables.

Lemma 4.11,1

Let $X_1,X_2,\ldots$ be exchangeable random variables and

$$P\{X_1=X_2\}=0. \tag{4.11.1}$$

Then record indicators ξ_1, ξ_2,... are independent and

$$P\{\xi_n=1\}=1/n\ ,\ n=1,2,\ldots\ .$$

Proof.

Firstly we will prove the second statement. Under conditions of the lemma,

$$P\{X_j=\max(X_1,\ldots,X_{j-1},X_{j+1},\ldots,X_n)\}$$

$$= P\{X_1=\max(X_2,\ldots,X_n)\}$$

$$\leq \sum_{j=2}^{n} P\{X_1=X_j\}= (n-1)P\{X_1=X_2\}=0$$

and hence

$$P\{X_j=\max(X_1,\ldots,X_{j-1},X_{j+1},\ldots,X_n)\}=0$$

for any $j \le n$. Then,

$$1=\sum_{j=1}^{n} P\{X_j>\max(X_1,\ldots,X_{j-1},X_{j+1},\ldots,X_n)\}. \tag{4.11.2}$$

Due to exchangeability of X's, the RHS of (4.11.2) coincides with the expression

$n\,P\{X_n>\max(X_1,\ldots,X_{n-1})\}$,

which evidently equals $nP\{\xi_n=1\}$. Hence,

$P\{\xi_n=1\}=1/n,\ n=1,2,\ldots.$

To prove the first statement we need to check that for any $2 \le r \le n$ and $1 \le m(1)<\ldots<m(r) \le n$ the following equality is true:

$$P\{\xi_{m(1)}=1,\ldots,\xi_{m(r)}=1\}=1/\prod_{j=1}^{r} m(j). \tag{4.11.3}$$

The exchangeability of X's implies that

$$P\{\xi_{m(1)}=1,\ldots,\xi_{m(r)}=1\}=\sum_{M\in\sigma} P\{X_{s(1)}<X_{s(2)}<\ldots<X_{s(n)}\}=N_\sigma\, P\{X_1<X_2<\ldots<X_n\}, \tag{4.11.4}$$

where the summation is taken over all arrangements $M==(s(1),\ldots,s(n))$, which provide inequalities

$X_{m(1)}>\max(X_1,\ldots,X_{m(1)-1}),\ldots,\ X_{m(r)}>\max(X_1,\ldots,X_{m(r)-1})$,

and N_σ denotes the number of such arrangements. The condition (4.11.4) and the equiprobability of all n! events

$\{X_{s(1)}<X_{s(2)}<\ldots<X_{s(n)}\}$

imply that

$P\{X_1<X_2<\ldots<X_n\}=1/n!$

as in the i.i.d. case. Since independent X's having a common continuous distribution function are also exchengeable they satisfy the relation (4.11.4). It means that both the sides of (4.11.4) have the same value for the i.i.d. case and for any exchangeable sequence of X's. From Renyi's lemma we know that

$$P\{\xi_{m(1)}=1,\dots,\xi_{m(r)}=1\}=1/\prod_{j=1}^{r} m(j)$$

for the classical model. Hence, the LHS of (4.11.4) is also equal to

$$\left(\prod_{j=1}^{r} m(j)\right)^{-1}$$

and

$$P\{\xi_{m(1)}=1,\dots,\xi_{m(r)}=1\}=\prod_{j=1}^{r} P\{\xi_{m(j)}=1\}$$

for exchangeable random variables $X_1,X_2,\dots$, satisfying (4.11.1). That completes the proof of the lemma. Additionally we get that

$$N_\sigma=n!/\prod_{j=1}^{r} m(j).$$

Remark 4.11.1.

Evidently, distributions of record times L(n), inter-record times Δ(n) and numbers of records N(n) are wholly determined by the joint distributions of record indicators $\xi_1,\xi_2,\dots$ .It means that all results for these random variables, which have been obtained for the classical model, can be reformulated without changing for the case of exchangeable X's satisfying (4.11.1). To study record times in sequences of exchangeable random variables one needs to recall the following de Finetti's representation.

Representation 4.11.1.

Let $X_1,X_2,\dots$ be an infinite sequence of exchangeable random variables. Then there exist a family of distribution functions F_θ, $-\infty<\theta<\infty$, and a distribution function T such that for any $n\geq 1$ and $x_1,\dots,x_n$ the following equality holds:

$$P\{X_1\leq x_1,\dots,X_n\leq x_n\}=\int_{-\infty}^{\infty} F_\theta(x_1)\dots F_\theta(x_n)T(d\theta). \tag{4.11.5}$$

The RHS of (4.11.5) can be regarded as a distribution of a mixture of vectors with independent identically distributed marginals. If we additionally suppose that F_θ is a continuous for almost all (with respect to the distribution T) values θ, then it follows from (4.11.5) and the corresponding results for i.i.d. random variables that

$$P\{X(n)<x\}=\int_{-\infty}^{\infty} P\{X_\theta(n)<x\}T(d\theta)=\int_{-\infty}^{\infty}\frac{1}{(n-1)!}\left(\int_0^{-\log(1-F_\theta(x))} u^{n-1}e^{-u}du\right)T(d\theta),\ n=1,2,\ldots, \tag{4.11.6}$$

where $X_\theta(n)$ denote record values in the sequences of independent random variables with the common distribution function F_θ .

Records in Markov sequences of random variables

Different properties of record times and record values in Markov sequences have been investigated in the main by Biondini and Siddiqui (1975) and Adke (1993).
In particular, it was proved that vectors (X(n),L(n)), n=1,2,..., generated by the Markov sequence $X_1, X_2,\ldots$, form a bivariate Markov sequence. It is interesting to mention that unlike {X(n)}, n=1,2,..., which is also a Markov sequence, record times L(1), L(2),... , in general, do not form a Markov chain.

Records in Archimedean copula processes

Following Ballerini (1994), who has initiated investigations of records in some special sequences of random variables with Archimedean dependence function, Nevzorova, Nevzorov and Balakrishnan (1997) considered more general Archimedean copula processes (AC processes).

Definition 4.11.1.

A sequence $X_1,X_2,\ldots$ with marginal distribution functions $F_1,F_2,\ldots$ is said to be an AC process if for any n=1,2,...,

$$P\{X_1<x_1,\ldots,X_n<x_n\}=B\left(\sum_{j=1}^{n} A(F_j(x_j))\right), \tag{4.11.7}$$

where the dependence function B is completely monotone, B(0)=1, B(∞)=1 and A is the inverse of B.

Ballerini has studied in detail the special case when

$$B(s)=\exp\{-s^{1/\gamma}\},\ \gamma\geq 1, \tag{4.11.8}$$

and

$F_j(x)=(F(x))^{\alpha(j)}$, $j=1,2,\ldots,$

where F is a continuous distribution function and $\alpha(j)$, $j=1,2,\ldots,$ are arbitrary positive constants. He called this situation as "dependent F^{α}-scheme", because marginal distribution functions coincide with those in the F^{α}-scheme. Note that Ballerini's scheme with $\gamma=1$ exactly presents the F^{α}-scheme. The most important result of Ballerini is the independence of record indicators $\xi_1,\xi_2,\ldots,\xi_n$ and maximal values $M_n=\max\{X_1,\ldots,X_n\}$, $n=1,2,\ldots$ .Following Nevzorova et al. (1997) we consider a more general record model than Ballerini's one. We take any complete monotone function B(s) such that $B(0)=1$ and $B(\infty)=0$. To characterize this set of dependence functions B, note that due to one of theorems of Bernstein, B(s) coincides with a Laplace transform of some proper distribution. Marginal distribution functions F_j are taken to be

$$F_j(x)=B(c(j)A(F(x))),\ j=1,2,\ldots \tag{4.11.9}$$

where F is an arbitrary continuous distribution function and $c(1),c(2),\ldots$ are positive constants. Let

$R(x)=A(F(x))$.

Then random variables $X_1,X_2,\ldots$ in the recent scheme have joint distribution functions given as follows:

$$H_n(x_1,\ldots,x_n)=P\{X_1<x_1,\ldots,X_n<x_n\}=B\Big(\sum_{j=1}^{n} c(j)R(x_j)\Big). \tag{4.11.10}$$

Indeed, if we take

$B(s)=\exp\{-s^{1/\gamma}\}$,

as Ballerini has done, then

$A(x)=(-\ln x)^{1/\gamma}$

and marginal distribution functions defined in (4.11.9) satisfy the relation

$F_j=F^{c(j)}$,

that is they are the same as in Ballerini's scheme. It means that Ballerini's situation is a partial case of the considered scheme. The following result is valid.

Theorem 4.11.1.

Let $X_1, X_2, \ldots$ be a sequence of random variables with joint distributions (4.11.10). Then record indicators $\xi_1, \xi_2, \ldots, \xi_n$ and maximal values $M_n = \max\{X_1, \ldots, X_n\}$ are independent for any $n=1,2,\ldots$.

The proof of the theorem can be found in Nevzorova et al (1997). Mention that it is based on the following useful formulae, where $b(x)=B'(x)$ and $\delta_n=c(1)+\ldots+c(n)$:

$$P\{M_n<x\}=H_n(x,\ldots,x)=B(\delta_n R(x)),\ n=1,2,\ldots,$$

and

$$P\{\xi_k=1\}=P\{M_{k-1}<X_k\}$$

$$= c_k \int_{-\infty}^{\infty} b(\delta_k R(x))dR(x)$$

$$= \frac{c(k)}{\delta(k)}(B(\delta_k R(\infty))-B(\delta_k R(-\infty)))$$

$$= \frac{c(k)}{\delta(k)}(B(0)-B(\infty))=\frac{c(k)}{\delta(k)},\ k=1,2,\ldots. \tag{4.11.11}$$

REFERENCES

1. Adke, S.R. (1993). Records generated by Markov sequences. *Statist. and Prob. Letters* **18**, 257-263.

2. Aggarwal, M.L. and Nagabhushanam, A. (1971). Coverage of a record value and related distribution problems. *Bull. Calcutta Stat. Assoc.* **20**, 99-103.

3. Ahsanullah, M. (1978). Record values and the exponential distribution *Ann. Inst. Statist. Math.* **30**, 429-433.

4. Ahsanullah, M.(1979). Characterization of the exponential distribution by record values. *Sankhyā* **B 41**, 116-121.

5. Ahsanullah, M. (1980). Linear prediction of record values for the two- parameter exponential distribution. *Ann. Inst. Statist. Math.* **32**, 363-368.

6. Ahsanullah, M. (1981a). On a characterization of the exponential distribution by weak homoscedasticity of record values. *Biom. J.* **23**, 715-717.

7. Ahsanullah, M. (1981b). Record values of exponentially distributed random variables. *Statist. Hefte* **22**, 121-127.

8. Ahsanullah, M. (1982). Characterizations of the exponential distribution by some properties of the record values. *Statist. Hefte* **23**, 326-332.

9. Ahsanullah, M.(1986a). Record values from a rectangular distribution. *Pakistan J. Statist.* **A2**, 1-5.

10. Ahsanullah, M. (1986b). Estimation of the parameters of a rectangular distribution by record values. *Comp. Stat. Quarterly* 2, 119- 125.

11. Ahsanullah, M. (1986b). Two Characterizations of the Exponential Distribution. Comm. Statist. Theory Meth. 16, 375-381.

12. Ahsanullah, M.(1987b). Record statistics and the exponential distribution. *Pakistan J. Statist.* **A3**, 17-40.

13. Ahsanullah, M. (1988). *Introduction to record statistics*. Needham Heights, MA: Ginn Press.

14. Ahsanullah, M. (1989). Estimation of the parameters of a power function distribution by record values. *Pakistan J. Statist.* **A5**, 189-194.

15. Ahsanullah, M. (1990a). Estimation of the parameters of the Gumbel distribution based on the m record values. *CSQ-Comput. Statist.Quart.* **5**, 231-239.

16. Ahsanullah, M. (1990b). Some characterizations of the exponential distribution by the first moment of record values. *Pakistan J. Statist.* **A6**, 183-188.

17. Ahsanullah, M. (1991a). Some characteristic properties of the record values from the exponential distribution. *Sankhyā* **B 53**, 403- 408.

18. Ahsanullah, M. (1991b). On record value from the generalized Pareto distribution. *Pakistan J. Statist.* **A7**, 126-129.

19. Ahsanullah, M. (1991c). Record values of the Lomax distribution. *Statist. Neerlandica* **41**, 21-29.

20. Ahsanullah, M. (1991d). Inference and prediction problems of the Gumbel distribution based on record values. *Pakistan J. Statist.* **A7**, 53-62.

21. Ahsanullah, M. (1992a).Inference and prediction problems of the generalized Pareto distribution based on record values. *Nonparametric Statistics. Order Statistics and Nonparametrics. Theory and Applications,* (Eds., P.K.Sen and I.A.Salama).Elsevier, 47-57.

22. Ahsanullah, M. (1992b). Record values of independent and identically distributed continuous random variables. *Pakistan J. Statist.* **A8**, 9-34.

23. Ahsanullah, M. (1995). *Record statistics.* Nova Science Publishers.

24. Ahsanullah, M. and Bhoj, D.S.(1997). Inference problems of modified power function distribution based on records. *J. Appl. Statist. Science* **5**, 143 -159.

25. Ahsanullah, M. and Holland, B. (1984). Record values and the geometric distribution. *Statist. Hefte* **25**, 319-327.

26. Ahsanullah, M. and Holland, B. (1987). Distributional properties of record values from the geometric distribution. *Statist. Neerl.* **41**, 129-137.

27. Ahsanullah, M. and Houchens, R.L. (1989).A note on the record values from a Pareto distribution. *Pakistan J. Statist.* **A5**, 51- 57.

28. Ahsanullah, M. and Kirmani, S.N.U.A. (1991). Characterizations of the exponential distribution through a lower record. *Commun. Statist.-Theory Meth.* **20**, 1293-1299.

29. Alpuim,M.T. (1985). Record values in populations with increasing or random dimension. *Metron* **43**, 145-155.

30. Andel, J. (1990). Records in an *AR(1)* process. *Ricerche Mat.* **39**, 327- 332.

31. Arnold, B.C., Balakrishnan, N. and Nagaraja, H.N. (1998). *Records* . New York: Wiley.

32. Bairamov , I.G. (1997). Some distribution free properties of statistics based on record values and characterizations of the distributions through a record. *J.Appl. Staist. Science* **5**, 17- 25.

33. Balakrishnan ,N. and Ahsanullah, M. (1995). Relations for single and product moments of record values from exponential distribution. *J. Appl. Statist. Science* **2**, 73-87.

34. Balakrishnan ,N., Ahsanullah, M. and Chan, P.S. (1992). Relations for single and product moments of record values from Gumbel distribution. *Statist. and Prob. Lett.* **15**, 223-227.

35. Balakrishnan ,N., Ahsanullah, M. and Chan, P.S. (1995). On the logistic record values and associated inference. *J. Appl. Statist. Science* **2**, 233-248.

36. Balakrishnan ,N. and Balasubramanian, K. (1995). A characterization of geometric distribution based on record values. *J. Appl. Statist. Science* **2**, 277-282.

37. Balakrishnan , N., Balasubramanian, K. and Panchapakesan, S. (1996). δ-Exceedance records . *J. Appl. Statist. Science* **4**, 123-132.

38. Balakrishnan , N.,Chan, P.S. and Ahsanullah, M. (1993). Recurrence relations for moments of record values from generalized extreme value distribution. *Comm. Statist.- Theory Meth.* **22**, 1471-1482.

39. Balakrishnan , N. and Nevzorov, V.B. (1997). Stirling numbers and records, *Advances in Combinatorial Methods and Applications to Probability and Statistics* (Ed. N.Balakrishnan), Burkhauser, Boston, 189-200.

40. Balakrishnan , N. and Nevzorov , V.B. (1998). Maxima and records in Archimedean copula processes. *Abstracts of communications of International conference " Asymptotic Methods in Probability and Mathematical Statistics",* St.-Petersburg, June 24-28, 1998, 25-30.

41. Ballerini, R. and Resnick, S. (1987). Records in the presence of a linear trend. *Adv. Appl. Prob.* **19**, 801-828.

42. Basak, P. and Bagchi P. (1990). Application of Laplace approximation to record values. *Commun. Statist.-Theory Meth.* **19**, 1875- 1888.

43. Berred, M. (1991). Record values and the estimation of the Weibull tail coefficient. *C.R. Acad. Sci. Paris* **312**, Serie 1, 943-946.

44. Berred, M. (1992). On record values and the exponent of a distribution with regularly varying upper tail. *J. Appl. Prob.* **29**, 575-586.

45. Berred, M. (1995). K-record values and the extreme-value index. *J. Stat. Planning and Inference* **45**, 49-64.

46. Blom, G. (1988). Om rekord. *Elementa* **71**, 67-69 .

47. Blom, G., Thorburn, D. and Vessey, T. (1990). The distribution of the record position and its applications. *American Statistician* **44**, 151-153.

48. Borovkov, K. and Pfeifer, D. (1995). On record indices and record times. *J. Stat. Planning and Inference* **45**, 65-80.

49. Browne, S. and Bunge, J. (1995). Random record processes and state dependent thinning. *Stoch. Processes and Appl.* **55**, 131-142.

50. Bruss, F.T. (1988). Invariant record processes and applications to best choice modelling. *Stoch. Proc. Appl.* **30**, 303-316.

51. Bruss, F.T., Mahiat, H.and Pierard, M. (1988). Sur une fonction generatrice du nombre de records d'une suite de variables aleatories de longueur aleatoire. *Ann. Soc. Sci. Bruxelles, ser.1,* **100**, 139-149 (in French).

52. Bruss, F.T.and Rogers, B. (1991). Pascal processes and their characterization. *Stoch. Proc. Appl.* **37**, 331-338.

53. Bunge, J.A. and Nagaraja, H.N. (1991). The distributions of certain record statistics from a random number of observations. *Stochast. Process. Appl.* **38**, 167-183.

54. Bunge, J.A. and Nagaraja, H.N. (1992a). Dependence structure of Poisson-paced records. *J. Appl. Prob.* **29**, 587-596.

55. Bunge, J.A. and Nagaraja, H.N. (1992b). Exact distribution theory for some point process record models. *Adv. Appl. Prob.* **24**, 20- 44.

56. Buzlaeva E. and Nevzorov, V. (1998). One new characterization based on correlations between records. *Abstracts of communications of International conference " Asymptotic Methods in Probability and Mathematical Statistics", St.-Petersburg, June 24-28, 1998,* 53-54.

57. Balabekyan, V.A. and Nevzorov, V.B. (1986). On number of records in a sequence of series of nonidentically distributed random variables. *Rings and Modules. Limit Theorems of Probability Theory* (Eds., Z.I.Borevich and V.V. Petrov), v.**1**, 147-153,Leningrad, Leningrad State University (in Russian).

58. Ballerini, R. (1994). A dependent F^{α}-scheme. *Statist. and Prob. Letters* **21**, 21-25.

59. Ballerini, R. and Resnick, S. (1985). Records from improving populations. *J. Appl. Prob.* **22**, 487-502.

60. Ballerini, R. and Resnick, S. (1987). Embedding sequences of successive maxima in extremal processes, with applications. *J. Appl. Prob.* **24**, 827-837.

61. Biondini, R. and Siddiqui, M.M. (1975). Record values in Markov sequences. *Statistical Inference and Rrelated Topics* **2**, 291-352. New York, Academic Press.

62. Chandler, K.N. (1952). The distribution and frequency of record values. *J. Roy. Statist. Soc.* **B14**, 220-228.

63. Cheng, Shi-hong (1987). Records of exchengeable sequences. *Acta Math.Appl. Sin.* **10**, 464-471 .

64. Dallas, A.C. (1981). Record values and the exponential distribution. *J. Appl. Prob.* **18**, 949-951.

65. Dallas, A.C. (1982). Some results on record values from the exponential and Weibull law. *Acta Math. Acad. Sci. Hungar.* **40**, 307-311.

66. Dallas, A.C. (1989). Some properties of record values coming from the geometric distribution. *Ann. Inst. Statist. Math.* **41**, 661-669.

67. De Haan, L. and Resnick, S.I. (1973). Almost sure limit points of record values. *J.Appl. Prob.* **10**, 528-542.

68. De Haan, L. and Verkade, E. (1987). On extreme-value theory in the presence of a trend. *J.Appl. Prob.* **24**, 62-76.

69. Deheuvels, P. (1982). Spacings, record times and extremal processes. *Exchangeability in Prob. and Statist. (Rome,1981)*. Amsterdam: North Holland/Elsevier, 233-243.

70. Deheuvels, P. (1983). The complete characterization of the upper and lower class of the record and inter-record times of an i.i.d. sequence. *Z. Wahrsch. Verw. Geb.* **62**, 1-6.

71. Deheuvels, P. (1984a). On record times associated with k-th extremes. *Proc. of the 3rd Pannonian Symp. on Math. Statist., Visegrad, Hungary, 13-18,Sept. 1982.* Budapest, 43-51.

72. Deheuvels, P. (1984b). The characterization of distributions by order statistics and record values- a unified approach. *J.Appl. Prob.* **21**, 326-334. (Correction: *J. Appl. Prob.* **22** (1985), 997).

73. Deheuvels, P. (1984c). Strong approximations of records and record times, *Statistical Extremes and Applications.* Proc. NATO Adv. Study Inst. . Dordrecht: Reidel, 491-496.

74. Deheuvels, P. (1988). Strong approximations of k-th records and k-th record times by Wiener processes. *Prob. Theor. Rel. Fields* **77**, 195-209.

75. Deheuvels, P., Gribkova N. and Nevzorov V.B. (1998). Bootstrapping order statistics and records. *Proceedings of the 3rd St- Petersburg Workshop on Simulation, St-Petersburg, June 28- July 3, 1998,* 349-354.

76. Deheuvels, P. and Nevzorov, V.B. (1993). Records in F^{α}-scheme.I. Martingale properties. *Zapiski Nauchn. Semin. POMI* **207**, 19-36 (in Russian). Translated version in *J.Math.Sci.***81** (1996), 2368-78.

77. Deheuvels, P. and Nevzorov, V.B. (1994). Records in F^{α}-scheme.II. Limit theorems. *Zapiski Nauchn. Semin. POMI* **216**, 42-51 (in Russian). Translated version in *J.Math.Sci.***88** (1998), 29-35.

78. Deheuvels, P. and Nevzorov V.B. (1994). Limit laws for K-record times. *J. Statist. Plann. Infer.* **38**, 279-308.

79. Deheuvels, P. and Nevzorov, V.B. (1999). Bootstrap for maxima and records. *Zapiski Nauchn. Semin. POMI* **260** (in Russian).

80. Devroye, L. (1988). Applications of the theory of records in the study of random trees. *Acta Informatica* **26**, 123-130.

81. Devroye, L. (1993). Records, the maximal layer, and uniform distributions in monotone sets. *Comput. Math. Appl.* **25**, 19- 31.

82. Dunsmore, J.R. (1983). The future occurrence of records. *Ann. Inst. Statist. Math.* **35, Part A**, 267-277.

83. Dziubdziela, W. (1977) Rozklady graniczne ekstremalnych statystyk pozycyjnych. *Roczniki Polsk. Tow. Mat.,ser.3,* **9**, 45-71 .

84. Dziubdziela, W. (1990) O czasach rekordowych i liczbie rekordow w ciagu zmiennych losowych. *Roczniki Polsk. Tow. Mat.,ser.2,* **29**, 57-70.

85. Dziubdziela, W. and Kopocinsky, B.(1976). Limiting properties of the k-th record values. *Zastos. Mat.* **15**, 187-190.

86. Embrechts, P. and Omey, E. (1983). On subordinated distributions and random record processes. *Math. Proc. Cam. Phil. Soc.* **93**, 339-353.

87. Engelen, R.,Tommassen, P. and Vervaat, W. (1988). Ignatov's theorem; a new and short proof. *J. Appl. Prob.* **25**, 229-236.

88. Ennadifi, G. (1995). Strong approximation of the number of renewal paced record times. *J. Stat. Planning and Inference* **45**, 113- 132.

89. Foster, F.G. and Teichroew, D. (1955). A sampling experiment on the powers of the records tests for trend in a time series. *J. Roy. Statist. Soc.* **B17**, 115-121.

90. Foster, F.G. and Stuart, A. (1954). Distribution free tests in time-series band on the breaking of records. *J. Roy. Statist. Soc.* **B16**, 1-22.

91. Franco, M. and Ruiz, J.M. (1997). On characterizations of distributions by expected values of order statistics and record values with gap. *Metrika* **45**, 107-119.

92. Freudenberg, W. and Szynal, D. (1976). Limit laws for a random number of record values. *Bull. Acad. Polon. Sci. Ser. Sci. Math. Astr. Phys.* **24**, 193-199.

93. Freudenberg, W. and Szynal, D. (1977). On domains of attraction of record value distributions. *Colloquium Math.* **38**, 129-139.

94. Galambos, J. and Seneta, E. (1975). Record times. *Proc. Amer. Math. Soc.* **50**, 383- 387.

95. Gajek, L. (1985). Limiting properties of difference between the successive k-th record values. *Prob. and Math. Statist.* **5**, 221- 224.

96. Gaver, D.P. (1976). Random record models. *J. Appl. Prob.* **13**, 538-547.

97. Gaver, D.P. and Jacobs, P.A. (1978). Non-homogeneously paced random records and associated extremal processes. *J.Appl. Prob.* **15**, 552-559.

98. Glick, N. (1978). Breaking records and breaking boards. *Amer. Math. Monthly* **85**, 2-26.

99. Goldie, Ch.M. (1982). Differences and quotients of record values. *Stoch. Process. Appl. 12*, 162.

100. Goldie, Ch.M. (1989). Records, permutations and greatest convex minorants. *Math. Proc. Cam. Phil. Soc.* **106**, 189-177.

101. Goldie, Ch.M. and Maller, R.A. (1996). A point-process approach to almost-sure behaviour of record values and order statistics. *Adv. Appl. Prob.* **28**, 426-462.

102. Goldie, Ch.M. and Resnick, S. (1987). Records in a partially ordered set. *Ann. Prob.* **17**, 678-699.

103. Goldie, Ch.M, and Rogers, L.C.G. (1984). The k-record processes are i.i.d.. *Z. Wahr. verw. Geb.* **67**, 197-211.

104. Grudzien, Z. (1979). On distribution and moments of i-th record statistic with random index. *Ann. Univ. Mariae Curie Sklodowska* **33A**, 89-108.

105. Grudzien, Z. and Szynal, D. (1985). On the expected values of k-th record values and associated characterizations of distributions. *Probability and Statistical Decision Theory*, Proc. 4th Pannonian Symp. on Math. Statist.,v.A, (F. Konecny, J.Mogyorodi and W.Wertz, Eds.). Budapest.

106. Grudzien, Z. and Szynal, D. (1997). Characterizations of continuous distributions via moments of the *K*-th record values with random indices. *J. Appl. Statist. Science* **5**, 259-266.

107. Gupta, R.C. (1984). Relationships between order statistics and record values and some characterization results. *J. Appl. Prob.* **21**, 425-430.

108. Gupta, R.C. and Kirmani, S.N.U.A. (1988). Closure and monotonicity properties of nonhomogeneous Poisson processes and record values. *Probability in the Engineering and Informational Sciences*, **v.2**, 475-484.

109. Gut, A. (1990a). Convergence rates for record times and the associated counting process. Stoch. Process. Appl. **36**, 135-152.

110. Gut, A. (1990b). Limit theorems for record times. *Probability theory and mathematical statistics,* **v.1** (Proceedings of the 5th Vilnius conference on probability theory and mathematical statistics), VSP/Mokslas, 490-503.

111. Guthree, G.L. and Holmes, P.T. (1975). On record and inter-record times for a sequence of random variables defined on a Markov chain. *Adv. Appl. Prob.* **7**, 195-214.

112. Haghighi-Talab, D. and Wright, C. (1973). On the distribution of records in a finite sequence of observations with an application to a road traffic problem. *J. Appl. Prob.* **10**, 556- 571.

113. Haiman, G. (1987). Almost sure asymptotic behavior of the record and record time sequences of a stationary Gaussian process. *Mathematical Statistics and Probability Theory*, **v. A**, (M.L.Puri, P.Revesz and W.Wertz, Eds). Dordrecht: Reidel, 105-120.

114. Haiman, G. (1996). Block records of some continuous time stationary 1-dependent processes. *Mathematical methods in stochastic simulation and experimental design* (Proceedings of the 2-nd St.-Petersburg Workshop on simulation, St. Petersburg, June 18-21, 1996. Eds. S.M.Ermakov, V.B. Melas). St. Petersburg, 287-293.

115. Haiman, G., Mayeur ,N., Nevzorov ,V. and Puri, M.L. (1998). Records and 2-block records of 1-dependent statuionary sequences under local dependence. *Ann. Inst. Henri Poincare* **34**, 481- 503.

116. Haiman, G. and Nevzorov, V. (1995). Stochastic ordering of the number of records. *Statistical theory and Applications: Papers in Honor of Herbert A.David* (eds. H.N.Nagaraja, P.K.Sen and D.F.Morrison). Springer Verlag, 105-116.

117. Holmes, P.T. and Strawderman, W. (1969). A note on the waiting times between record observations. *J. Appl. Prob.* **6**, 711-714.

118. Huang, W.J. and Li, S.H. (1993). Characterization results based on record values. *Statist. Sinica* **3**, 583-599.

119. Hungary, 1978 (P.Bartfai and J.Tomko,Eds). Amsterdam: North Holland, 109-116.

120. Ignatov, Z. (1981). Point processes generated by order statistics and their applications. *Colloquia Mathematica Societatis Janos Bolyai* **24**, Point Processes and Queueing Problems ,Keszthely,

121. Ignatov, Z. (1986). Ein von der variationsreihe erzeugter poissonscher punkt-prozess. *Annuaire Univ. Sofia Fac. Math. Mec.* **71** (1976-77), part 2, 79-94

122. Imlahi, A. (1993). Functional laws of the iterated logarithm for records *J. Stat. Planning and Inference* **45**, 215-224.

123. Iwinska, M. (1986). On the characterizations of the exponential distribution by record values. *Fasc. Math.* **15**, 159-164.

124. Iwinska, M. (1986). On the characterizations of the exponential distribution by order statistics and record values. *Fasc. Math.* **16**, 101-107.

125. Kamps, U. (1992). Identities for the difference of moments of successive order statistics and record values. *Metron* **50**, 179- 180.

126. Kamps, U. (1994). Reliability properties of record values from non-identically distributed random variables. *Comm. Statist.- Theory Meth.* **23**, 2101-2112.

127. Kamps, U. (1995). *A Concept of Generalized Order Statistics* (Teubner Scripten zur Mathematischen Stochastik), Stuttgart: Teubner.

128. Kamps, U. (1995b). Recurrence relations for moments of record values. *J. Stat. Planning and Inference* **45**, 225-234.

129. Katzenbeisser, W. (1990). On the joint distribution of the number of upper and lower records and the number of inversions in a random sequence. *Adv. Appl. Prob.* **22**, 957-960.

130. Kinoshita, K. and Resnick, S.I. (1989). Multivariate records and shape, *Lectures Notes Statist.* **51**, Extreme value theory, Oberwolfach, December 6-12, 1987 (J. Husler and R.D. Reiss, Eds). Berlin: Springer Verlag, 222-233.

131. Kirmani, S.N.U.A. and Beg, M.I. (1984). On characterization of distribution by expected records. *Sankhyā* **A46**, part 3, 463- 465.

132. Korwar, R.M. (1984). On characterizing distributions for which the second record value has a linear regression on the first. *Sankhyā* **B46**, 108-109.

133. Koshar, S.C. (1990). Some partial ordering results on record values. *Comm. Statist.- Theory Meth.* **19**, 299-306

134. Lin, G.D. (1987). On characterizations of distributions via moments of record values. *Prob. Theory Rel. Fields* **74**, 479-483.

135. Lin, G.D. and Huang, J.S. (1987). A note on the sequence of expectations of maxima and of record values. *Sankhyā* **A 49**, part 2, 272-273.

136. Malov, S.V. (1997). Sequential τ- ranks. *J. Appl. Statist. Science* **5**, 211--224.

137. Malov, S.V., Nevzorov, V.B. and Nikulin, M.S. (1996). Sequential ranks and related topics, *Mathematical methods in stochastic simulation and experimental design* (Proceedings of the 2nd St.-Petersburg Workshop on simulation, St.

Petersburg, June 18-21, 1996. Eds.: S.M.Ermakov, V.B. Melas). St.-Petersburg, 294-299.

138. Malov , S.V.and Nevzorov, V.B (1997). Characterizations using ranks and order statistics. *Advances in the theory and practice of statistics: A volume in honor of Samuel Kotz* (Eds.N.L.Johnson and N.Balakrishnan). NY: Wiley, 479-489.

139. Mohan, N.R. and Nayak, S.S. (1982). A characterization based on the equidistribution of the first two spacings of record values. *Z. Wahr. verw. Geb.* **60**, 219-221.

140. Nagaraja, H.N. (1977). On a characterization based on record values. *Austral. J. Statist.* **19**, 70-73.

141. Nagaraja, H.N. (1978). On the expected values of record values. *Austral. J. Statist.* **20,** No.2, 176-182.

142. Nagaraja, H.N. (1982). Record values and extreme value distributions. *J. Appl. Prob.* **19**, 233-259.

143. Nagaraja, H.N. (1988a). Record values and related statistics-a review. *Comm. Statist. Theory Meth., ser.A, 17*, 2223-2238.

144. Nagaraja, H.N. (1988b). Some characterizations of continuous distributions based on regressions of adjacent order statistics and record values. *Sankhyā* **A 50**, part 1, 70-73.

145. Nagaraja, H.N. and Nevzorov, V.B. (1996). Correlations between functions of records may be negative. *Statistics and Probability Letters* **29**, 95-100.

146. Nagaraja, H.N. and Nevzorov, V.B. (1997). On characterizations based on record values and order statistics. *J. Stat. Plann. And Inference* **63**, 271-284.

147. Nagaraja, H.N., Sen, P. and Srivastava, R.C. (1989). Some characterizations of geometric tail distributions based on record values. *Statist. Papers* **30**, 147- 159.

148. Nayak, S.S. (1981). Characterizations based on record values. *J. Indian. Stat. Assoc.* **19**, 123-127.

149. Nayak, S.S. (1984). Almost-sure limit points of and the number of boundary crossings related to SLLN and LIL for record times and the number of record values. *Stoch. Proc. Appl.* **17**, 167- 176.

150. Nayak, S.S. (1985). Record values for and partial maxima of a dependent sequence. *J. Indian. Stat. Assoc.* **23**, 109-125.

151. Nayak, S.S. (1989). On the tail behaviour of record values. *Sankhyā* **A 51**, 390-401.

152. Nayak, S.S. and Wali, K.S. (1992). On the number of boundary crossings related to LIL and SLLN for record values and partial maxima of i.i.d. sequences and extremes of uniform spacings. *Stoch. Process. Appl.* **43**, 317-329.

153. Neuts, M.F. (1967). Waiting times between record observations. *J. Appl. Prob.* **4**, 206-208.

154. Nevzorov, V.B.(1981). Limit theorems for order statistics and record values. *Abstracts of the Third Vilnius conference on probability theory and mathematical statistics,*V.**2**, 86-87.

155. Nevzorov, V.B. (1984). Record times in the case of nonidentically distributed random variables. *Theory Probab. Appl.* **29**, 808-809.

156. Nevzorov,V.B. (1985). Record and interrecord times for sequences of non-identically distributed random variables. *Zapiski Nauchn. Semin. LOMI* **142**, 109-118 (in Russian). Translated version in *J.Soviet.Math.* **36** (1987),510-516.

157. Nevzorov, V.B. (1986a).K^{th} record times and their generalizations. *Zapiski Nauchn. Semin. LOMI* **153**, 115-121 (in Russian). Translated version in *J.Soviet.Math.* **44** (1989), 510-515.

158. Nevzorov, V.B. (1986b). Record times and their generalizations. *Theory Probab. Appl.* **31**, 629-630 .

159. Nevzorov, V.B. (1986c). Two characterizations using records. *Stability Problems for Stochastic Models* (V.V.Kalashnikov, B.Penkov and V.M. Zolotarev, Eds.), *Lecture Notes in Math.* **1233**, 79-85. Berlin: Springer Verlag.

160. Nevzorov, V.B. (1987a). Moments of some random variables connected with records. *Vestnik of the Leningrad Univ.* **8**, 33-37 (in Russian).

161. Nevzorov, V.B. (1987b). Records. *Theory Probab. Appl.* **32**, 219-251.

162. Nevzorov, V.B. (1987c). Distributions of kth- record values in the discrete case. *Zapiski Nauchn. Semin. LOMI* **158**, 133-137 (in Russian). Translated version in *J.Soviet.Math.* **43** (1988), 2830-2833.

163. Nevzorov, V.B. (1988). Centering and normalizing constants for extrema and for records. *Zapiski Nauchn. Semin. LOMI* **166**, 103-111 (in Russian). Translated version in *J.Soviet.Math.* **52** (1990), 2935-2941.

164. Nevzorov, V.B. (1989). Martingale methods of investigation of records. *Statistics and Control Random Processes* , 156-160 (in Russian).

165. Nevzorov, V.B. (1990a). Generating functions for k^{th} record values- a martingale approach. *Zapiski Nauchn. Semin. LOMI* **184**, 208-214 (in Russian). Translated version in *J.Math. Sci.* **68** (1994),545-550.

166. Nevzorov, V.B. (1990b). Records for nonidentically distributed random variables. *Proc. of 5th Vilnius Conf. on Prob. & Statist.*, v.2. VSP/Mokslas, 227-233.

167. Nevzorov, V.B. (1992). A characterization of exponential distributions by correlations between records. *Mathematical Methods of Statistics* **1**, 49-54, Allerton Press.

168. Nevzorov, V.B. (1993). Characterizations of certain non-stationary sequences by properties of maxima and records. *Rings and Modules. Limit Theorems of Probability Theory* (Eds., Z.I.Borevich and V.V. Petrov), v.**3**, 188-197, St.-Petersburg, St. Petersburg State University (in Russian).

169. Nevzorov, V.B. (1995). Asymptotic distributions of records in non-stationary schemes. *J. Statist. Plann. Infer.* **44**, 261-273.

170. Nevzorov, V.B. (1997). One limit relation between order statistics and records. *Zapiski Nauchn. Semin. POMI* **244**, 218-226 (in Russian).

171. Nevzorov , V.B and Ahsanullah, M. (2000). Some distributions of induced records. Biometrical Journal, 42, 1066-1081.

172. Nevzorov , V.B. and Balakrishnan, N. (1998). Record of records. *Handbook of Statistics* **16**. North Holland, Amsterdam, 515-570.

173. Nevzorov, V.B. and Rannen, M. (1992). On record times in sequences of nonidentically distributed discrete random variables. *Zapiski Nauchn. Semin. POMI* **194**, 124-133 (in Russian). Translated version in *J.Math. Sci.*

174. Nevzorov, V.B. and Stepanov, A.V. (1988). Records: martingale approach to finding of moments. *Rings and Modules. Limit Theorems of ProbabilityTheory* (Eds., Z.I.Borevich and V.V. Petrov), v.**2**, 171-181, St.-Petersburg, St. Petersburg State University (in Russian).

175. Nevzorova , L.N. and Nevzorov, V.B. (1999). Ordered random variables. *Acta Applicandae Mathematica,* **58**, 1-3, 217-229.

176. Nevzorova, L.N., Nevzorov, V.B. and Balakrishnan , N.(1997). Characterizations of distributions by extremes and records in Archimedean copula processes. *Advances in the Theory and Practice of Statistics: A volume in honor of Samuel Kotz* (Eds.N.L.Johnson and N.Balakrishnan), 469-478, NY: Wiley.

177. Pfeifer, D. (1979). *Record values in einem stochastischen modell mit nicht-identischen verteilungen.* (Ph.D. Thesis, RWTH, Aachen,).

178. Pfeifer, D. (1981). Asymptotic expansions for the mean and variance of logarithmic inter-record times. *Meth. Operat. Res.* **39**, 113-121.

179. Pfeifer, D. (1982). Characterizations of exponential distributions by independent non- stationary record increments. *J. Appl. Prob.* **19**, 127-135 (correction: **19**, 906).

180. Pfeifer, D. (1984a). Limit laws for inter-record times from non-homogeneous record values. *J. Organizat. Behav. and Statist.* **1**, 69-74.

181. Pfeifer, D. (1984b). A note on moments of certain record statistics. *Z. Wahr. verw. Geb.* **B66**, 293-296.

182. Pfeifer, D. (1985). On a relationship between record values and Ross's model of algorithm efficiency. *Adv. Appl. Prob.* **27**, 470-471.

183. Pfeifer, D. (1986). Extremal processes, record times and strong approximation. *Publ. Inst. Statist. Univ. Paris* **31**, 47-65.

184. Pfeifer, D. (1987). On a joint strong approximation theorem for record and inter-record times. *Prob. Theory Rel. Fields* **75**, 213-221.

185. Pfeifer, D. (1989). Extremal processes, secretary problems and the 1/e law. *J. Appl. Prob.* **26**, 722-733.

186. Pfeifer, D. (1991). Some remarks on Nevzorov's record model. *Adv. Appl. Prob.* **23**, 823-834.

187. Pfeifer, D. and Zhang, Y.C. (1989). A survey on strong approximation techniques in connection with records, *Lectures Notes Statist.* **51**, *Extreme value theory*, Oberwolfach, December 6-12, 1987 (J. Husler and R.D. Reiss, Eds). Berlin: Springer Verlag, 50-58.

188. Rahimov, I. (1995). Record values of a family of branching processes. *I MA volumes in mathematics and its applications* **84**. Springer, 285- 296.

189. Rannen, M.M. (1991). Records in sequences of series of nonidentically distributed random variables. *Vestnik of the Leningrad State University* **24**, 79-83.

190. Renyi, A. (1962). Theorie des elements saillants d'une suite d'observations. *Colloquim on Combinatorial Methods in Probability Theory* (August 1-10, 1962), Math. Inst., Aarhus Univ., Aarhus, Danmark, 104-117 . See also: On the extreme elements of observations. *Selected Papers of Alfred Renyi*, v. **3** (1976), 50-65. Budapest: Akademiai Kiado.

191. Resnick, S.I. (1973a). Limit laws for record values. *Stoch. Process. Appl.* **1**, 67-82.

192. Resnick, S.I. (1973b). Extremal processes and record value times. *J.Appl. Prob.* **10**, 864-868.

193. Resnick, S.I. (1973c). Record values and maxima. *Ann. Prob.* **1**, 650- 662.

194. Rogers, L.C.G. (1989). Ignatov's theorem: an abbreviation of the proof of Engelen, Tommassen and Vervaat. *Adv. Appl. Prob.* **21**, 933-934.

195. Roy, D. (1990). Characterization through record values. *J. Indian Stat. Assoc.* **28**, 99-103.

196. Samaniego, F.J. and Whitaker, L.R. (1986). On estimating population characteristics from record-breaking observations. I. Parametric results. *Nav. Res. Log. Quart.* **33**, 531-543.

197. Samaniego, F.J. and Whitaker, L.R. (1988). On estimating population characteristics from record-breaking observations. II. Nonparametric results. *Nav. Res. Log.Quart.* **35,** 221-236.

198. Samaniego, F.J. and Kaiser, L.D. (1978). Estimating value in a uniform auction. *Naval Res. Log. Quart.* **25**, 621-632.

199. Shiganov, I.S. (1982). On sharpening of upper bound for constant in reminder term of central limit theorem. *Stability Problems for Stochastic Models,* 109-115.

200. Shorrock, R.W. (1972). A limit theorem for inter-record times. *J.Appl. Prob.* **9**, 219-223.

201. Shorrock, R.W. (1972). On record values and record times. *J. Appl. Prob.* ***9***, 316-326

202. Shorrock, R.W. (1973). Record values and inter-record times. *J. Appl. Prob.* **10**, 543-555.

203. Shorrock, R.W. (1974). On discrete time extremal processes. *Adv. Appl. Prob.* **6**, 580-592.

204. Siddiqui, M.M. and Biondini, R.W. (1975). The joint distribution of record values and inter-record times. *Ann.Prob.* **3**, 1012- 1013.

205. Smith, R.L. (1988). Forecasting records by maximum likelihood. *J. Amer. Statist. Assoc.* **83**, 331-338.

206. Smith, R.L. and Miller, J.E. (1986). A non-gaussian state space model and application to prediction of records. *J. Roy. Statist. Soc.* **B48**, 79-88.

207. Srivastava, R.C. (1979). Two characterizations of the geometric distribution by record values. *Sankhyā* **B 40**, 276-278.

208. Srivastava, R.C. (1981a). On some characterizations of the geometric distribution *Statistical Distribution s in Scientific work.* V.4 (C. Tallie, g.P. Patil and B.A.Baldessari, Eds.). Dordrecht: Reidel, 349-355.

209. Srivastava, R.C. (1981b). Some characterizations of the exponential distribution based on record values. *Statistical Distributions in Scientific Work,* **v.4**, (C.Taillie, G.P.Patil and B.A.Baldessari, Eds.). Dordrecht: Reidel, 411-416.

210. Srivastava, R.C. and Bagchi ,S.N. (1985). On some characterizations of the univariate and multivariate geometric distributions. *J. Indian. Statist. Assoc.* **23**, 27-33.

211. Stam, A.I. (1985). Independent Poisson processes generated by record values and inter-record times. *Stoch. Process. Appl.* **19**, 315- 325.

212. Stepanov, A.V. (1987). On logarithmic moments for inter-record times. *Theory Probab. Appl.* **32**, 708-710.

213. Stepanov, A.V. (1989). Characterizations of geometric class of distributions. *Teoriya Veroyatnostey i Matematicheskaya Statistika* **41**, 133-136 (in Russian). Translated version in *Theory Probability and Mathematical Statistics* **41** (1990).

214. Stepanov A.V. (1992). Limit theorems for weak records. *Theory Probab. Appl.* **37**, 586-590.

215. Stuart, A. (1954). Asymptotic relative efficiencies of distribution-free tests of randomness against normal alternatives. *J. Amer. Stat. Assoc.* **49**, 147-157.

216. Stuart, A. (1956). The efficiencies of tests of randomness against normal regression. *J. Amer. Statist. Assoc.* **51**, 285-287.

217. Stuart, A. (1957). The efficiency of the records test for trend in normal regression. *J. Roy. Statist. Soc.* **B 19**, 149-153.

218. Strawderman, W.E. and Holmes, P.T. (1970). On the law of the iterated logarithm for inter-record times. *J. Appl. Probab.*7, 432-439.

219. Tallie, C. (1981). A note on Srivastava's characterization of the exponential distribution based on record values. *Statistical Distributions in Scientific Work,* **v.4**, (C.Tallie, G.P.Patil and B.A.Baldessari, Eds.). Dordrecht: Reidel, 417-418.

220. Tata, M.N. (1969). On outstanding values in a sequence of random variables. *Z. Wahr. Verw. Geb.* **B12**, 9-20.

221. Teugels, J.L. (1984). On successive record values in a sequence of Independent identically distributed random variables. *Statist. Extremes and Appl.* (Tiago de Oliveira, Ed.). Dordrecht: Reidel, 639-650.

222. Too, Y.H. and Lin, G.D. (1989). Characterizations of uniform and exponential distributions. *Statist. and Prob. Lett.* **7**, 357-359.

223. Tryfos, P. and Blackmore, R. (1985). Forecasting records. *J. Amer. Statist. Assoc.* **80**, 46-50.

224. Vervaat, W. (1973). Limit theorems for records from discrete distributions. *Stoch. Process. Appl.* **1**, 317-334.

225. Westcott, M. (1977). A note on record times. *J. Appl. Prob.* **14**, 637-639.

226. Wilks, S.S. (1959). Recurrence of extreme observations. *J. Amer. Math. Soc.* **1**, 106- 112.

227. Westcott, M. (1977). The random record model. *Proc. Roy. Soc. London* **A356**, 529-547.

228. Westcott, M. (1979). On the tail behavior of record-time distributions in a random record process. *Ann. Prob.* **7**, 868-873.

229. Wilks, S.S. (1959). Recurrence of extreme observations. *J. Amer. Math. Soc.* **1**, 106- 112.

230. Witte, H.J. (1988). Some characterizations of distributions based on the integrated Cauchy functional equation. *Sankhyā* **A 50**, 59-63.

231. Witte, H.J. (1990). Characterizations of distributions of exponential or geometric type by the integrated lack of memory property and record values. *Comput. Statist. and Data Analysis* **10**, 283-288.

232. Witte, H.J. (1993). Some characterizations of exponential or geometric distributions in a nonstationary record value model. *J.Appl. Prob.* **30**, 373-381.

233. Williams, D. (1973). On Renyi's record problem and Engel's series. *Bull. London Math. Soc.* **5**, part 2, 235-237.

234. Vervaat, W., (1973). Success epochs in a sequence of Bernoulli trials. *Adv. Appl. Prob.* **5**, 35-36.

235. Yakimiv , A.L. (1995). Asymptotics of kth record values. *Theory Probab. Appl.* **40**, 925-928.

236. Yanushkevichius, R. (1993). Stability of characterization by record properties. *Lecture Notes in Math.* **1546**, *Stability problems for stochastic models*, Suzdal, 1991, Berlin: Springer Verlag, 189- 196.

237. Yang, M.C.K. (1975). On the distribution of the inter-record times in an increasing population. *J. Appl. Prob.* **12**, 148-154.

238. Zahle, U. (1989). Self-similar random measures, their carrying dimension and application to records. *Lectures Notes Statist.* **51**, *Extreme value theory*, Oberwolfach, December 6-12, 1987 (J. Husler and R.D. Reiss, Eds). Berlin: Springer Verlag, 59-68.

Chapter 5

INDUCED ORDER STATISTICS AND RECORD VALUES

Let (X,Y), (X_1,Y_1), $(X_2,Y_2),\ldots,(X_n, Y_n)$ be two-dimensional random vectors with the common bivariate distribution function F(x,y). For the sake of simplicity we suppose that

$$P\{X_k=X_m\}=0$$

for any $k\neq m$. Let D(1),...,D(n) be antiranks based on X's. It means that D(k) is equal to an index of such X in the initial sequence $X_1,\ldots, X_n$, which coincides with $X_{k,n}$, i.e. the following relation for events holds:

$$\{D(r)=m\}=\{X_{r,n}=X_m\},\ 1\leq r\leq n,\ 1\leq m\leq n\ .$$

Now let us order vectors

$$(X_k,Y_k),\ k=1,\ldots,n,$$

by increasing values of X's. We get a sequence of dependent vectors, the first components of which form a variational series $X_{1,n}\leq X_{2,n}\leq\ldots\leq X_{n,n}$:

$$(X_{k,n}, Y_{D(k)}),\ k=1,2,\ldots,n.$$

The second components $Y_{D(k)}$, k=1,2,...,n, of the ordered vectors are said to be induced order statistics (Bhattacharya (1974)) or concomitants of order statistics (David (1973)). Note that both the denominations have appeared independently from each other. More often one uses symbols $Y_{[k,n]}$ or $Y_{[k:n]}$ instead of $Y_{D(k)}$ to show that these statistics correspond to a two-dimensional sample of a fixed size n. Induced order statistics arise in different selection procedures, when one can choose the best (with respect to some feature X) k of n objects to investigate the second characteristic Y. Say, X's and Y's are weights and heights of individuals under investigation respectively, or X's are results of sportsmen in a preliminary round of some competition and $Y_{[1,n]},\ldots,Y_{[k,n]}$ are results of k best sportsmen in the final part of the competition.

The situation described above can be easily generalized for multinomial induced order statistics (Barnett (1976)). Let us have (m+1)-variate vectors

$$(X_1,Y_{1,1},\ldots,Y_{m,1}),\ldots,(X_n,Y_{1,n},\ldots,Y_{m,n}).$$

If to order these vectors by values of X's, the rest components form multinomial concomitants

$$(Y_{1,D(1)},\ldots,Y_{m,D(1)}),\ldots,(Y_{1,D(n)},\ldots,Y_{m,D(n)})$$

corresponding to order statistics $X_{1,n},\ldots,X_{n,n}$ respectively. Another generalization of induced order statistics was suggested by Egorov and Nevzorov (1981). They have ordered random vectors

$$(Y_{1,1},\ldots,Y_{m,1}),\ldots,(Y_{1,n},\ldots,Y_{m,n})$$

by the values of

$$X_r= f(Y_{1,r},\ldots,Y_{m,r}),\ r=1,2,..,n,$$

$f(y_1,\ldots,y_m)$ being a real-valued function, and they called the corresponding orderings as f-induced order statistics. Indeed, we can obtain the same f-induced order statistics applying Barnett's approach with random vectors $(X_r,Y_{1,r},\ldots,Y_{m,r})$, the first components X_r of which coincide initially with

$$f(Y_{1,r},\ldots,Y_{r,n}),\ r=1,2,\ldots,n.$$

Observe that the case r=1 and f(x)=x corresponds to the usual order statistics. If r=1 and f(x)=|x|, then the so-called absolute order statistics arise (see, for, example, Egorov and Nevzorov (1975)). An interesting generalizations of absolute order statistics can be obtain if we consider any r>1 and

$$f(x_1,\ldots,x_r)=(x_1^2+\ldots+x_r^2)^{1/2}$$

or

$$f(x_1,\ldots,x_r)=(x_1+\ldots+x_r).$$

The detailed surveys of results related to induced order statistics are given in Bhattacharya (1984) and David and Nagaraja (1998). These authors give also a detailed review of applications of induced order statistics in different statistical procedures.

5.1. ABSOLUTE ORDER STATISTICS

Suppose that X, X_1, X_2,... be independent random variables with a common cumulative distribution function (cdf) F and a probability density function (pdf) p. Let $X_{[1,n]},\ldots,X_{[n,n]}$, n=1,2,..., denote absolute order statistics , i.e. $X_{[1,n]},\ldots,X_{[n,n]}$ are obtained by ordering of random variables $X_1, X_2,\ldots X_n$ according to increasing values of their modulus. It means that

$$|X_{[1,n]}|\leq\ldots\leq|X_{[n,n]}|.$$

Statistical procedures based on absolute order statistics are useful in many situations. For example, let a statistician know that a sample from some population distribution probably includes one or more outliers, which come from another distribution with heavy tails. In this case X's having the maximal absolute values (that is absolute order statistics $X_{[n,n]},X_{[n-1,n]},\ldots$) are the most appropriate to be the outliers. The most suitable cases to apply absolute order statistics are connected with symmetric distributions. Below we will list some useful results for absolute order statistics based on samples from symmetric distributions.

It is easy to prove that any symmetric random variable X satisfies the following equality in probability:

$$X\stackrel{d}{=}\xi|X|, \tag{5.1.1}$$

where the r.v. ξ takes the two values −1 and 1 with equal probabilities and does not depend on |X|. Moreover, if we denote $Y_k=|X_k|$, then the following generalization of (5.1.1) is true for symmetric random variables:

$$\{X_k\}_{k=1}^{n} \stackrel{d}{=} \{\xi_k Y_k\}_{k=1}^{n}, \tag{5.1.2}$$

where ξ_k, k=1,2,..., take values −1 and 1 with probabilities ½ , and all ξ's and Y's are independent. From (5.2.1) we immediately obtain the next result.

Representation 5.1.1 (Egorov and Nevzorov (1975)).

Let $X_{[1,n]},\ldots, X_{[n,n]}$ be absolute order statistics based on symmetric random variables $X_1,\ldots,X_n$. Then

$$\{X_{[k,n]}\}_{k=1}^{n} \stackrel{d}{=} \{\xi_k Y_{k,n}\}_{k=1}^{n}, \tag{5.1.3}$$

where $Y_{1,n}\leq\ldots\leq Y_{n,n}$ are order statistics, based on the random variables $Y_k=|X_k|$, k=1,2,...,n,

ξ_k, k=1,2,...,n, are independent random variables, taking values –1 and 1 with probabilities ½, and the vectors $(\xi_1,...,\xi_n)$ and $(Y_{1,n},...,Y_{n,n})$ are independent.

Let us call X a quasi-symmetric random variable if there exists such $0 \le p \le 1$ that

$$pP\{X<-x\}=(1-p)P\{X>x\}. \quad (5.1.4)$$

for any positive x.
Note that p=1/2 corresponds to symmetric random variables, p=0 in (5.1.4) means that $P\{X \le 0\}=1$ and p=1 if $P\{X \ge 0\}=1$.

Remark 5.1.1.

If we consider quasi-symmetric random variables $X_1,...,X_n$ then (5.1.2) and (5.1.3) stay to be true with random variables ξ_k, k=1,2,...,n, taking values 1 and -1 with probability p and 1-p respectively.

A question arises in connection with representations (5.1.1) and (5.1.3): what nondegenerate random variables X may be represented in the form

$$X \stackrel{d}{=} \xi|X|,$$

where ξ and |X| are independent? As it turns out, X has to be quasi-symmetric and ξ has to assume two values only. Indeed, from (5.1.1) it follows that

$$|X| \stackrel{d}{=} |\xi||X|,$$

where random variables on the right side are independent. From the relations

$$\alpha=P\{X=0\}=P\{X=0\}+P\{\xi=0, X\neq 0\}=\alpha+(1-\alpha)P\{\xi=0\}$$

we obtain (on eliminating the degenerate case, when $\alpha=P\{X=0\}=1$) that $P\{\xi=0\}=0$.
With no loss of generality, we may assume that X does nor take the value zero (otherwise we consider the conditional distribution of X, given that $X\neq 0$). We have now

$$\log|X| \stackrel{d}{=} \log|\xi|+\log|X|.$$

Using characteristic functions we conclude that $\log|\xi|=0$ with probability 1, i.e.

$P\{|\xi|=1\}=1$. It means that

$P\{\xi=1\}=1-P\{\xi=-1\}=p$

for some p, $0 \le p \le 1$. If p equals 0 or 1, we immediately get that $P\{X\le 0\}=1$ or $P\{X\le 0\}=0$. Finally, if $0<p<1$, then it follows from (5.1.1) that

$$P\{X<-x\}=(1-p)P\{|X|>x\}$$

and

$$P\{X>x\}=pP\{|X|>x\},$$

i.e. X is a quasi-symmetric random variable.

Representation 5.1.1 enables us to express the distribution of absolute order statistics as a mixture of distributions of the classical order statistics. In fact, for absolute order statistics based on quasi-symmetric random variables we have the following equalities for $x>0$:

$$P\{X_{[k,n]}<-x\}=P\{\xi_k Y_{k,n}<-x\}=(1-p)P\{-Y_{k,n}<-x\} \qquad (5.1.5)$$

and

$$P\{X_{[k,n]}<x\}=1-p+pP\{Y_{k,n}<x\}. \qquad (5.1.6)$$

Let $V=-|X|$, $V_k=-|X_k|$ and $V_{1,n}\le\ldots\le V_{n,n}$ be the order statistics based on random variables $V_1,\ldots,V_n$. Note that

$$-Y_{k,n}\overset{d}{=}V_{n-k+1,n}.$$

Then it follows from (5.1.5) and (5.1.6) that

$$P\{X_{[k,n]}<x\}=(1-p)P\{V_{n-k+1,n}<x\}+pP\{Y_{k,n}<x\}. \qquad (5.1.7)$$

for any x.

Egorov and Nevzorov (1976) obtained the following result for the classical order statistics.

Theorem 5.1.1.

Let $X_1,X_2,\ldots,X_n$ be independent random variables with a common cdf F and a pdf p(x) such that

$$\sup_{x:p(x)>0} |p'(x)| \le M.$$

Then

$$\sup_x |P\{(X_{k,n}-F^{-1}(\frac{k}{n+1}))p(F^{-1}(\frac{k}{n+1}))<x\beta_2\}-\Phi(x)|\leq C(k^{-1/2}+(n-k+1)^{-1/2}+M\beta_2/p^2(F^{-1}(\frac{k}{n+1}))), \quad (5.1.8)$$

where C is an absolute constant, F^{-1} is the inverse of F, Φ is the cdf of the standard normal law and

$$\beta_2=k^{1/2}(n-k+1)^{1/2}/(n+1)(n+2)^{1/2}.$$

We can apply the statement of theorem 5.1.1 to the RHS of 5.1.7. Let G(x) and g(x) be the cdf and pdf of |X|. Easy to see that

$$g(x)=p(x)+p(-x)=p(x)/p=p(-x)/(1-p),\ x>0,$$

and

$$G(x)=(F(x)-(1-p))/p,\ x>0,$$

for the quasi-symmetric random variable defined in (5.1.4) Analogously, the pdf of V=-|X| has the form

$$v(x)=g(-x),\ x<0,$$

and the cdf of V is given as

$$V(x)=(1-F(-x))/p=F(x)/(1-p),\ x<0.$$

Note also that

$$-V^{-1}(\frac{n-k+1}{n+1})=G^{-1}(\frac{k}{n+1})=F^{-1}(1-p(\frac{n-k+1}{n+1})).$$

Denote

$$x_{p,k,n}=F^{-1}(1-p(\frac{n-k+1}{n+1})).$$

On combining relations (5.1.7) and (5.1.8), under conditions of theorem 5.1.1 we obtain that

$$\sup_x|P\{X_{[k,n]}<x\}-(1-p)\Phi\left(\frac{x-V^{-1}(\frac{n-k+1}{n+1})}{\beta_2}v(V^{-1}(\frac{n-k+1}{n+1}))\right)-$$

$$p\,\Phi\left(\frac{x-G^{-1}(\frac{k}{n+1})}{\beta_2}g(G^{-1}(\frac{k}{n+1}))\right)|$$

$$\leq C(k^{-1/2}+(n-k+1)^{-1/2}+M(1-p)\max\{\frac{1}{p},\frac{1}{1-p}\}\beta_2/v^2(V^{-1}(\frac{n-k+1}{n+1}))$$

$$+Mp\max\{\frac{1}{p},\frac{1}{1-p}\}\beta_2/g^2(G^{-1}(\frac{k}{n+1}))). \qquad (5.1.9)$$

In (5.1.9) we used the evident inequality

$$|v'(x)|=|g'(x)|\leq M\max\{\frac{1}{p},\frac{1}{1-p}\}.$$

One can rewrite (5.1.9) in terms of the cdf $F(x)$ and the pdf $p(x)$:

$$\sup_x|P\{X_{[k,n]}<x\}-(1-p)\Phi\left(\frac{x+x_{p,k,n}}{p\beta_2}p(x_{p,k,n})\right)-p\,\Phi\left(\frac{x-x_{p,k,n}}{p\beta_2}p(x_{p,k,n})\right)|$$

$$\leq C(k^{-1/2}+(n-k+1)^{-1/2}+M\,p^2\max\{\frac{1}{p},\frac{1}{1-p}\}\beta_2/p^2(x_{p,k,n})). \qquad (5.1.10)$$

As a consequence of (5.1.10) we get for $k=\alpha n$ that

$$\sup_x|P\{X_{[\alpha n,n]}<x\}-(1-p)\Phi\left(\frac{x+x_{p,\alpha}}{p\alpha^{1/2}(1-\alpha)^{1/2}}p(x_{p,\alpha})n^{1/2}\right)-$$

$$p\Phi\left(\frac{x-x_{p,\alpha}}{p\alpha^{1/2}(1-\alpha)^{1/2}}p(x_{p,\alpha})n^{1/2}\right)|\leq C(M,p,x_{p,\alpha})/n^{1/2}, \qquad (5.1.11)$$

where

$$x_{p,\alpha}=F^{-1}(1-p(1-\alpha))$$

and the constant on the RHS of (5.1.11) depends on M, p and $p(x_{p,\alpha})$ only. Now we obtain from (5.1.11) that

$$\lim P\{ X_{[\alpha n,n]}<x\}=P\{\eta_{\alpha,p}<x\}, \quad (5.1.12)$$

as $n\to\infty$, where the random variable $\eta_{\alpha,p}$ takes values $-x_{p,\alpha}$ and $x_{p,\alpha}$ with probabilities (1-p) and p correspondingly.

We have used representation 5.1.1 to prove formulae (5.1.10)-(5.1.12) for quasi-symmetric random variables. Egorov and Nevzorov (1982) obtained the similar results for absolute order statistics in a more+ general situation. They have compared the following two equalities (the first of them is given for absolute order statistics $X_{[k,n]}$ and the second is formulated for the classical order statistics based on random variables $Y_k=|X_k|$):

$$P\{ X_{[k,n]}<x\}= \int_0^\infty P\{X<x|\ |X|=u\}\beta_{k,n-k+1}(G(u))dG(u)$$

$$= \int_0^\infty \frac{p(-u)}{p(-u)+p(u)} P\{X<x|\ X=-u\}\beta_{k,n-k+1}(G(u))dG(u)$$

$$+ \int_0^\infty \frac{p(u)}{p(-u)+p(u)} P\{X<x|\ X=u\}\beta_{k,n-k+1}(G(u))dG(u) \quad (5.1.13)$$

and

$$P\{Y_{k,n}<x\}= \int_0^\infty P\{Y<x|\ Y=u\}\beta_{k,n-k+1}(G(u))dG(u), \quad (5.1.14)$$

where

$$\beta_{k,n-k+1}(x)= (n!/(k-1)!(n-k)!)x^{k-1}(1-x)^{n-k},\ 0\le x\le 1,$$

is the density function of Beta distribution with parameters k and n-k+1.

Denote

$$\lambda_{k,n}=p(G^{-1}(k/(n+1))),$$
$$\mu_{k,n}= p(-G^{-1}(k/(n+1)))$$

and

$$L_{k,n}(x) = \frac{\lambda_{k,n}}{\lambda_{k,n}+\mu_{k,n}}\Phi(\frac{x-G^{-1}(k/(n+1))}{\beta_2}(\lambda_{k,n}+\mu_{k,n}))$$

$$+\frac{\mu_{k,n}}{\lambda_{k,n}+\mu_{k,n}}\Phi(\frac{x+G^{-1}(k/(n+1))}{\beta_2}(\lambda_{k,n}+\mu_{k,n})).$$

The following result is valid.

Theorem 5.1.2.

Suppose that

$$\sup_{x:0<G(x)<1}|p'(x)| \leq M<\infty$$

and

$$\gamma=\inf_n (\lambda_{k,n}+\mu_{k,n})>0.$$

Then the equality

$$\sup_x |P\{X_{[k,n]}<x\}-L_{k,n}(x)| \leq C(k^{-1/2}+(n-k+1)^{-1/2})+C(\gamma,M)\,\beta_2 \qquad (5.1.15)$$

holds for all n and $k=k(n)\leq n$, where C is an absolute constant and $C(\gamma,M)$ depends on γ and M only.

The next result follows from theorem 5.1.2.

Theorem 5.1.3 (Egorov and Nevzorov (1982)).

Suppose that

$$k/n\to\alpha,\ 0<\alpha<1, \text{ as } n\to\infty,$$

$$p_1=p(G^{-1}(\alpha))>0,\ p_2=p(-G^{-1}(\alpha))>0$$

and

$$p'(\pm G^{-1}(\alpha))<\infty.$$

Then

$P\{X_{[k,n]}<x\}\to L_\alpha(x)$, as $n\to\infty$, where

$$L_\alpha(x)=\begin{cases}0, & if \quad x\le G^{-1}(\alpha),\\ p_1/(p_1+p_2), & if \quad -G^{-1}(\alpha)<x\le G^{-1}(\alpha),\\ 1, & if \quad x>G^{-1}(\alpha).\end{cases}$$

The following sharpening of theorem 5.1.3 was obtained by Egorov and Nevzorov (1982).

Theorem 5.1.4.

Let the sequence $k=k(n)$ be chosen in such a way that

$k/(n+1)=\alpha$, $0<\alpha<1$.

Suppose that

$$\lim_{x\downarrow\alpha} p(-G^{-1}(x))=p_1,\ \lim_{x\uparrow\alpha} p(-G^{-1}(x))=p_2,$$

$$\lim_{x\uparrow\alpha} p(G^{-1}(x))=p_3,\ \lim_{x\downarrow\alpha} p(G^{-1}(x))=p_4,$$

where $p_1+p_4>0$ and $p_2+p_3>0$, and

$$\lim_{x\uparrow\alpha}\ p'(\pm G^{-1}(x))<\infty,\ \lim_{x\downarrow\alpha}\ p'(\pm G^{-1}(x))<\infty.$$

Then

$P\{X_{[k,n]}<x\}\to \tilde{L}_\alpha(x)$,

$$\text{as } n\to\infty, \text{ where } \tilde{L}_\alpha(x)=\begin{cases}0, & if \quad x\le G^{-1}(\alpha),\\ \dfrac{p_1}{2(p_1+p_4)}+\dfrac{p_2}{2(p_2+p_3)}, & if \quad -G^{-1}(\alpha)<x\le G^{-1}(\alpha),\\ 1, & if \quad x>G^{-1}(\alpha).\end{cases}$$

Egorov and Nevzorov (1975) studied convergence to the normal law of linear combinations

$$S_{n,k}=\sum_{j=1}^{n-k} a_j X_{[j,n]}$$

of absolute order statistics based on a sequence of independent symmetric i.i.d. random variables. To formulate their main result we need some additional notation. Let, as above, the initial random variables have a common continuous distribution function F and let

$G(x)=F(x)-F(-x),\ x>0,$
be the cdf of $|X|$. For any $s>0$ introduce new random variables $X^{(s)}, X_1^{(s)}, X_2^{(s)},\ldots$, which are independent and have a common cdf

$$U_s(x)=P\{X^{(s)}<x\}=P\{X<x \mid |X|\le s\}=(F(x)-F(-s))/(F(s)-F(-s)),\ |x|\le s. \tag{5.1.16}$$

We suppose that X's have a symmetric distribution. Then $X^{(s)}$, for any $s>0$, also has a symmetric distribution. Denote absolute order statistics, based on $X_1^{(s)},\ldots, X_n^{(s)}$, as

$X^{(s)}_{[k,n]}$, $k=1,2,\ldots,n,\ n=1,2,\ldots.$

It follows from this definition and Representation 5.1.1 that

$EX^{(s)}_{[k,n]}=0.$

Let

$$s_0=G^{-1}\left(\frac{n-k}{n}\right),$$

where G^{-1} is the inverse of the cdf G. Denote

$$B^2_{n,k}=\sum_{j=1}^{n-k} a^2_{j,n} E(X^{(s_0)}_{j,n-k})^2. \tag{5.1.17}$$

Note that

$E(X^{(s_0)}_{j,n-k})^2$, $j=1,2,\ldots,n-k,$

can be regarded as expected values of the classical order statistics corresponding to a sample of size (n-k) and the cdf

$$\mathbf{H(x)=(F(\sqrt{x})-F(-\sqrt{x}))/(F(\sqrt{s_0})-F(-\sqrt{s_0})),\ 0<x\le s_0.}$$

Now we are ready to formulate the following theorem, the proof of which can be found in Egorov and Nevzorov (1975).

Theorem 5.1.5.

Suppose that the relation

$$0<\liminf \frac{k(n)}{n} \leq \liminf \frac{k(n)}{n} <1 \qquad (5.1.18)$$

holds, as $n \to \infty$. Then

$$\Delta_n = \sup_x |P\{S_{n,k(n)}<xB_{n,k(n)}\} - \Phi(x)| = O\left(\sqrt{\frac{\log n}{n}}\right), \qquad (5.1.19)$$

where $B_{n,k}$ is defined in (5.1.17) and Φ is the cdf of the standard normal law.

5.2. *F*-INDUCED ORDER STATISTICS

A generalization of absolute order statistics was suggested in Egorov and Nevzorov (1981, 1982). They have introduced the so-called *f*-induced order statistics.

Consider a sequence of i.i.d. random variables $X_1,\ldots,X_n$ and some function f. Now arrange $X_1,\ldots,X_n$ in order of growth of the values

$$Y_k=f(X_k),\ k=1,\ldots,n.$$

The random variables obtained as a result of such ordering are said to be f-induced order statistics . For the sake of simplicity we will denote them as absolute order statistics : $X_{[1,n]},\ldots,X_{[n,n]}$. Note also that absolute order statistics are a partial case ($f(x)=|x|$) of *f*-induced order statistics.

Thus, let $X, X_1,\ldots,X_n$ be a sequence of independent random variables having a common density function $p(x)$ and let $X_{[1,n]},\ldots,X_{[n,n]}$ be *f*-induced order statistics, generated by a function f. Denote

$$G(x)=P\{f(X)<x\}.$$

It is not hard to see (compare with 5.1.13) that

$$P\{X_{[k,n]}<x\}=\int_0^\infty P\{X<x|\ f(X)=u\}\beta_{k,n-k+1}(G(u))dG(u)=$$

$$\int_0^1 P\{X<x|\ f(X)=G^{-1}(u)\}\ \beta_{k,n-k+1}(u)du, \qquad (5.2.1)$$

where

$$\beta_{k,n-k+1}(x)=(n!/(k-1)!(n-k)!)x^{k-1}(1-x)^{n-k},\ 0\le x\le 1,$$

and G^{-1} is the inverse of the cdf G.

Since $\beta_{k,n-k+1}$-measures asymptotically concentrate in neighborhoods of points $k/(n+1)$, the asymptotic distribution of $X_{[k,n]}$ is determined by the solutions of the equation

$$f(x)=G^{-1}(\frac{k}{n+1}).$$

In the case $f(x)=|x|$, there are two solutions $\pm\ G^{-1}(\frac{k}{n+1})$ of this equation and theorem 5.1.2 reflects the corresponding situation. Now we consider a more general case.

Let

$$\Delta_{\varepsilon,n}=\{z:\ |G^{-1}(\frac{k}{n+1})-z|<\varepsilon\}$$

denote the ε-neighborhood of the point $G^{-1}(\frac{k}{n+1})$. The following result belongs to Egorov and Nevzorov (1982).

Theorem 5.2.2.

Suppose that a positive ε exists, for which in ε-neighborhoods $\Delta_{\varepsilon,n}$ of the points $G^{-1}(\frac{k}{n+1})$ the equation $f(x)=z$ has s continuous monotone decreasing or increasing non-intersecting solutions $f_1(z),\ldots,f_s(z)$ with bounded derivatives in these neighborhoods:

$$0<m_1\le|f_i'(z)|\le m_2<\infty,\ 1\le i\le s.$$

Suppose in addition that

$$\lambda=\inf_n(\lambda_{k,n}^{(1)}+\ldots+\lambda_{k,n}^{(s)})>0,$$

where

$$\lambda_{k,n}^{(j)}=p(f_j(G^{-1}(\frac{k}{n+1}))),\ 1\le j\le s,$$

and

$$\sup_{x:0<G(x)<1} |p'(x)| \le M<\infty.$$

Then , for all n and k(n),

$$\sup_x | P\{X_{[k,n]}<x\}-L_{k,n}(x)| \le C(k^{-1/2}+(n-k+1)^{-1/2})+C(m_1,m_2,M,s)\ \beta_2, \tag{5.2.2}$$

where

$$L_{k,n}(x)=\sum_{i=1}^{s} \frac{\lambda_{k,n}^{(i)}}{\lambda_{k,n}^{(1)}+\ldots+\lambda_{k,n}^{(s)}}\Phi(\frac{x-f_i(G^{-1}(k/(n+1)))}{\beta_2}(\lambda_{k,n}^{(1)}+\ldots+\lambda_{k,n}^{(s)})),$$

$$\beta_2=k^{1/2}(n-k+1)^{1/2}/(n+1)(n+2)^{1/2},$$

C is an absolute constant , $C(m_1,m_2,M,s)$ depends on m_1,m_2, M and s only and $\Phi(x)$ is the cdf of the standard normal law.

Let now the sequence k=k(n) be chosen in such a way that

$$k/(n+1)\to\alpha,\ 0<\alpha<1,$$

as $n \to \infty$, and let the following limits exist:

$$\lim_{z\to G^{-1}(\alpha)} p(f_k(z))=p_k>0,\quad \lim_{z\to G^{-1}(\alpha)} p'(f_k(z))<\infty.$$

Denote

$$f_k=f_k(G^{-1}(\alpha)).$$

Then the limiting distribution function of $P\{X_{[k,n]}<x\}$ is given as

$$L_\alpha(x)=\sum_{k: f_k<x} \frac{p_k}{p_1+\ldots+p_s}. \tag{5.2.3}$$

Indeed, the asymptotic cdf from theorem 5.1.3 is a particular case of (5.2.3).

As an illustration of the results for *f*-induced order statistics, given above, we consider the following example.

Example 5.2.1.

Let f(x) be the fractional part of x, i.e. f(x)=x-[x], and let the initial random variables have the standard exponential density function p(x)=exp(-x), x≥0. Consider the asymptotic behavior of induced medians $X_{[n+1,2n+1]}$. In this case,

$$G(x)=P\{f(X)<x\}=P\{X-[X]<x\}=(1-e^{-x})/(1-e^{-1}),\ 0\le x<1,$$

$$k/(n+1)=\alpha=1/2,\ G^{-1}(1/2)=\log\frac{2e}{e+1},$$

$$f_j(x)=j+x,\ 0\le x<1,\ j=0,1,2,\ldots,$$

$$p_0==(e+1)/2e,$$

$$p_j=p(f_j(G^{-1}(1/2)))=\exp(-j)p_0,\ j=1,2,\ldots,$$

$$\sum_{j=0}^{\infty} p_j=p_0/(1-e^{-1})$$

and then relations, analogous to (5.2.2) and (5.2.3), have the following form:

$$\sup_x|\ P\{X_{[n+1,2n+1]}<x\}-\sum_{j=0}^{\infty}\ \exp(-j)(1-e^{-1})\ \Phi((x-j-\log\frac{2e}{e+1})\frac{e+1}{e-1}\sqrt{2n}\ )|\le Cn^{-1/2},$$

where C is an absolute constant, and

$$L_\alpha(x)=\sum_{k:k<x-\log\frac{2e}{e+1}}\exp(-k)(1-e^{-1}).$$

We can see that the asymptotic distribution function $L_\alpha(x)$ corresponds to the geometric distribution with the weights

$$r_k=\exp(-k)(1-e^{-1}),$$

concentrated in the points

$$x_k=k+\log\frac{2e}{e+1},\ k=0,1,2,\ldots.$$

Multi-dimensional generalizations of *f*-induced order statistics may be also considered. In this case one takes p-dimensional random vectors

$$X_k=(X_k^{(1)},\ldots,X_k^{(r)}),\ k=1,2,\ldots,n,$$

and some function f(x), where $x=(x^{(1)},\ldots,x^{(r)})$. The random vectors X_k are ordered by increasing of values

$$f(X_1^{(1)},\ldots,X_1^{(r)}),\ldots,\ f(X_n^{(1)},\ldots,\ X_n^{(r)}).$$

The corresponding ordering gives us the so-called *f*-ordered random vectors

$$X_{[1,n]}=(X_{[1.n]}^{(1)},\ldots,X_{[1.n]}^{(r)}),\ldots,\ X_{[n,n]}=(X_{[n.n]}^{(1)},\ldots,X_{[n.n]}^{(r)}),$$

such that

$$f(X_{[1.n]}^{(1)},\ldots,X_{[1.n]}^{(r)})\le\ldots\le f(X_{[n.n]}^{(1)},\ldots,X_{[n.n]}^{(r)}).$$

Of the most interest are *f*-induced order vectors corresponding to the functions $f(x)=||x||$, where $||x||$ denotes some norm of the vector $x=(x^{(1)},\ldots,x^{(r)})$, $f(x)=x^{(1)}+\ldots+x^{(r)}$ and $f(x)=x^{(k)}$, where $x^{(k)}$ is the kth component of the vector x. If r=2 and $f(x)=x^{(1)}$ then *f*-induced order statistics coincide with the classical induced order statistics (concomitants of order statistics), which were suggested independently by David (1973) and Bhattacharya (1974). The approach of David and Bhattacharya will be studied in the next section. To complete this section we mention that Egorov and Nevzorov (1981) considered normalized sums

$$S_n=B_n\left(\sum_{j=1}^{n-k} X_{[i,n]}-M_n\right)$$

of *f*-induced order vectors, where M_n is a sequence of centering vectors and B_n is a sequence of normalizing positive-definite matrices, under the condition that

$$\lim_{n\to\infty}\frac{1}{n}=\alpha$$

and

$$\lim_{n\to\infty} \frac{n-k}{n} = \beta \ (0 \le \alpha < \beta \le 1).$$

They have found sufficient conditions, which provide the asymptotic normality of the vectors S_n, as $n\to\infty$.

5.3. CONCOMITANTS OF ORDER STATISTICS

Now we describe the classical approach to induced order statistics, which was suggested independently by David (1973) and Bhattacharya (1974). Note that David has used another term (concomitants of order statistics) for induced order statistics. Up to now both the terms are used in statistics. For example, the list of references in the recent review " Concomitants of order statistics" prepared by David and Nagaraja (1998) includes 12 papers with the term "induced order statistics" in their titles and 26 papers with the term "concomitant". We also will use both the terms.

Let (X,Y), (X_1,Y_1), (X_2,Y_2),..., (X_n, Y_n) be independent bivariate random vectors with a common joint distribution function $F(x,y)$. To simplify the exposition of the material we will assume that (X,Y) has a density distribution function $f(x,y)$.

Denote marginal distribution functions and density functions of components X and Y as $G(x)$ and $g(x)$ (for X) , $H(x)$ and $h(x)$ (for Y).

Let $D(1)$,...,$D(n)$ be antiranks based on X's. It means that $D(k)$, $k=1,2,\ldots,n$, are defined by equalities

$$\{D(r)=m\}=\{X_{r,n}=X_m\},\ 1\le r\le n,\ 1\le m\le n .$$

Since in our situation X's are i.i.d. random variables,

$$P\{X_{r,n}=X_m\}=1/n$$

for any $1\le r\le n$, $1\le m\le n$, and each antirank $D(r)$ takes values $1,2,\ldots,n$ with equal probabilities.

Now let us order vectors (X_k,Y_k), $k=1,\ldots,n$, by increasing values of X's. We get a sequence of dependent vectors, the first components of which form a variational series $X_{1,n}\le X_{2,n}\le\ldots\le X_{n,n}$:

$$(X_{k,n}, Y_{D(k)}),\ k=1,2,\ldots,n.$$

The second components

$Y_{D(k)}$, k=1,2,...,n,

of the ordered vectors are said to be induced order statistics (concomitants of order statistics).We will use symbols $Y_{[k,n]}$ instead of $Y_{D(k)}$ to show that these statistics correspond to a two-dimensional sample of a fixed size n.

Distributions of concomitants

It is easy to find using the standard methods of order statistics that the joint density function of random variables $X_{1,n},\ldots,X_{n,n}$ and $Y_{[1,n]},\ldots,Y_{[n,n]}$ has the following form:

$$f_{X_{1,n},\ldots,X_{n,n},Y_{[1,n]},\ldots,Y_{[n,n]}}(x_1,\ldots,x_n,y_1,\ldots,y_n)=n!\prod_{j=1}^{n} f(x_j,y_j), \quad (5.3.1)$$

if $x_1<x_2<\ldots<x_n$,

and

$$f_{X_{1,n},\ldots,X_{n,n},Y_{[1,n]},\ldots,Y_{[n,n]}}(x_1,\ldots,x_n,y_1,\ldots,y_n) = 0. \text{ otherwise.}$$

Now we get that the density function of concomitants $Y_{[1,n]},\ldots,Y_{[n,n]}$ is given as

$$f_{Y_{[1,n]},\ldots,Y_{[n,n]}}(y_1,\ldots,y_n)=n! \int\ldots\int_{-\infty<x_1<\ldots<x_n<\infty} \prod_{j=1}^{n} f(x_j,y_j)dx_1\ldots dx_n. \quad (5.3.2)$$

Let us find the conditional density function of concomitants given that

$X_{1,n}=x_1,\ldots,X_{n,n}=x_n$.

Since the density function of all the set of order statistics $X_{1,n},\ldots,X_{n,n}$ has the following form:

$$f_{X_{1,n},\ldots,X_{n,n}}(x_1,\ldots,x_n)=n!\prod_{j=1}^{n} g(x_j),\ -\infty<x_1<\ldots<x_n<\infty,$$

we obtain that

$$f_{Y_{[1,n]},\ldots,Y_{[n,n]}|X_{1,n},\ldots,X_{n,n}}(y_1,\ldots,y_n|x_1,\ldots,x_n)$$

$$= f_{X_{1,n},\ldots,X_{n,n},Y_{[1,n]},\ldots,Y_{[n,n]}}(x_1,\ldots,x_n,y_1,\ldots,y_n)/\, f_{X_{1,n},\ldots,X_{n,n}}(x_1,\ldots,x_n)$$

$$= \prod_{j=1}^{n} (f(x_j,y_j)/g(x_j)). \tag{5.3.3}$$

The latest equality implies that the concomitants are conditionally independent given that order statistics $X_{1,n},\ldots,X_{n,n}$ are fixed. Since the initial sequence of X's uniquely determines order statistics, we can add that the concomitants are also conditionally independent given that $X_1,\ldots,X_n$ are fixed. These results belong to Bhattacharya (1974). Note that

$$r(x,y)=f(x,y)/g(x)$$

is the conditional density function of Y given X=x. It follows from (5.3.3) that for any

$$-\infty<x_1<\ldots<x_n<\infty,$$

$$P\{Y_{[1,n]}<y_1,\ldots,Y_{[n,n]}<y_n|X_{1,n}=x_1,\ldots,X_{n,n}=x_n\}=\prod_{j=1}^{n} R(x_j,y_j), \tag{5.3.4}$$

where

$$R(x,y)=P\{Y<y|X=x\}.$$

Moreover, for any k=1,2,..., and $1\le m(1)<m(2)<\ldots<m(k)\le n$, the following equality holds:

$$P\{Y_{[m(1),n]}<y_{m(1)},\ldots,Y_{[m(k),n]}<y_{m(k)}|X_{1,n}=x_1,\ldots,X_{n,n}=x_n\}$$

$$= P\{Y_{[m(1),n]}<y_{m(1)},\ldots,Y_{[m(k),n]}<y_{m(k)}|X_{m(1),n}=x_{m(1)},\ldots,X_{m(k),n}=x_{m(k)}\}$$

$$= \prod_{j=1}^{k} R(x_{m(j)},y_{m(j)}). \tag{5.3.5}$$

It follows from (5.3.5) that

$$P\{Y_{[m(1),n]}<y_{m(1)},\ldots,Y_{[m(k),n]}<y_{m(k)}\}$$

$$=E\prod_{j=1}^{k} R(X_{m(j),n},y_{m(j)})=E\prod_{j=1}^{k} P\{Y<y|X_{m(j),n}\}$$

$$=\int\limits_{-\infty<x_{m(1)}<\ldots<x_{m(k)}<\infty}\!\!\!\cdots\!\!\!\int \prod_{j=1}^{k} R(x_{m(j)},y_{m(j)})\, f_{X_{m(1),n},\ldots,X_{m(k),n}}(x_{m(1)},\ldots,x_{m(k)})dx_{m(1)}\ldots dx_{m(k)}, \tag{5.3.6}$$

where

$f_{X_{m(1),n},\ldots,X_{m(k),n}}(x_{m(1)},\ldots,x_{m(k)})=$

$$\frac{n!}{(m(1)-1)!(m(2)-m(1)-1)!\cdots(m(k)-m(k-1)-1)!(n-m(k))!}(G(x_{m(1)}))^{m(1)-1}$$

$$(G(x_{m(2)})-G(x_{m(1)}))^{m(2)-m(1)-1}\ldots(G(x_{m(k)})-G(x_{m(k-1)}))^{m(k)-m(k-1)-1}g(x_{m(1)})\cdots g(x_{m(k)})$$

is the density function of order statistics $X_{m(1),n},\ldots,X_{m(k),n}$.

In particular,

$$P\{Y_{[m,n]}<y\}=\frac{n!}{(m-1)!(n-m)!}\int_{-\infty}^{\infty} R(x,y)(G(x))^{m-1}(1-G(x))^{n-m}g(x)dx. \tag{5.3.7}$$

Evidently, the corresponding density functions are given as follows:

$f_{Y_{[m(1),n]},\ldots,Y_{[m(k),n]}}(y_{m(1)},\ldots,y_{m(k)})=$

$$\int\limits_{-\infty<x_{m(1)}<\ldots<x_{m(k)}<\infty}\!\!\!\cdots\!\!\!\int \prod_{j=1}^{k}\frac{f(x_{m(j)},y_{m(j)})}{g(x_{m(j)})}\, f_{X_{m(1),n},\ldots,X_{m(k),n}}(x_{m(1)},\ldots,x_{m(k)})dx_{m(1)}\ldots dx_{m(k)} \tag{5.3.8}$$

and

$$f_{Y_{[m,n]}}(y)=\int_{-\infty}^{\infty}\frac{f(x,y)}{g(x)}\, f_{X_{m,n}}(x)dx=$$

$$f(x,y)(G(x))^{m-1}(1-G(x))^{n-m}dx. \qquad (5.3.9)$$

Marginal transformations of bivariate distributions

We have onsidered above concomitants $Y_{[k,n]}$, $k=1,2,\ldots n$, $n=1,2,\ldots$, based on independent random vectors (X_1,Y_1), (X_2,Y_2), ... with a common cdf $F(x,y)$ and continuous marginal pdf's $G(x)$ and $H(y)$. It is evident that distributions of antiranks $D_1,\ldots,D_n$ of order statistics $X_{1,n},\ldots,X_{n,n}$ do not change under some monotone transformations of X's. Hence such transformations of X's save the distributions of concomitants $Y_{[k,n]}=Y_{D(k)}$. Let $S(x)$ be a strictly increasing real function and we consider a sequence of vectors $(S(X_1),Y_1)$, $(S(X_2), Y_2)$,.... The new vectors have the common cdf $F(S^{-1}(x),y)$, S^{-1} being the inverse function of S. The given transformation does not change the ordering of the first components and hence the distributions of concomitants will be the same. Moreover, to guarantee this property we need the strict increase of the monotone increasing function S only for such points x which belong to the support of X's The most important for us is the case when $S(x)=G(x)$. Note that for any continuous cdf G, the probability integral transformation $U=G(X)$ gives us a random variable having the standard uniform distribution. Hence without loss of generality we can restrict ourselves by consideration of the initial vectors (X_n,Y_n), the first components of which have the standard uniform distribution.

Let us now consider vectors $(X_1, R(Y_1))$, $(X_2, R(Y_2))$,..., where $R(y)$ be an increasing real function. These vectors have the common cdf $F(x,R^{-1}(y))$, where R^{-1} is the inverse of the function R. It is not difficult to show that concomitants based on the given transformed vectors have the same distribution as the random variables $R(Y_{[1.n]}),\ldots,$ $R(Y_{[n,n]})$, where $Y_{[k,n]}$, $k=1,2,\ldots,n$, are concomitants corresponding to the initial sequence $(X_1, Y_1),\ldots, (X_n, Y_n)$. Here the most interesting is also the case $R(y)=H(y)$, which enables us to work with the uniformly distributed Y's.

Hence, if we compare vectors $(X_{k,n},Y_{[k,n]})$, corresponding to the cdf $F(x,y)$ with continuous marginal cdf's G and H, and vectors $(U_{k,n},V_{[k,n]})$, corresponding to the cdf $F_1(x,y)= F(G^{-1}(x), H^{-1}(y))$ with the standard uniform marginals, we will come to the following relations:

$$(U_{k,n},V_{[k,n]})\overset{d}{=}(G(X_{k,n}), H(Y_{[k,n]})), \qquad (5.3.10)$$

or

$$(X_{k,n}, Y_{[k,n]})\overset{d}{=} (G^{-1}(U_{k,n}), H^{-1}(V_{[k,n]})), \qquad (5.3.11)$$

where the notation

$$W\overset{d}{=}Z$$

means that the vectors (or random variables) W and Z have the same distribution.

Example 5.3.1.

Very often in the testing of independence one accepts (or rejects) the hypothesis

$$F(x,y)= G(x)H(y)$$

against one of the following alternatives:

$$F_\theta(x,y)=G(x)H(y)+\theta\Omega(G(x),H(y)),$$

where θ is a parameter and $\Omega(x,y)$ be a fixed real valued function. Consider the bivariate cdf $F_\theta(x,y)$. Let us need to investigate distributions and some properties of the corresponding record values X(n) and concomitants Y[n]. In this situation it is essentially easier for us to deal with record values U(n) and concomitants V[n] corresponding to the cdf

$$F_\theta(G^{-1}(x),H^{-1}(y))= xy+\theta\Omega(x,y),$$

and then to use the equality

$$(X(n),Y[n]) \stackrel{d}{=} (G^{-1}(U(n)), H^{-1}(V[n])).$$

Note that some properties of concomitants for the family of Farlie-Gumbel-Morgenstern distributions (the case $\Omega(x,y)=x(1-x)y(1-y)$) are discussed in Houchens (1984) and Arnold, Balakrishnan, Nagaraja (1998).

The dependence structure of concomitants

We know that $Y_{[k,n]}=Y_{D(k)}$, where D(k) is the antirank of $X_{k,n}$. If components X and Y are independent then vectors (D(1),...,D(n)) and $(Y_1,\ldots,Y_n)$ are also independent and it implies that concomitants $Y_{[1,n]},\ldots,Y_{[n,n]}$ and order statistics $X_{1,n},\ldots,X_{n,n}$ are independent. Moreover, it is easy to show in this case that

$$(Y_{[1,n]},\ldots,Y_{[n,n]}) \stackrel{d}{=} (Y_1,\ldots,Y_n).$$

Hence, the studying of concomitants is essential only for the case when X and Y are dependent. In some situations the dependence structure of X and Y can be inherited by vectors $(X_{k,n},Y_{[k,n]})$. Let us consider some important cases.

Suppose that components X_k and Y_k , k=1,2,...,n, are linked by the linear regression model

$$Y_k=a+bX_k+\xi_k,$$

where X_k and ξ_k are mutually independent. Then

$$Y_{[k,n]}=Y_{D(k)}=a+bX_{D(k)}+\xi_{D(k)}.$$

Since ξ's and antiranks D(1),...,D(n) are independent, then

$$(\xi_{D(1)},\ldots,\xi_{D(n)})\overset{d}{=}(\xi_1,\ldots,\xi_n)$$

and vectors $(\xi_{D(1)},\ldots,\xi_{D(n)})$ and

$$(Y_{D(1)},\ldots,Y_{D(n)})=(Y_{1,n},\ldots,Y_{n,n})$$

are independent. Thus, concomitants $Y_{[k,n]}$ can be represented as

$$Y_{[k,n]}=a+bX_{k,n}+\mu_k,\ k=1,2,\ldots, \tag{5.3.12}$$

where the vector $(\mu_1,\ldots,\mu_n)$ has the same distribution as $(\xi_1,\ldots,\xi_n)$ and is independent with $(X_{1,n},\ldots,X_{n,n})$.

We can consider a more general model with

$$Y_k=g(X_k,\xi_k),$$

where X's and ξ's are independent and g(x,y) is arbitrary function. The similar arguments provide the following relation for concomitants:

$$Y_{[k,n]}=g(X_{k,n},\mu_k),\ k=1,2,\ldots,n, \tag{5.3.13}$$

where the vectors $(\mu_1,\ldots,\mu_n)$ and $(X_{1,n},\ldots,X_{n,n})$. are independent and

$$(\mu_1,\ldots,\mu_n)\overset{d}{=}(\xi_1,\ldots,\xi_n).$$

The conditional independence of concomitants given that X's are fixed enables us to formulate some martingale statements. Let m(x)=E(Y|X=x) be the regression of Y on X and

$$S_{nk}=\sum_{r=1}^{k} c_r(Y_{[r,n]}-m(X_{r,n})),\ k=1,2,\ldots,n,$$

where $c_1,c_2,\ldots$ be arbitrary constants. Sen (1976) (see also Bhattacharya (1984)) proved that for each n, $S_{n,1},\ldots,S_{n,n}$ is a martingale. The simple arguments based on the conditional independence of the concomitants $Y_{[r,n]}$ enable us to show that a sequence

$$T_{nk}= S_{nk}^2-\sum_{r=1}^{k} c_r^2 E(Y_{[r,n]}-m(X_{r,n}))^2,\ k=1,2,\ldots,n,$$

also forms a martingale .

Markov structure of concomitants

As we know from (5.3.1), the joint density function of random variables $X_{1,n},\ldots,X_{n,n}$ and $Y_{[1,n]},\ldots,Y_{[n,n]}$ is given as

$$f_{X_{1,n},\ldots,X_{n,n},Y_{[1,n]},\ldots,Y_{[n,n]}}(x_1,\ldots,x_n,y_1,\ldots,y_n)=n!\prod_{j=1}^{n} f(x_j,y_j),$$

if $x_1<x_2<\ldots<x_n$. Then the similar expression for $X_{1,n},\ldots,X_{k,n}$ and $Y_{[1,n]},\ldots,Y_{[k,n]}$ is given as follows:

$$f_{X_{1,n},\ldots,X_{k,n},Y_{[1,n]},\ldots,Y_{[k,n]}}(x_1,\ldots,x_k,y_1,\ldots,y_k)=$$

$$\frac{n!}{(n-k)!}(1-G(x_k))^{n-k}\prod_{j=1}^{k} f(x_j,y_j), \tag{5.3.14}$$

if $x_1<\ldots<x_k$, and

$$f_{X_{1,n},\ldots,X_{k,n},Y_{[1,n]},\ldots,Y_{[k,n]}}(x_1,\ldots,x_k,y_1,\ldots,y_k)= 0,\text{otherwise}.$$

Now we can use (5.3.14) to find the conditional density function of the vector $(X_{k+1,n},Y_{[k+1,n]})$ given that

$$X_{j,n}=x_j,\ Y_{[j,n]}=y_j,\ j=1,2,\ldots,k.$$

It is given as follows:

$$f_{X_{k+1,n},Y_{[k+1,n]}|X_{1,n},\ldots,X_{k,n},Y_{[1,n]},\ldots,Y_{[k,n]}}(x_{k+1},y_{k+1}|\,x_1,\ldots,x_k,y_1,\ldots,y_k)=$$

$$f_{X_{1,n},\ldots,X_{k+1,n},Y_{[1,n]},\ldots,Y_{[k+1,n]}}(x_1,\ldots,x_{k+1},y_1,\ldots,y_{k+1})/\,f_{X_{1,n},\ldots,X_{k,n},Y_{[1,n]},\ldots,Y_{[k,n]}}(x_1,\ldots,x_k,y_1,\ldots,y_k)=$$

$$(n-k)(1-G(x_{k+1}))^{n-k-1}f(x_{k+1},y_{k+1})/(1-G(x_k))^{n-k},\ 1\le k\le n-1. \tag{5.3.15}$$

Now (5.3.15) implies that

$$\mathbf{f}_{X_{k+1,n},Y_{[k+1,n]}|X_{1,n},\ldots,X_{k,n},Y_{[1,n]},\ldots,Y_{[k,n]}}(\mathbf{x_{k+1},y_{k+1}|\,x_1,\ldots,x_k,y_1,\ldots,y_k})=$$

$$f_{X_{k+1,n},Y_{[k+1,n]}|X_{k,n},Y_{[k,n]}}(x_{k+1},y_{k+1}|x_k,y_k).$$

Thus , we have obtained that vectors

$$(X_{1,n},Y_{[1,n]}),\ldots,(X_{n,n},Y_{[n,n]})$$

form a bivariate Markov chain. It is known that the Markov property is valid for order statistics $X_{1,n},\ldots,X_{n,n}$, which corresponds to a continuous distribution. For many situations, say such as given in (5.3.13), the Markov property holds also for concomitants $Y_{[1,n]},\ldots,Y_{[n,n]}$, but this fact can not be guaranteed for concomitants in the general case.

Distributions of maximal values of concomitants

Since $Y_{[1,n]},\ldots,Y_{[n,n]}$ and order statistics $Y_{1,n},\ldots,Y_{n,n}$ are results of different orderings of random variables $Y_1,\ldots,Y_n$, one obtains that order statistics based on concomitants $Y_{[1,n]},\ldots,Y_{[n,n]}$ coincide with order statistics $Y_{1,n},\ldots,Y_{n,n}$. In particular,

$$\min\{Y_{[1,n]},\ldots,Y_{[n,n]}\}=Y_{n,n}$$

and

$$\max\{Y_{[1,n]},\ldots,Y_{[n,n]}\}=Y_{n,n}.$$

Useful statistics based on maximal values of concomitants can be applied in the following situation. Let X denote the score on a screening test , while Y is the score on a later test. As a result of screening test the best k out of n performers are selected for the later test. Their preliminary scores are $X_{n,n}, X_{n-1,n},\ldots,X_{n-k+1,n}$. Then

$$V_{k,n}=\max\{Y_{[n-k+1,n]},\ldots,Y_{[n,n]}\}$$

presents the best score on the later test. The ratio $EV_{k,n}/EY_{n,n}$ may be chosen as a measure of effectiveness of the screening procedure. Indeed, statistics

$$\max\{Y_{[1,n]},\ldots,Y_{[k,n]}\},\ \min\{Y_{[n-k+1,n]},\ldots,Y_{[n,n]}\},\ \min\{Y_{[1,n]},\ldots,Y_{[k,n]}\}$$

can be also applied in the analogous procedures. The extreme values of concomitants of selected order statistics were discussed in Feinberg and Huber (1996), Nagaraja (1992), Nagaraja and David (1994), Joshi and Nagaraja (1995). Below we will find the distribution of $V_{k,n}$.

Analogously (5.3.14), the joint density function of random variables $X_{n-k+1,n},\ldots,X_{n,n}$ and $Y_{[n-k+1,n]},\ldots,Y_{[n,n]}$ is given as

$$f_{X_{n-k+1,n},\ldots,X_{n,n},Y_{[n-k+1,n]},\ldots,Y_{[n,n]}}(x_{n-k+1},\ldots,x_n,y_{n-k+1},\ldots,y_n)=$$

$$\frac{n!}{(n-k)!}(G(x_k))^{n-k}\prod_{j=n-k+1}^{n} f(x_j,y_j),\ x_1<\ldots<x_n. \tag{5.3.16}$$

Then

$$P\{V_{k,n}<y\}=$$

$$\frac{n!}{(n-k)!}\int_{-\infty}^{y}\cdot\int_{-\infty}^{y}\ \int\ldots\int_{-\infty<x_{n-k+1}<\ldots<x_n<\infty}(G(x_k))^{n-k}\prod_{j=n-k+1}^{n} f(x_j,y_j)dx_{n-k+1}\ldots dx_n dy_{n-k+1}\ldots dy_n$$

$$=\frac{n!}{(n-k)!}\int_{-\infty}^{y}\int_{-\infty}^{\infty}(G(x_{n-k+1}))^{n-k} f(x_{n-k+1},y_{n-k+1})$$

$$\left(\int\ldots\int_{x_{n-k+1}<\ldots<x_n<\infty}\prod_{j=n-k+2}^{n}\left(\int_{-\infty}^{y} f(x_j,y_j)dy_j\right)dx_{n-k+2}\ldots dx_n\right)dx_{n-k+1}dy_{n-k+1}. \tag{5.3.17}$$

Using the symmetry argument we get that

$$\int\ldots\int_{x_{n-k+1}<\ldots<x_n<\infty}\prod_{j=n-k+2}^{n}\left(\int_{-\infty}^{y} f(x_j,y_j)dy_j\right)dx_{n-k+2}\ldots dx_n$$

$$=\frac{1}{(k-1)!}\prod_{j=n-k+2}^{n}\left(\int_{x_{n-k+1}}^{\infty}\int_{-\infty}^{y} f(x_j,y_j)dy_j=\frac{1}{(k-1)!}(P\{X>x_{n-k+1},Y<y\})^{k-1}\right.$$

and then

$P\{V_{k,n}<y\}$

$$=\frac{n!}{(n-k)!(k-1)!}\int_{-\infty}^{\infty}(G(x))^{n-k}(P\{X>x,Y<y\})^{k-1}\int_{-\infty}^{y} f(x,v)\,dvdx$$

$$=-\frac{n!}{(n-k)!k!}\int_{-\infty}^{\infty}(G(x))^{n-k}d_x(S^k(x,y)), \tag{5.3.18}$$

where

$S(x,y)=P\{X>x,Y<y\}=H(y)-F(x,y).$

The RHS of (5.3.18) coincides with

$S^n(-\infty,y)-S^n(\infty,y)=H^n(y),$

if $k = n$, and it can be rewritten as

$$\frac{n!}{k!(n-k-1)!}\int_{-\infty}^{\infty} S^k(x,y)\,(G(x))^{n-k-1}dG(x), \tag{5.3.19}$$

if $k<n$.

Note that if components X and Y are independent, then

$S(x,y)=(1-G(x))H(y)$

and the expression (5.3.19) can be transformed as follows:

$$\frac{n!}{k!(n-k-1)!}\int_{-\infty}^{\infty} S^k(x,y)\,(G(x))^{n-k-1}dG(x)$$

$$= \frac{n!}{k!(n-k-1)!} H^k(y) \int_{-\infty}^{\infty} (1-G(x))^k (G(x))^{n-k-1}\, dG(x)$$

$$= \frac{n!}{k!(n-k-1)!} H^k(y) \int_0^1 u^{n-k-1}(1-u)^k du = H^k(y) = P\{\max(Y_{n-k+1},\ldots,Y_n)<y\}.$$

Number of records in sequences of concomitants

Let N(n) denote a number of records amongst n concomitants

$Y_{[k,n]}$, k=1,2,...,n,

and let

$E_n = EN(n)$.

We know that independence of the components X and Y implies that $Y_{[1,n]},\ldots, Y_{[n,n]}$ are independent and have the common distribution function H(y). In this situation one can use Lemma 4.2.1, the statement of which enables us to find that record indicators

$\xi_k = 1_{\{Y_{[k,n]} \text{ is a record value}\}}$, k=1,2,...,n,

are independent and

$P\{\xi_k=1\}=1- P\{\xi_k=0\}=1/k$.

Then

$N(n)= \xi_1+\ldots+\xi_n$

and

$E_n=1+1/2+\ldots+1/n$, n=1,2, ... (5.3.20)

The next theorem gives a general formula for EN(n).

Theorem 5.3.1.

For any n=1,2,...,

$$E_n=n\int_{-\infty}^{\infty}\int_{-\infty}^{\infty} f(x,y)(1-G(x)+F(x,y))^{n-1}dydx. \tag{5.3.21}$$

Proof.

It is clear that

$$E_n=\sum_{k=1}^{n} P\{\xi_k=1\}.$$

We need now to find probabilities

$p_k= P\{\xi_k=1\}$, $k=1,2,\ldots,n$.

We obtain from (5.3.14) that

$p_k=P\{Y_{[k,n]}>\max(Y_{[1,n]},\ldots,Y_{[k-1,n]})\}=$

$$\frac{n!}{(n-k)!}\int\cdots\int_{-\infty<x_1<\ldots<x_k<\infty}(1-G(x_k))^{n-k}\left(\int_{-\infty}^{\infty} f(x_k,y_k)\left(\prod_{j=1}^{k-1}\int_{-\infty}^{y_k} f(x_j,y_j)dy_j\right)dy_k dx_1\ldots dx_k\right)=$$

$$\frac{n!}{(n-k)!}\int_{-\infty}^{\infty}(1-G(x_k))^{n-k}\int_{-\infty}^{\infty} f(x_k,y_k)\int\cdots\int_{-\infty<x_1<\ldots<x_k}\left(\prod_{j=1}^{k-1}\int_{-\infty}^{y_k} f(x_j,y_j)dy_j)dx_1\ldots dx_{k-1}\right)dy_k dx_k. \tag{5.3.22}$$

Since

$$\int\cdots\int_{-\infty<x_1<\ldots<x_k}\left(\prod_{j=1}^{k-1}\int_{-\infty}^{y_k} f(x_j,y_j)dy_j\right)dx_1\ldots dx_{k-1}$$

$$=\frac{1}{(k-1)!}\prod_{j=1}^{k-1}\int_{-\infty}^{x_k}\int_{-\infty}^{y_k} f(x_j,y_j)dx_j dy_j$$

$$=\frac{1}{(k-1)!}(F(x_k,y_k))^{k-1},$$

the RHS of (5.3.22) can be simplified and we obtain that

$$p_k = \frac{n!}{(k-1)!(n-k)!} \int_{-\infty}^{\infty} (1-G(x))^{n-k} \int_{-\infty}^{\infty} f(x,y)\, (F(x,y))^{k-1} dydx. \tag{5.3.23}$$

Some natural simplifications enable us to get the required result :

$$E_n = \sum_{k=1}^{n} p_k = n \int_{-\infty}^{\infty} \int_{-\infty}^{\infty} f(x,y) \sum_{k=1}^{n} \binom{n-1}{k-1} (1-G(x))^{n-k} (F(x,y))^{k-1} dydx =$$

$$n \int_{-\infty}^{\infty} \int_{-\infty}^{\infty} f(x,y) \sum_{k=0}^{n-1} \binom{n-1}{k} (1-G(x))^{n-1-k} (F(x,y))^{k} dydx$$

$$= n \int_{-\infty}^{\infty} \int_{-\infty}^{\infty} f(x,y)(1-G(x)+F(x,y))^{n-1} dydx.$$

The proof is completed.

Note that if the components X and Y are independent then

$F(x,y)=G(x)H(y)$, $f(x,y)=g(x)h(y)$

and we obtain from (5.3.23) that

$$P\{\xi_k=1\} = \frac{n!}{(k-1)!(n-k)!} \int_{-\infty}^{\infty} g(x)(G(x))^{k-1} (1-G(x))^{n-k} dx \int_{-\infty}^{\infty} h(y)\, (H(y))^{k-1} dy =$$

$$\frac{n!}{(k-1)!(n-k)!} \int_{0}^{1} u^{k-1}(1-u)^{n-k} dx \int_{0}^{1} v^{k-1} dv = 1/k,$$

which is equivalent to (5.3.20).

Ranks of concomitants

Let us consider two orderings of the second components of independent identically distributed vectors $(X_1,Y_1),\ldots,(X_n,Y_n)$, results of which are the classical order statistics $Y_{k,n}$, $k=1,2,\ldots$, and induced order statistics (concomitants) $Y_{[k,n]}$, $k=1,2,\ldots$. Indeed, the arrangement of $Y_{[k,n]}$ in order of growth of their values gives the sequence $Y_{1,n},\ldots,Y_{n,n}$. It is interesting to compare the location of Y's in both the sequences. Note that for any m and k, $1\le m\le n$, $1\le k\le n$,

$$P\{Y_m=Y_{[k,n]}\}=P\{X_m=X_{k,n}\}=P\{Y_m=Y_{k,n}\}=1/n.$$

One can see that ranks

$$R(Y_m)=\sum_{j=1}^{n} 1_{\{Y_j\le Y_m\}}$$

and

$$R(X_m)=\sum_{j=1}^{n} 1_{\{X_j\le X_m\}}$$

are uniformly distributed on the set $\{1,2,\ldots,n\}$. Below we will find expressions for probabilities

$$p_{m,s,t}=P\{r(X_m)=s,\ R(Y_m)=t\}=P\{X_m=X_{s,n}, Y_m=Y_{t,n}\}=P\{X_m=X_{s,n}, Y_{[m,n]}=Y_{t,n}\}. \quad (5.3.24)$$

This enables us to calculate ranks

$$\pi_{s,n}=\sum_{j=1}^{n} 1_{\{Y_{[j,n]}\le Y_{[s,n]}\}}=\sum_{j=1}^{n} 1_{\{Y_j\le Y_{[s,n]}\}}$$

of concomitants, since

$$P\{\pi_{s,n}=t\}=\sum_{m=1}^{n} p_{m,s,t}. \quad (5.3.25)$$

It is not difficult to verify that

$$p_{m,s,t}=\int_{-\infty}^{\infty}\int_{-\infty}^{\infty} P\{A_{x,y}^{s-1,t-1}\}f(x,y)dxdy, \tag{5.3.26}$$

where f(x,y) is a bivariate density function of the vector (X_m, Y_m) and

$A_{x,y}^{s-1,t-1}=$

{exactly s-1 of r.v.'s $X_1,\ldots,X_{m-1},X_{m+1},\ldots,X_n$ are less than x and exactly t-1 of r.v.'s $Y_1,\ldots,Y_{m-1},Y_{m+1},\ldots,Y_n$ are less than y}.

There are four possible options for each vector (X_j,Y_j):

$\{X_j<x,Y_j<y\}$, $\{X_j<x,Y_j\geq y\}$, $\{X_j\geq x,Y_j<y\}$, $\{X_j\geq x, Y_j\geq y\}$.

Denote

$p_1=P\{X<x,Y<y\}$, $p_2=P\{X<x,Y\geq y\}$, $p_3=P\{X\geq x, Y<y\}$

and

$p_4=P\{X\geq x, Y\geq y\}=1-p_1-p_2-p_3$.

Then probabilities of events $A_{x,y}^{s-1,t-1}$ are given as follows:

$$P\{A_{x,y}^{s-1,t-1}\}=\sum_{\alpha_1,\alpha_2,\alpha_3,\alpha_4} C(\alpha_1,\alpha_2,\alpha_3,\alpha_4)p_1^{\alpha_1}\, p_2^{\alpha_2}\, p_3^{\alpha_3}\, p_4^{\alpha_4}, \tag{5.3.27}$$

where

$C(\alpha_1,\alpha_2,\alpha_3,\alpha_4)=(n-1)!/\ \alpha_1!\alpha_2!\alpha_3!\alpha_4!$

and summation in (5.3.27) is taken over all negative integers $\alpha_1,\alpha_2,\alpha_3,\alpha_4$ such that

$\alpha_1+\alpha_2+\alpha_3+\alpha_4=n-1$, $\alpha_1+\alpha_2=s-1$, $\alpha_1+\alpha_3=t-1$.

From (5.3.26) and (5.3.27) we obtain the necessary expressions for probabilities $p_{m,s,t}$ and find that they do not depend on n. Then distributions of ranks $\pi_{s,n}$ are given as follows:

$$P\{\pi_{s,n}=t\}=n\int_{-\infty}^{\infty}\int_{-\infty}^{\infty} P\{A_{x,y}^{s-1,t-1}\}f(x,y)dxdy, \tag{5.3.28}$$

with $P\{A_{x,y}^{s-1,t-1}\}$ defined in (5.3.27). For example, we find from (5.3.28) that

$$P\{\pi_{n,n}=n\}=n\int_{-\infty}^{\infty}\int_{-\infty}^{\infty} F^{n-1}(x,y)f(x,y)dxdy. \tag{5.3.28}$$

The RHS of (5.3.28) gives the probability that $Y_{[n,n]}=Y_{n,n}$. Symmetrically we can write that

$$P\{\pi_{1,n}=1\}=P\{Y_{[1,n]}=Y_{1,n}\}=n\int_{-\infty}^{\infty}\int_{-\infty}^{\infty} (1-P\{X>x,Y>y\})^{n-1} f(x,y)dxdy.$$

The similar arguments enable us to obtain the expression for

$P\{\pi_{1,n}=1,\ \pi_{n,n}=n\}$.

Easy to see that

$P\{\pi_{1,n}=1,\ \pi_{n,n}=n\}=P\{Y_{[1,n]}=Y_{1,n},\ Y_{[n,n]}=Y_{n,n}\}=$

$$n(n-1)\int_{-\infty}^{\infty}\int_{-\infty}^{\infty}\int_{x}^{\infty}\int_{y}^{\infty} (P\{x<X<u,y<Y<v\})^{n-2}f(x,y)f(u,v)dvdudydx. \tag{5.3.29}$$

For some important distributions such as bivariate normal there are tables of the probabilities $P\{\pi_{s,n}=t\}$ (see, for example, David, O'Connell and Yang (1977)).

Moments of concomitants

It follows from (5.3.9) that

$$E(Y_{[m,n]})^{\alpha}=\int_{-\infty}^{\infty} y^{\alpha}f_{Y_{[m,n]}}(y)=\int_{-\infty}^{\infty} \frac{f(x,y)}{g(x)}\, f_{X_{m,n}}(x)dx=$$

$$\int_{-\infty}^{\infty} \frac{1}{g(x)} f_{X_{m,n}}(x) \int_{-\infty}^{\infty} y^{\alpha} f(x,y)dydx=$$

$$\frac{n!}{(m-1)!(n-m)!} \int_{-\infty}^{\infty} (G(x))^{m-1}(1-G(x))^{n-m} \int_{-\infty}^{\infty} y^{\alpha} f(x,y)dydx. \tag{5.3.30}$$

In particular,

$$EY_{[m,n]}= \frac{n!}{(m-1)!(n-m)!} \int_{-\infty}^{\infty} (G(x))^{m-1}(1-G(x))^{n-m} \int_{-\infty}^{\infty} yf(x,y)dydx. \tag{5.3.31}$$

Earlier we have introduced the regression of Y on X:

$$m(x)=E(Y|X=x)= \int_{-\infty}^{\infty} yf(x,y)/g(x)dy. \tag{5.3.32}$$

Let also

$$\sigma^2(x)=Var(Y|X=x)= \int_{-\infty}^{\infty} (y-m(x))^2 f(x,y)/g(x)dy \tag{5.3.33}$$

be the residual variance.

We can give another form of (5.3.31):

$$E(Y_{[m,n]})=E(E(Y_{[m,n]}|X_{m,n}))=Em(X_{m,n})).$$

On comparing formulae (5.3.31)-(5.3.33) one can see that the analogous expression for variances $Var(Y_{[m,n]})$ is given in terms of order statistics $X_{m,n}$ as follows:

$$Var(Y_{[m,n]})=E(\sigma^2(X_{m,n}))+Var(X_{m,n}).$$

The conditional independence of concomitants $Y_{[r,n]}$ and $Y_{[s,n]}$ given that order statistics $X_{[r,n]}$ and $X_{[s,n]}$ are fixed enables us to obtain the following useful relations for covariances:

$$cov(X_{[r,n]}, Y_{[s.n]})=cov(X_{r,n}, m(X_{s,n})),$$

$cov(Y_{[r,n]}, Y_{[s.n]})=cov(m(X_{r,n}), m(X_{s,n})), r\neq s.$

Characteristic functions of concomitants

We know from (5.3.4) that

$$P\{Y_{[1,n]}<y_1,\dots,Y_{[n,n]}<y_n|X_{1,n}=x_1,\dots,X_{n,n}=x_n\}=\prod_{j=1}^{n} R(x_j,y_j),$$

where

$$R(x,y)=P\{Y<y|X=x\}.$$

It enables us to write an expression for the joint characteristic functions of concomitants $Y_{[1,n]},\dots,Y_{[n,n]}$:

$$E\exp\{i\sum_{k=1}^{n} t_k X_{k,n}\}$$

$$=\int_{-\infty<x_1<\dots<x_n<\infty}\!\!\dots\int \left(\prod_{k=1}^{n}\int_{-\infty}^{\infty} \exp(it_k y_k)R\{x_k,dy_k\}\right)$$

$$f_{X_{1,n},\dots,X_{n,n}}(x_1,\dots,x_n)dx_1\dots dx_n$$

$$=n!\int_{-\infty<x_1<\dots<x_n<\infty}\!\!\dots\int \left(\prod_{k=1}^{n}\int_{-\infty}^{\infty} \exp(it_k y_k)\, f(x_k,y_k)dy_k\right) dx_1\dots dx_n. \qquad (5.3.34)$$

Then the characteristic function of the vector $(Y_{[1,n]},\dots,Y_{[m,n]})$ has the form

$$E\exp\{i\sum_{k=1}^{m} t_k X_{k,n}\}$$

$$=n!\int_{-\infty<x_1<\dots<x_n<\infty}\!\!\dots\int \left(\prod_{k=1}^{m}\int_{-\infty}^{\infty} \exp(it_k y_k)\, f(x_k,y_k)dy_k \prod_{k=m+1}^{n}\int_{-\infty}^{\infty} f(x_k,y_k)dy_k\right) dx_1\dots dx_n$$

$$= \frac{n!}{(n-m)!} \int \ldots \int_{-\infty < x_1 < \ldots < x_m < \infty} \left(\prod_{k=1}^{m} \int_{-\infty}^{\infty} \exp(it_k y_k)\, f(x_k, y_k) dy_k\right)(1-G(x_m))^{n-m}\, dx_1 \ldots dx_m,$$

m =1,2,...,n. (5.3.35)

Analogously one obtains that a characteristic function

$f_{m,n}(t)=E\exp(itY_{[m,n]})$, m=1,2,..., n,

is given as follows

$$f_{m,n}(t)= \int_{-\infty}^{\infty} \int_{-\infty}^{\infty} \exp(ity) R\{x,dy\}\, f_{X_{m,n}}(x)dx$$

$$= \frac{n!}{(m-1)!(n-m)!} \int_{-\infty}^{\infty} \int_{-\infty}^{\infty} \exp(ity)\, f(x,y)(G(x))^{m-1}(1-G(x))^{n-m} dydx. \quad (5.3.36)$$

Let us denote for any fixed x a characteristic function of a random variable with the cdf

$R(x,y) = P\{Y<y|X=x\}$

as $f^{(x)}(t)$. Then (5.3.35) can be rewritten as follows:

$$f_{m,n}(t)=Ef^{(X_{m,n})}(t). \quad (5.3.37)$$

From (5.3.35) one can find characteristic functions of sums

$S_{m,n}=Y_{[1,n]}+\ldots+Y_{[m,n]}$, m=1,2,...,n-1.

Using the symmetry argument we obtain that

$E\exp\{itS_{m,n}\}$

$$= \frac{n!}{(n-m)!} \int \ldots \int_{-\infty < x_1 < \ldots < x_m < \infty} \left(\prod_{k=1}^{m} \int_{-\infty}^{\infty} \exp(ity_k)\, f(x_k, y_k) dy_k\right) (1-G(x_m))^{n-m}\, dx_1 \ldots dx_m$$

$$= \frac{n!}{(n-m)!} \int_{-\infty}^{\infty} (1\text{-}G(x_m))^{n-m} \Big(\int_{-\infty<x_1<\ldots<x_m<\infty} \cdots \int \prod_{k=1}^{m-1} \int_{-\infty}^{\infty} \exp(ity_k)\, f(x_k,y_k)dy_k)\, dx_1\ldots dx_{m-1}\Big)$$

$$\int_{-\infty}^{\infty} \exp(ity_m)\, f(x_m,y_m)dy_m\, dx_m$$

$$= \frac{n!}{(m-1)!(n-m)!} \int_{-\infty}^{\infty} (1\text{-}G(x_m))^{n-m} \prod_{k=1}^{m-1} \int_{-\infty}^{\infty} \int_{-\infty}^{x_m}$$

$$\exp(ity_k)\, f(x_k,y_k)dy_k dx_k) \int_{-\infty}^{\infty} \exp(ity_m)\, f(x_m,y_m)dy_m\, dx_m$$

$$= \frac{n!}{(m-1)!(n-m)!} \int_{-\infty}^{\infty} \Big(\int_{-\infty}^{\infty} \int_{-\infty}^{x_m} \exp(ity)\, f(x,y)dydx\Big)^{m-1}$$

$$\int_{-\infty}^{\infty} \exp(ity)\, f(x_m,y)dy\, (1\text{-}G(x_m))^{n-m}\, dx_m. \tag{5.3.38}$$

On noting that

$$m\Big(\int_{-\infty}^{\infty} \int_{-\infty}^{x_m} \exp(ity)\, f(x,y)dydx\Big)^{m-1} \int_{-\infty}^{\infty} \exp(ity)\, f(x_m,y)dydx_m =$$

$$d_{x_m} \Big(\int_{-\infty}^{\infty} \int_{-\infty}^{x_m} \exp(ity)\, f(x,y)dydx\Big)^{m},$$

we can transform the RHS of (5.3.38) as

$$\frac{n!}{m!(n-m-1)!} \int_{-\infty}^{\infty} \chi^m(t,x)G^m(x)(1\text{-}G(x))^{n-m-1} g(x)dx,$$

where

$$\chi(t,x)=\int_{-\infty}^{\infty}\exp(ity)\,\frac{1}{G(x)}\int_{-\infty}^{x}f(u,y)dudy \qquad (5.3.39)$$

for any fixed x is a characteristic function, which corresponds to the cdf

$H(y,x)=P\{Y<y|X<x\}$.

All the saying enables us to represent the characteristic functions of sums $S_{m,n}$ in the following form:

$$E\exp\{itS_{m,n}\}=\int_{-\infty}^{\infty}\chi^m(t,x)f_{X_{m+1,n}}(x)dx=E\chi^m(t,X_{m+1,n}),\ m=1,\ldots n-1. \qquad (5.3.40)$$

If m = n, then

$$S_{n,n}=Y_{[1,n]}+\ldots+Y_{[n,n]}=Y_1+\ldots+Y_n$$

and

$$E\exp\{it\,S_{n,n}\}=(E\exp\{itY\})^n.$$

Asymptotic distributions of concomitants and their sums

Consider the asymptotic behavior of concomitants $Y_{[\alpha n,n]}$, $0<\alpha<1$, as $n\to\infty$. As above, we suppose that the vector (X,Y) has a joint density function f(x,y), and its components have marginal densities g(x) and h(y).

Suppose additionally that g(x) is bounded away from 0 and continuous in a neighborhood of

$$\mu_\alpha=G^{-1}(\alpha).$$

Then

$$X_{\alpha n,n}\to\mu_\alpha$$

in probability. From (5.3.37) we obtain that

$$E\exp\{itY_{[\alpha n,n]}\}=E\,f^{(X_{\alpha n,n})}(t)\to f^{\mu_\alpha}(t), \qquad (5.3.41)$$

where $f^{\mu_\alpha}(t)$ is a characteristic function , corresponding to cdf

$$R_\alpha(y)=R(\mu_\alpha,y) = P\{Y<y|X= G^{-1}(\alpha)\}.$$

It follows from (5.3.41) that then

$$P\{Y_{[\alpha n,n]}<y\}\to R_\alpha(y), \qquad (5.3.42)$$

as $n\to\infty$. The similar arguments show that if g(x) is bounded away from 0 and continuous in neighborhoods of quantiles

$$\mu_{\alpha(k)}=G^{-1}(\alpha(k)),\ k=1,2,\ldots,r,\ 0<\alpha(1)<\ldots<\alpha(r)<1,$$

then

$$P\{Y_{[\alpha(1)n,n]}<y_1,\ldots, Y_{[\alpha(r)n,n]}<y_r\}\to R_{\alpha(1)}(y_1)\cdots R_{\alpha(r)}(y_r). \qquad (5.3.43)$$

From (5.3.43) we find that concomitants of sample quantiles are asymptotically independent.

Now let us consider concomitants of extreme order statistics. We will study the asymptotic behavior of $Y_{[n-k+1,n]}$ under fixed $k=1,2,\ldots$, as $n\to\infty$.
We need to recall (5.3.7), which in our situation has the following form:

$$P\{Y_{[n-k+1,n]}<y\}=\frac{n!}{(n-k)!(k-1)!}\int_{-\infty}^{\infty} R(x,y)(G(x))^{n-k}(1-G(x))^{k-1}g(x)dx=$$

$$\int_{-\infty}^{\infty} R(x,y)dP\{X_{n-k+1,n}<x\}. \qquad (5.3.44)$$

It is known from the theory of extreme order statistics that if

$$P\{X_{n,n}-a_n<xb_n\}= (G(a_n+b_nx))^n\to T_1(x),\ n\to\infty,$$

for some sequences of normalizing constants a_n and b_n, then for any fixed $k=1,2,\ldots$,

$$P\{X_{n-k+1,n}-a_n<xb_n\} \to T_k(x)=\sum_{m=0}^{k-1} T_1(x)\{-\ln T_1(x)\}^m/m!.$$

Moreover, for any k there are only three types of possible nondegenerate limit cdf's $T_k(x)$. They are:

$$T_{1,k}(x)=\frac{1}{(k-1)!}\int_{\exp(-x)}^{\infty} e^{-t}t^{k-1}\,dt,\ -\infty<x<\infty;$$

$$T_{2,k,\alpha}(x)=\frac{1}{(k-1)!}\int_{x^{-\alpha}}^{\infty} e^{-t}t^{k-1}\,dt,\ x>0,\ \alpha>0,$$

and

$$T_{3.k,\alpha}=\frac{1}{(k-1)!}\int_{(-x)^{\alpha}}^{\infty} e^{-t}t^{k-1}\,dt,\ x<0,\ \alpha>0.$$

Assume that the distribution function G belongs to the domain of attraction of some limit cdf T_1. It means that there exist such sequences $\{a_n\}$ and $\{b_n\}$, that

$$P\{X_{n,n}-a_n<xb_n\}=(G(a_n+b_nx))^n\to T_1(x),\ n\to\infty,$$

and hence

$$P\{X_{n-k+1,n}-a_n<xb_n\}\to T_k(x)=\sum_{m=0}^{k-1} T_1(x)\{-\ln T_1(x)\}^m/m!.$$

We will suppose additionally that for some sequences $\{A_n\}$ and $\{B_n\}$ there exists such function $\vec{R}(x,y)$ that

$$R(a_n+b_nx,\ A_n+B_ny)=P\{Y<A_n+B_ny|X<a_n+b_nx)\to \vec{R}(x,y). \tag{5.3.45}$$

Then (5.3.44) and (5.3.45) imply that

$$P\{Y_{[n-k+1,n]}-A_n<yB_n\}=\int_{-\infty}^{\infty} R(x,A_n+B_ny)dP\{X_{nk+1,n}<x\}=$$

$$\int_{-\infty}^{\infty} R(a_n+b_nx,A_n+B_ny)dP\{X_{n-k+1,n}<a_n+b_nx\}\to\int_{-\infty}^{\infty}\vec{R}(x,y)dT_k(x).$$

To study the asymptotic distribution of sums

$S_{m,n}= Y_{[1,n]}+\ldots+Y_{[m,n]}$

we can use the relation (5.3.40) for the characteristic function of $S_{m,n}$:

$$E\exp\{itS_{m,n}\}=\int_{-\infty}^{\infty} \chi^{m}(t,x) f_{X_{m+1,n}}(x)dx=E\chi^{m}(t,X_{m+1,n}),$$

where $\chi(t,x)$, given in (5.3.39) is a characteristic function, which corresponds to the cdf

$H_x(y)=P\{Y<y|X<x\}=F(x,y)/G(x)$

and the density function

$$h_x(y)=\frac{1}{G(x)}\int_{-\infty}^{x} f(u,y)du.$$

From this relation we obtain that the distribution of $S_{m,n}$ coincides with a mixture of distributions of sums of independent identically distributed random variables:

$$P\{S_{m,n}<y\}=\int_{-\infty}^{\infty} P\{V_1(x)+\ldots+V_m(x)<y\}\, f_{X_{m+1,n}}(x)dx, \tag{5.3.46}$$

where $V_1(x),V_2(x),\ldots$ (for each fixed x) are independent and have a common cdf $H(y,x)$.

To explain how one can use (5.3.46) we consider the quantile case, when

$m=[\alpha n]+1,\ 0<\alpha<1.$

In this situation

$(X_{m,n}-\mu_\alpha)\, g(\mu_\alpha)n^{1/2}\, \alpha^{-1/2}(1-\alpha)^{-1/2},$

where

$\mu_\alpha=G^{-1}(\alpha),$

is asymptotically normal, as well as random variables

$(V_1(x)+\ldots+V_m(x)-mEV_1(x))/ (\mathrm{var}V_1(x))^{1/2}\, m^{1/2}$.

It enables us to approximate the RHS of (5.3.46) by the expression

$$\int_{-\infty}^{\infty} \Phi((y-n\alpha e(x))/\sigma(x)\, \alpha^{1/2}n^{1/2})d\Phi((x-\mu_\alpha)g(\mu_\alpha)n^{1/2}\alpha^{-1/2}(1-\alpha)^{-1/2}), \tag{5.3.47}$$

where

$$e(x)=EV_1(x)=\int_{-\infty}^{\infty} yh_x(y)=\frac{1}{G(x)}\int_{-\infty}^{\infty} y\int_{-\infty}^{x} f(u,y)dudy,$$

$$\sigma^2(x)=\mathrm{Var}\ V_1(x)=\frac{1}{G(x)}\int_{-\infty}^{\infty} y^2\int_{-\infty}^{x} f(u,y)dudy - e^2(x)$$

and Φ is the distribution function of the standard normal law.

The expression (5.3.47) is transformed as

$$\int_{-\infty}^{\infty} \Phi((y-n\alpha e(t(x)))/\sigma(t(x))\, \alpha^{1/2}n^{1/2})d\Phi(x), \tag{5.3.48}$$

where

$t(x)=\mu_\alpha+x\alpha^{1/2}(1-\alpha)^{1/2}/g(\mu_\alpha)n^{1/2}$.

Upon expanding

$\Phi((y-n\alpha e(t(x)))/\sigma(t(x))\, \alpha^{1/2}n^{1/2})$

in a Taylor series around the point $x^*=\mu_\alpha$, we approximate (5.3.48) by the expression

$\Phi((y-An)/Bn^{1/2})$,

where A and B do not depend on n. It implies that

$$(S_{[\alpha n]+1,n}-An)/Bn^{1/2} \tag{5.3.49}$$

is asymptotically normal. The given approach was used by Egorov and Nevzorov (1981) in a more general situation -they got the asymptotic normality of f-induced order vectors.

Bhattacharya (1974,1976) introduced random processes, based on sums $S_{m,n}$:

$$S_n(t)=S_{[nt],n}=\sum_{j=0}^{[nt]} Y_{[j,n]},\ 0\le t\le 1.$$

Let

$$S(t)=\int_{-\infty}^{G^{-1}(t)} m(x)dG(x)=\int_{-\infty}^{G^{-1}(t)} E(Y|X=x)dG(x),$$

$$s(t)=m(G^{-1}(t)),$$

$$\psi(t)=\int_{-\infty}^{G^{-1}(t)} \sigma^2(x)dG(x)=\int_{-\infty}^{G^{-1}(t)} Var(Y|X=x)dG(x),$$

$W(t)$ is a standard Brownian motion and $W^0(t)$ is a Brownian bridge independent of $B(t)$. The process $S_n(t)$ was represented as a sum of two asymptotically independent processes,

$$S_{n,1}(t)=\sum_{j=0}^{[nt]} (Y_{[j,n]}-m(X_{j,n}))$$

and

$$S_{n,2}(t)=\sum_{j=0}^{[nt]} m(X_{j,n}),$$

and Bhattacharya has proved that under some conditions the sequence of processes

$$n^{-1/2}(S_n(t)-nS(t))$$

converges weakly to a limit process, which has the form

$$W(\psi(t))+\int_0^t W^0(u)ds(u).$$

In particular, if we take $t=\alpha$, $0<\alpha<1$, then the result of Bhattacharya implies, that asymptotically

$$(S_{[n\alpha],n} - nS(\alpha))/n^{1/2}$$

has the normal

$$N(0, \psi(\alpha)+\mathrm{Var}(\int_0^{\alpha} W^0(u)ds(u)))$$

distribution. Coming back to (5.3.49), we see that this expression is valid with $A= S(\alpha)$ and

$$B=(\psi(\alpha)+\mathrm{Var}(\int_0^{\alpha} W^0(u)ds(u)))^{1/2}.$$

We have listed here only a few limit theorems related to induced order statistics and their sums. A detailed review of other limit theorems for concomitants is given by David and Nagaraja (1998).

5.4. Concomitants of Record Values

Following the definition of induced order statistics (concomitants of order statistics) let us introduce induced record values (concomitants of record values).

Consider a sequence

$$T^{(i)}=\{ X_i , Y_i \}, i = 1,2,\ldots,$$

of independent bivariate random vectors, having a common distribution function $F(x,y)$, and let $L(1)<L(2)<\ldots$ and $X(n)$, $n=1,2,\ldots$, be record times and record values, based on the sequence of random variables $X_1, X_2,\ldots$. Select the subsequence

$$T^{(L(n))} =\{X_{L(n)},Y_{L(n)}\}=\{X(n),Y_{L(n)}\} , n=1,2,\ldots,$$

from the sequence $T^{(n)}$, $n=1,2,\ldots$. Then the second components

$$Y[n]=Y_{L(n)}, n=1,2,\ldots,$$

of the selected vectors $T^{(L(n))}$ are said to be induced record values or concomitants of record values.

Indeed, if the components Y's and X's are independent, i.e. $F(x,y)=G(x)H(y)$, where $G(x)=P\{X<x\}$ and $H(y)=P\{Y<y\}$, we easily find that induced record values $Y[1],Y[2],\ldots, Y[n]$ as well as induced order statistics $Y_{[1,n]},Y_{[2,n]},\ldots,Y_{[n,n]}$ coincide in distribution with the initial random variables $Y_1,Y_2,\ldots,Y_n$ .More interesting and important case of concomitants is connected with dependent components X and Y. Some properties of concomitants Y[k] were discussed in Houchens (1984), Ahsanullah (1994, 1995), Arnold, Balakrishnan, Nagaraja (1998).

Distributions of concomitants

Let

$$T=(X,Y),\ T^{(1)}=(X_1,Y_1),\ T^{(2)}=(X_2,Y_2),\ \ldots$$

be a sequence of independent bivariate random vectors having a common c.d.f. $F(x,y)$. For the sake of convenience we assume that $F(x,y)$ is absolutely continuous and $f(x,y)$ will denote the corresponding probability density function . Let also $G(x)$ and $g(x)$ denote c.d.f. and p.d.f. of X's while $H(y)$ and $h(y)$ are c.d.f. and p.d.f. for the second components $Y_1,Y_2,\ldots$.

At first we will find a joint density function $f_n(x_1,\ldots,x_n,y_1,\ldots,y_n)$ of random variables $X(1),\ldots,X(n)$ and $Y[1],Y[2],\ldots,Y[n]$.

It is easy to see that

$$f_1(x_1,y_1)=f(x_1,y_1),$$

since

$$(X(1),Y[1])=(X_1,Y_1).$$

To find $f_2(x_1, x_2, y_1, y_2)$ note that for $x_1<x_2$ the following relation holds:

$$f_2(x_1, x_2, y_1, y_2)=\lim P\{A(\delta_1,\delta_2, \delta_3, \delta_4)\}/ \delta_1 \delta_2 \delta_3 \delta_4 , \tag{5.4.1}$$

as $\delta_k\to 0$, k=1,2,3,4, where

$$A(\delta_1,\delta_2, \delta_3, \delta_4)=\{x_1\le X_1<x_1+\delta_1,\ y_1\le Y_1<y_1+\delta_2,\ x_2\le X(2)<x_2+\delta_3,\ y_2\le Y[2]<y_2+\delta_4\}.$$

One can see that

$$P\{A(\delta_1,\delta_2,\delta_3,\delta_4)\}=\sum_{k=0}^{\infty} P\{x_1\le X_1<x_1+\delta_1,\ \max\{X_2,\ldots,X_{k+1}\}\le X_1,\ x_2\le X_{k+2}<x_2+\delta_3,$$

$$y_1\le Y_1<y_1+\delta_2,\ y_2\le Y_{k+2}<y_2+\delta_4\}. \tag{5.4.2}$$

The probability on the LHS of (5.4.2) can be estimated as follows:

$$\sum_{k=0}^{\infty} P\{x_1\le X_1<x_1+\delta_1,\ \max\{X_2,\ldots,X_{k+1}\}\le x_1,\ x_2\le X_{k+2}<x_2+\delta_3,\ y_1\le Y_1<y_1+\delta_2,$$

$$y_2\le Y_{k+2}<y_2+\delta_4\}\le P\{A(\delta_1,\delta_2,\delta_3,\delta_4)\}\le$$

$$\sum_{k=0}^{\infty} P\{x_1\le X_1<x_1+\delta_1,\ \max\{X_2,\ldots,X_{k+1}\}\le x_1+\delta_1,\ x_2\le X_{k+2}<x_2+\delta_3,$$

$$y_1\le Y_1<y_1+\delta_2,\ y_2\le Y_{k+2}<y_2+\delta_4\}. \tag{5.4.3}$$

Since vectors (X_i,Y_i) are independent and

$$P\{\max\{X_2,\ldots,X_{k+1}\}\le v\}=G^k(v),$$

we obtain from (5.4.3) that

$$\frac{P\{x_1\le X<x_1+\delta_1,\ y_1\le Y<y_1+\delta_2\}P\{x_2\le X<x_2+\delta_3,\ y_2\le Y<y_2+\delta_4\}}{1-G(x_1)}\le$$

$$P\{A(\delta_1,\delta_2,\delta_3,\delta_4)\}\le$$

$$\frac{P\{x_1\le X<x_1+\delta_1,\ y_1\le Y<y_1+\delta_2\}P\{x_2\le X<x_2+\delta_3,\ y_2\le Y<y_2+\delta_4\}}{1-G(x_1+\delta_1)}. \tag{5.4.4}$$

Relations (5.4.1) and (5.4.4) imply that

$$f_2(x_1, x_2, y_1, y_2)=f(x_1,y_1)f(x_2,y_2)/(1-G(x_1)), \tag{5.4.5}$$

if $x_1<x_2$. Evidently, $f_2(x_1, x_2, y_1, y_2)=0$, if $x_1\ge x_2$.
Analogously, we can obtain that

$$f_n(x_1,\ldots,x_n,y_1,\ldots,y_n)=\prod_{k=1}^{n-1}\frac{f(x_k,y_k)}{1-G(x_k)}\,f(x_n,y_n), \tag{5.4.6}$$

if $x_1<x_2<\ldots<x_n$, and $f_n(x_1,\ldots,x_n,y_1,\ldots,y_n)=0$ otherwise.

From (5.4.6) we get that the joint pdf $h_n(y_1,y_2,\ldots,y_n)$ of concomitants Y[1],Y[2],...,Y[n] is given by the expression

$$h_n(y_1,y_2,\ldots,y_n)=\int\limits_{-\infty<x_1<\ldots<x_n<\infty}\!\!\cdots\!\int\ \prod_{k=1}^{n-1}\frac{f(x_k,y_k)}{1-G(x_k)}\,f(x_n,y_n)\,dx_1\ldots dx_n. \tag{5.4.7}$$

It is known that the joint p.d.f. $g_n(x_1,x_2,\ldots,x_n)$ of record values X(1),...,X(n) has the following form:

$$g_n(x_1,x_2,\ldots,x_n)=\prod_{k=1}^{n-1}\frac{g(x_k)}{1-G(x_k)}\,g(x_n),\ x_1<x_2<\ldots<x_n.$$

Hence, the conditional pdf of concomitants Y[1],Y[2],...,Y[n], given that

$X(1)=x_1,\ X(2)=x_2,\ldots,X(n)=x_n\ ,x_1<x_2<\ldots<x_n,$

can be found as follows

$f_{Y[1],\ldots,Y[n]|X(1),\ldots,X(n)}\,(y_1,\ldots,y_n|x_1,\ldots,x_n)=$

$f_n(x_1,\ldots,x_n,y_1,\ldots,y_n)/\ g_n(x_1,x_2,\ldots,x_n)=$

$$\prod_{k=1}^{n}\frac{f(x_k,y_k)}{g(x_k)}=\prod_{k=1}^{n}\ h^{(x_k)}(y_k), \tag{5.4.8}$$

where

$h^{(x)}(y)=f(x,y)/g(x)$

is the conditional p.d.f. of Y given that X=x.
Equality (5.4.8) implies that concomitants Y[1],Y[2],... are conditionally independent and

$$P\{Y[1]<y_1,\ldots,Y[n]<y_n|X(1)=x_1,\ldots,X(n)=x_n\}=\prod_{k=1}^{n}\ P\{Y_k<y_k|X_k=x_k\}. \tag{5.4.9}$$

From (5.4.8) we have also that for any r=1,2,... and $1 \le k(1)<k(2)<\ldots<k(r)$ the following equality is valid:

$$f_{Y[k(1)],\ldots,Y[k(r)]|X(k(1)),\ldots,X(k(r))}(y_1,\ldots,y_r|x_1,\ldots,x_r)=\prod_{l=1}^{r} h^{(x_k)}(y_k),\ x_1<x_2<\ldots<x_n, \tag{5.4.10}$$

and the joint p.d.f. $h_{k(1),k(2),\ldots,k(r)}$ of concomitants Y[k(1)],Y[k(2)],....Y[k(r)] has the form

$$h_{k(1),\ldots,k(r)}(y_1,\ldots,y_r)=$$

$$\int\limits_{-\infty<x_1<\ldots<x_n<\infty}\!\!\ldots\!\int \prod_{l=1}^{r} h^{(x_l)}(y_l)g_{k(1),\ldots,k(r)}(x_1,\ldots,x_r)dx_1\ldots dx_r=$$

$$E\prod_{l=1}^{r} h^{(X(k(l)))}(y_l), \tag{5.4.11}$$

where $g_{k(1),\ldots,k(r)}$ is the joint p.d.f. of record values X(k(1)),X(k(2)),...,X(k(r)).
In particular, since the pdf of X(n) has the expression

$$g_{X(n)}(x)=g(x)(-\ln(1-G(x)))^{n-1}/\Gamma(n),$$

we obtain from (5.4.11) that for any n=1,2,... the p.d.f. of Y[n] is given as follows:

$$h_{Y[n]}(y)=Eh^{(X(n))}(y)=\int_{-\infty}^{\infty} f(x,y)\frac{\{-\ln(1-G(x))\}^{n-1}}{\Gamma(n)}dx. \tag{5.4.12}$$

From (5.4.6) we can get a joint pdf of random variables X(n) and Y[n]. It has the form

$$f_{X(n),Y[n]}(x,y)=f(x,y)\frac{\{-\ln(1-G(x))\}^{n-1}}{\Gamma(n)}. \tag{5.4.13}$$

Marginal transformations of bivariate distributions

Above we considered concomitants Y[n], based on independent random vectors $T^{(1)}=(X_1,Y_1)$, $T^{(2)}=(X_2,Y_2)$, ... with a common cdf F(x,y) and continuous marginal pdf's G(x) and H(y). It seems natural that distributions of Y[n], n=1,2,..., do not change under some monotone transformations of X's. In fact, let S(x) be a strictly increasing real function and we consider a sequence of vectors $(S(X_1),Y_1)$, $(S(X_2),Y_2)$,.... The new vectors have the common c.d.f. $F(S^{-1}(x),y)$, S^{-1} being the inverse function of S. The given transformation does not change the ordering of the first components and hence the distributions of concomitants will be the same. The most important is the case when S(x)=G(x). Indeed, the distribution function G can be non-strictly increasing function, but really we need the strict increase of S only for such points x which belong to the support of X's. The probability integral transformation U=G(X) gives us a random variable having the standard uniform distribution. Hence without loss of generality we can restrict ourselves by consideration of the initial vectors $T^{(1)}$, $T^{(2)}$,..., the first components of which have the standard normal distribution. Let us now consider vectors (X_1, R(Y_1)), (X_2,R(Y_2)),..., where R(y) be an increasing real function. These vectors have the common cdf $F(x,R^{-1}(y))$, where R^{-1} is the inverse of the function R. It is not difficult to show that the concomitants based on the given transformed vectors have the same distribution as the random variables R(Y[1]), R(Y[2]),..., where Y[n], n=1,2,..., are concomitants corresponding to the initial sequence $T^{(1)}$, $T^{(2)}$,.... Here again the most interesting case is R(y)=H(y), which provides us by the uniformly distributed Y's.

Hence, if we compare the vectors (X(n),Y[n]), corresponding to the cdf F(x,y) with marginal cdf 's G and H, and vectors (U(n),V[n]), corresponding to the cdf $F(G^{-1}(x), H^{-1}(y))$, we will come to the following relation:

$$\mathbf{(U(n),V[n]) \stackrel{d}{=} (G(X(n)), H(Y[n]).} \qquad (5.4.14)$$

Distribution of maximal and minimal concomitants

Now let us find distributions of maximal concomitants

M(n)=max{Y[1],...,Y[n]}.

In view of (5.4.9) and (5.4.10) we get that

$$P\{M(n)<y|X(1)=x_1,\ldots,X(n)=x_n\}=\prod_{k=1}^{n} P\{Y_k<y|X_k=x_k\}=$$

$$\prod_{k=1}^{n} \int_{-\infty}^{y} \frac{f(x_k, v)}{g(x_k)} dv$$

and

$$P\{M(n)<y\}= \int\ldots\int_{-\infty<x_1<\ldots<x_n<\infty} \left(\prod_{k=1}^{n} \int_{-\infty}^{y} \frac{f(x_k, v)}{g(x_k)} dv\right) g_n(x_1,x_2,\ldots,x_n)dx_1\ldots dx_n$$

$$= \int_{-\infty<x_1<\ldots<x_n<\infty}\!\!\!\ldots\int \left(\prod_{k=1}^{n}\int_{-\infty}^{y}\frac{f(x_k,v)}{g(x_k)}dv\right)\prod_{k=1}^{n-1}\frac{g(x_k)}{1-G(x_k)}g(x_n)dx_1\ldots dx_n$$

$$= \int_{-\infty<x_1<\ldots<x_n<\infty}\!\!\!\ldots\int \left(\prod_{k=1}^{n-1}\int_{-\infty}^{y}\frac{f(x_k,v)}{1-G(x_k)}dv\right)\int_{-\infty}^{y}f(x_n,v)dv\, dx_1\ldots dx_n$$

$$= \int_{-\infty}^{\infty}\left(\int_{-\infty}^{y}f(x_n,v)dv\right)\left(\int_{-\infty<x_1<\ldots<x_{n-1}<x_n}\!\!\!\ldots\int \left(\prod_{k=1}^{n-1}\int_{-\infty}^{y}\frac{f(x_k,v)}{1-G(x_k)}dv\right)dx_1\ldots dx_{n-1}\right)dx_n. \quad (5.4.15)$$

Taking into account the symmetry of the expression in the RHS of (5.4.15) with respect to $x_1,x_2,\ldots,x_{n-1}$, we come to the following equalities:

$$P\{M(n)<y\}=\frac{1}{(n-1)!}\int_{-\infty}^{\infty}\prod_{k=1}^{n-1}\left(\int_{-\infty}^{x_n}\left(\int_{-\infty}^{y}\frac{f(x_k,v)}{1-G(x_k)}dv\right)dx_k\right)\int_{-\infty}^{y}f(x_n,v)dvdx_n$$

$$=\frac{1}{(n-1)!}\int_{-\infty}^{\infty}\left(\int_{-\infty}^{u}\left(\int_{-\infty}^{y}\frac{f(x,v)}{1-G(x)}dv\right)dx\right)^{n-1}\int_{-\infty}^{y}f(u,v)dvdu \quad (5.4.16)$$

Denote

$$R(u,y)=\int_{-\infty}^{u}\int_{-\infty}^{y}\frac{f(x,v)}{1-G(x)}dvdx.$$

It is easy to see that

$0\le R(u,y)\le R(u,\infty)=-\ln(1-G(u))$

and for any $n=1,2,\ldots$, the following relation hold:

$\lim\,(1-G(u))R^n(u,y)=0,$

as $u\to-\infty$ or $u\to\infty$.

Now (5.4.16) can be rewritten as

$$P\{M(n)<y\}=\frac{1}{n!}\int_{-\infty}^{\infty}(1-G(u))d(R^n(u,y))=\frac{1}{n!}\int_{-\infty}^{\infty}R^n(u,y)g(u)du. \quad (5.4.17)$$

Note that if components X and Y are independent, then

$$f(x,y)=g(x)h(y),\ R(u,y)=H(y)(-\ln(1-G(y))$$

and we get the natural relation for the distribution of the maximal value:

$$P\{M(n)<y\}=\frac{1}{n!}H^n(y)\int_{-\infty}^{\infty}(-\ln(1-G(y)))^n g(u)du=$$

$$\frac{1}{n!}H^n(y)\int_{0}^{1}(-\ln(1-u))^n du=H^n(y).$$

Evidently, the minimal value for Y[1],Y[2],…,Y[n] coincides with –M*(n), where M*(n)=max{Y*[1],…,Y*[n]) is the maximal concomitant, based on vectors $(X_1,-Y_1),(X_2,-Y_2),\ldots$. Thus,

$$P\{\min\{Y[1],\ldots,Y[n]\}<y\}=1-P\{M^*(n)>-y\}=1-\frac{1}{n!}\int_{-\infty}^{\infty}(R^*(u,-y))^n g(u)du,$$

where

$$R^*(u,v)=\int_{-\infty}^{u}\int_{-\infty}^{v}\frac{f(x,-v)}{1-G(x)}dv=\int_{-\infty}^{u}\int_{-v}^{\infty}\frac{f(x,v)}{1-G(x)}dv,$$

and

$$R^*(u,-y)=\int_{-\infty}^{u}\int_{y}^{\infty}\frac{f(x,v)}{1-G(x)}dv=-\ln(1-G(u))-R(u,y).$$

Record values in a sequence of concomitants

Consider a sequence of concomitants Y[1],Y[2],…. Record values V(1), V(2),… in this sequence we define analogously X(n) in (4.1.8). Let us find distributions of V(n), n=1,2,….

Evidently, V(1)=Y[1] and the p.d.f. of V(1) is given as

$$f_{V(1)}(v)=\int_{-\infty}^{\infty}f(x,v)dv.$$

To find the joint p.d.f. of V(1) and V(2) we will use the following equality for $v_1<v_2$:

$$f_{V(1),V(2)}(v_1,v_2)=\sum_{k=2}^{\infty}\int_{-\infty}^{v_1}\cdots\int_{-\infty}^{v_1} h_{1,2,\ldots,k}(v_1,y_2,\ldots,y_{k-1},v_2)dy_2\ldots dy_{k-1}, \qquad (5.4.18)$$

where $h_{1.2....,k}$ is the joint pdf of concomitants Y[1],…,Y[k] and (see (5.4.11))
$h_{1.2....,k}(v_1,y_2,\ldots,y_{k-1},v_2)=$

$$\int\ldots\int_{-\infty<x_1<\ldots<x_k<\infty} \frac{f(x_1,v_1)}{1-G(x_1)}\frac{f(x_2,y_2)}{1-G(x_2)}\cdots\frac{f(x_{k-1},y_{k-1})}{1-G(x_{k-1})} f(x_k,v_2)dx_1\ldots dx_k.$$

It is possible to simplify the RHS of (5.4.18) and we get that

$$f_{V(1),V(2)}(v_1,v_2)=\sum_{k=2}^{\infty}\int_{-\infty}^{\infty}\frac{f(x_1,v_1)}{1-G(x_1)}\int_{x_1}^{\infty} f(x_k,v_2)$$

$$(\int\ldots\int_{x_1<\ldots<x_k}(\int_{-\infty}^{v_1}\frac{f(x_2,y_2)}{1-G(x_2)}dy_2\cdots\int_{-\infty}^{v_1}\frac{f(x_{k-1},y_{k-1})}{1-G(x_{k-1})}dy_{k-1})dx_2\ldots dx_{k-1})dx_k dx_1=$$

$$\sum_{k=2}^{\infty}\frac{1}{(k-2)!}\int_{-\infty}^{\infty}\frac{f(x_1,v_1)}{1-G(x_1)}\int_{x_1}^{\infty} f(x_k,v_2)\,(\int_{x_1}^{x_k}\int_{-\infty}^{v_1}\frac{f(x_2,y_2)}{1-G(x_2)}dy_2dx_2$$

$$\cdots\int_{x_1}^{x_k}\int_{-\infty}^{v_1}\frac{f(x_{k-1},y_{k-1})}{1-G(x_{k-1})}dy_{k-1}dx_{k-1})dx_kdx_1$$

$$=\int_{-\infty}^{\infty}\frac{f(s,v_1)}{1-G(s)}\int_{s}^{\infty} f(t,v_2)\,(\sum_{k=2}^{\infty}\frac{1}{(k-2)!}(\int_{s}^{t}\int_{-\infty}^{v_1}\frac{f(x,y)}{1-G(x)}dydx)^{k-2})dtds. \qquad (5.4.19)$$

Denote

$$I(s,t,v)=\int_{s}^{t}\int_{-\infty}^{v}\frac{f(x,y)}{1-G(x)}dydx\,.$$

Then we finally have from (5.4.19) that

$$f_{V(1),V(2)}(v_1,v_2)=\int_{-\infty}^{\infty}\frac{f(s,v_1)}{1-G(s)}\int_{s}^{\infty} f(t,v_2)\exp\{ I(s,t,v_1)\}\, dtds$$

$$=\iint_{-\infty<s<t<\infty}\frac{f(s,v_1)}{1-G(s)} f(t,v_2)\exp\{ I(s,t,v_1)\}\, dtds, \tag{5.4.20}$$

for any $v_1<v_2$.

Analogously we can get that the joint p.d.f. of V(1),…,V(n) is given as follows:

$f_{V(1),\dots V(n)}(v_1,\dots,v_n)$

$$=\int_{-\infty<x_1<\dots<x_n<\infty}\!\!\dots\int\frac{f(x_1,v_1)\,f(x_2,v_2)\cdots f(x_n,v_n)}{(1-G(x_1))\cdots(1-G(x_{n-1}))}\exp\{\sum_{r=1}^{n-1} I(x_r,x_{r+1},v_r)\}dx_1\dots dx_n, \tag{5.4.21}$$

if $v_1<v_2<\dots<v_n$, and

$f_{V(1),\dots V(n)}(v_1,\dots,v_n)= 0$, otherwise.

Markov property of concomitants

As we know (see (5.4.6)), the joint p.d.f. of random variables X(1),…,X(n), Y[1],…, Y[n] has the form

$$f_n(x_1,\dots,x_n,y_1,\dots,y_n)=\prod_{k=1}^{n-1}\frac{f(x_k,y_k)}{1-G(x_k)} f(x_n,y_n),\ -\infty<x_1<\dots<x_n<\infty.$$

Let us find the conditional pdf

$f_{X(n),Y[n]|X(1),\dots,X(n-1),Y[1],\dots,Y[n-1]}$

of the vector $\widetilde{T}(n)= (X(n),Y[n])$ given that

$\widetilde{T}(k)= (X(k),Y[k])=(x_k,y_k)$, $k=1,2,\dots,n-1$.

We obtain that

$f_{X(n),Y[n]|X(1),\dots,X(n-1),Y[1],\dots,Y[n-1]}(x_n,y_n|x_1,\dots,x_{n-1},y_1,\dots,y_{n-1})=$

$f_n(x_1,\ldots,x_n,y_1,\ldots,y_n)/ f_{n-1}(x_1,\ldots,x_{n-1},y_1,\ldots,y_{n-1})=$

$$f(x_n,y_n)/(1-G(x_{n-1})),\ -\infty<x_1<\ldots<x_n<\infty, \qquad (5.4.22)$$

depends on x_{n-1},x_n and y_n only. The given relation implies that the bivariate sequence

$$\widetilde{T}(k)= (X(k),Y[k]),\ k=1,2,\ldots$$

forms a Markov chain.

Note that if the cdf F is continuous, then it is known from the theory of records that sequences of record values X(1),X(2),... also are Markovian. One can expect that the sequence Y[1],Y[2],... inherits the Markov property of the sequences $\widetilde{T}(k)$, but the following example shows that there is no necessity for concomitants to form a Markov chain.

Example 5.4.1

Consider the partial case of Farlie-Gumbel-Morgenstern family of bivariate distributions

$$F_\theta(x,y)=G(x)H(y)+\theta\Omega(G(x),H(y))$$

with the uniform U([-1/2,1/2]) marginals and $\theta=1$. Then the joint pdf of the components X and Y has the form

$$f(x,y)=1+4xy,\ |x|\leq 1/2,\ |y|\leq 1/2.$$

The corresponding evaluations give the following pdf $h_{1,2,\ldots,k}$ of concomitants Y[1],...,Y[k] under k=2 and k=3:

$$h_{1,2}(y_1,y_2)=1+y_2+\frac{2}{3}y_1y_2$$

and

$$h_{1,2,3}(y_1,y_2,y_3)=1+y_2+\frac{3}{2}y_3+\frac{2}{3}y_1y_2+\frac{1}{3}y_1y_3+\frac{17}{9}y_2y_3+\frac{10}{9}y_1y_2y_3,$$

if $|y_k|\leq 1/2$, k=1,2,3.

Now one can find that the conditional pdf of Y[3] given that Y[1]=y_1 and Y[2]=y_2 are of the form

$$h_{Y[3]|Y[1],Y[2]}(y_3|y_1,y_2)=1+y_3+y_3(9+6y_1+16y_2+8y_1y_2)/6(3+3y_2+2y_1y_2) \qquad (5.4.23)$$

and the RHS of (5.4.23) depends on y_1. It means that Y[1],Y[2],Y[3] do not form a Markov chain.

Moments of concomitants

From (5.4.12) we can obtain the moments of Y[n]:

$$E(Y[n])^{\alpha}=\int_{-\infty}^{\infty}\frac{\{-\ln(1-G(x))\}^{n-1}}{\Gamma(n)}\int_{-\infty}^{\infty} y^{\alpha} f(x,y)\,dy\,dx. \qquad (5.4.24)$$

If we denote

$$m_{\alpha}(x)=E(Y^{\alpha}|X=x)=\int_{-\infty}^{\infty} y^{\alpha}\frac{f(x,y)}{g(x)}dy,$$

then (5.4.24) can be rewritten as

$$E(Y[n])^{\alpha}=\int_{-\infty}^{\infty} g_{X(n)}(x)\, m_{\alpha}(x)dx, \qquad (5.4.25)$$

Where

$$g_{X(n)}(x)=g(x)(-\ln(1-G(x)))^{n-1}/\Gamma(n)$$

is the p.d.f. of the record value X(n), and hence

$$E(Y[n])^{\alpha}=Em_{\alpha}(X(n)). \qquad (5.4.26)$$

In particular,

$EY[n]=Em_1(X(n))$,
$E(Y[n])^2=Em_2(X(n))$
and

$$Var(Y(n))=Em_2(X(n))-(Em_1(X(n)))^2.$$

If we introduce the function

$$\sigma^2(x)=Var(Y|X=x)=\int_{-\infty}^{\infty} y^2\frac{f(x,y)}{g(x)}dy-(\int_{-\infty}^{\infty} y\frac{f(x,y)}{g(x)}dy)^2$$

and use the equality

$$Var(Y)=Var(m_1(X))+E\sigma^2(X),$$

which can be easily proved, we get one more expression for variances of concomitants:

$$Var(Y[n])=Var(m_1(X(n)))+E\sigma^2(X(n)).$$

Analogously one can get the joint moments of concomitants. For example, for any $r=1,2,\ldots$ and $1\le k(1)<k(2)<\ldots<k(r)$ the following relation is valid:

$$EY[k(1)]Y[k(2)]\cdots Y[k(r)]= E(m_1(X(k(1)))\, m_1(X(k(2)))\cdots m_1(X(k(r)))). \qquad (5.4.27)$$

It follows from (5.4.27) that

$$Cov(Y[k(1)],Y[k(2)])=Cov(m_1(X(k(1))),m_1(X(k(2)))).$$

Characteristic functions of concomitants

The formula for characteristic functions of the concomitants also can be obtained from (5.4.13). We get that

$$E\exp\{itY(n)\}=\int_{-\infty}^{\infty}\frac{\{-\ln(1-G(x))\}^{n-1}}{\Gamma(n)}\int_{-\infty}^{\infty} e^{ity}f(x,y)dydx. \qquad (5.4.28)$$

The RHS of (5.4.28) can be represented as $EH(t,X(n))$, where

$$H(t,x)=E(\exp(itY)|X=x). \qquad (5.4.29)$$

Let the components X and Y of the vector $T=(X,Y)$ be tied by the relation $Y=s(X)+Z$, where random variables X and Z are independent. The expression for the characteristic function given above immediately gives us the following equality:

$$E\exp\{itY[n]\}=E\exp(itZ)E\exp\{its(X(n))\},$$

and it means that

$$Y[n] \stackrel{d}{=} s(X(n)) + Z,$$

where X(n) and Z are independent

References

1. Ahsanullah, M. (1994). Record values, random record models and concomitants. *J.Statist.Res.* **28**, 89-109.

2. Ahsanullah, M. (1995). *Record statistics*. Nova Science Publishers, Commack, NY.

3. Arnold, B.C., Balakrishnan, N. and Nagaraja, H.N. (1998). *Records.* John Wiley, NY.

4. Barnett, V. (1976). The ordering of multivariate data (with discussion). J. Roy. Statist. Soc. **A 139**, 318-354.

5. Barnett, V., Green, P.J. and Robinson A. (1976). Concomitants and correlation estimates. Biometrika **45**, 431-435.

6. Bhattacharya, P.K. (1974). Convergence of sample paths of normalized sums of induced order statistics. *Ann.Statist.* **2**, 1034-1039.

7. Bhattacharya, P.K. (1976). An invariance principle in regression analysis. *Ann. Statist.* **4**, 621-624.

8. Bhattacharya, P.K. (1984). Induced order statistics. Theory and applications. In: P.R. Krishnaiah and P.K. Sen, eds., Handbook of Statistics Vol.4. Elsevier, Amsterdam, 383-403.

9. Chanda, K.C. and Ruymgaart, F.H. (1992). Asymptotic normality of linear combinations of functions of concomitant order statistics. *Commun. Statist.-Theory Meth.* **21**, 3247-3254.

10. David, H.A. (1973). Concomitants of order statistics. *Bull. Internat. Statist. Inst.* **45**, 295-300.

11. David, H.A. (1994). Concomitants of extreme order statistics. In: J. Galambos et al., eds, *Extreme Value Theory and Applications,* Kluwer, Dordrecht, 211-224.

12. David, H.A. and Galambos, J. (1974). The asymptotic theory of concomitants of order statistics. *J. Appl. Probab.* **11**, 762-770.

13. David, H.A. and Nagaraja, H.N. (1998). Concomitants of order statistics. *Handbook of statistics* **16.** Eds: N. Balakrishnan and C.R. Rao, Elsevier, Amsterdam, 487-513.

14. David, H.A., O'Connell, M.J. and Yang, S.S. (1977). Distribution and expected value of the rank of a concomitant of an order statistic. *Ann. Statist.* **5**, 216-223.

15. Egorov, V.A. and Nevzorov, V.B. (1975a). Asymptotic expansions of the distribution function of sums of absolute order statistics. *Vestnik of Leningrad Univ.* **19**, 18-25.

16. Egorov, V.A. and Nevzorov, V.B. (1975b). On the rate of convergence to the normal law of linear combinations of absolute order statistics . *Theory Probab. Appl.* **20**, 207-215.

17. Egorov, V.A. and Nevzorov, V.B. (1976). Limit theorems for linear combinations of order statistics. *lectures Notes in Mathematics* **550**, 63-79. Springer Verlag, New York.

18. Egorov, V.A. and Nevzorov, V.B. (1981). Rate of convergence to the normal law of sums of induced order statistics. *Notes of Sci. Semin. LOMI* **108**, 45-56 (in Russian). English version: *J. Soviet Math.* **25** (1984), 1139-1146.

19. Egorov, V.A. and Nevzorov, V.B. (1982). Some theorems on induced order statistics. *Theory Probab. Appl.* 27, 633-639.

20. Feinberg, F.M. and Huber, J. (1996). A theory of cutoff formation under imperfect information. *Mgmt. Sci.* **42**, 65-48.

21. Goel, P.K. and Hall, P. (1994). On the average difference between concomitants and order statistics. *Ann. Probab.* **22**, 126-144.

22. Gomes, M.I. (1984). Concomitants in a multidimensional extreme model. In: J. Tiago de Oliveira, ed., *Statistical Extremes and Applications.* Reidel, Holland, 353- 364.

23. Harrell, F.E., and P.K.Sen (1979). Statistical inference for censored bivariate normal distributions based on induced order statistics. *Biometrika* **66**, 293-298.

24. Houchens, R.L. (1984). Record value theory and inference. *Ph.D. Dissertation,* University of California, Riverside, California

25. Joshi, S.N. and Nagaraja, H.N. (1995). Joint distribution of maxima of concomitants of selected order statistics. *Bernoulli* **1**, 245-255.

26. Jha, V.D. and Hossein, M.G. (1986). A note on concomitants of order statistics. *J. Ind. Soc. Agric. Statist.* **38**, 417-420.

27. Kim, S.H. and David, H.A. (1990). On the dependence structure of order statistics and concomitants of order statistics. *J. Statist. Plann. Infer.* **24**, 363-368.

28. Krishnaiah and P.K. Sen, eds, *Handbook of Statistics*, Vol.**4**. Elsevier, Amsterdam, 383-403

29. Mehra, K.L. and Upadrasta, S.P. (1992). Asymptotic normality of linear combinations induced order statistics with double weights. *Sankhya* **A 54**, 332-350.

30. Nagaraja, H.N. (1992). Some asymptotic results for the induced selection differential. *J. Appl. Probab.* **19**, 253-261.

31. Nagaraja, H.N. and H.A. David (1994). Distribution of the maximum of concomitants of selected order statistics. *Ann. Statist.* **22**, 478-494.

32. Nevzorov, V.B. and Ahsanullah, M. (2000). Some distributions of Induced Records. Biometrical Journal, 42, 1069-1081.

33. Nevzorov, V.B. and Balakrishnan, N. (1998). A record of records. *Handbook of statistics* **16.** Eds: N. Balakrishnan and C.R. Rao, Elsevier, Amsterdam, 515- 570.

34. O'Connell, M.J. and David, H.A. (1976). Order statistics and their concomitants in some double sampling situations. In S.Ikeda et al.,eds, *Essays in Probability and Statistics.* Shinko Tsusho, Tokyo. 451-466.

35. Sandstrom, A. (1987). Asymptotic normality of linear functions of concomitants of order statistics. *Metrika* **34**, 129-142.

36. Sen, P.K. (1976). A note on invariance principles for induced order statistics. *Ann. Prob.* 4. 474-479.

37. Sen, P.K. (1981). Some invariance principles for mixed rank statistics and induced order statistics and some applications. *Commun. Statist.- Theory Meth.* **10**, 1691-1718.

38. Veraverbeke, N. (1992). Asymptotic results for U-functions of concomitants of order statistics. *Statistics* **23**, 257-264.

39. Yang, S.S. (1977). General distribution theory of the concomitants of order statistics. *Ann. Statist.* **5**, 996-1002.

40. Yang, S.S. (1981). Linear functions of concomitants of order statistics with application to nonparametric estimation of a regression function. *J. Amer. Statist. Assoc.* **76**, 658-662.

41. Yang, S.S. (1981). Linear combinations of concomitants of order statistics with application to testing and nonparametric estimation of a regression function. *Ann. Inst. Statist. Math.* **33**, 463-470.

Chapter 6

GENERALIZED ORDER STATISTICS

In section 6.1 we will consider some of the basic properties of the generalized order statistics. In section 6.2 we will present different relationships between order statistics, record values and generalized order statistics and show that generalized order statistics can be considered as a "bridge" between the classical order statistics and records. Some characterizations of distributions based on properties of generalized order statistics will be given in section 6.3.

6.1. DEFINITIONS AND PROPERTIES OF GENERALIZED ORDER STATISTICS

Definitions and detailed theory of the so-called generalized order statistics (gos) were given in Kamps (1995). In this section we will give only basic properties of gos. Note that the classical order statistics, record values and sequential order statistics are special cases of generalized order statistics.
Let $\{X(1,n,m_1,k), ..., X(n,n,m_n,k); F\}$, where $k \geq 1$, $m_1, \ldots, m_n$ are real numbers and F is a distribution function, denote n generalized order statistics, which correspond to the c.d.f. F. Let also $\{U(1,n,m_1,k), ..., U(n,n,m_n,k)\}$ be a vector of the uniform gos (the case , when $F(x)=x$, $0<x<1$) and $\{Z(1,n,m_1,k), ..., Z(n,n,m_n,k)\}$ be a vector of the exponential gos, which correspond to the case $F(x)=\max\{0, 1-e^{-x}\}$. The simplest way to define the joint distribution of gos $X(1,n,m_1,k), ..., X(n,n,m_n,k)$ in the general case is to give the joint distribution of the uniform gos $U(1,n,m_1,k), ..., U(n,n,m_n,k)$ and then to use the following definition:

$$\{X(1,n,m_1,k), ..., X(n,n,m_n,k); F\} \overset{d}{=} \{G(U(1,n,m_1,k)), ..., G(U(n,n,m_n,k))\},$$

where $G(x)=F^{-1}(x)$ is the inverse of the cdf F. For the sake of simplicity we will consider gos only for absolutely continuous distributions with a cdf F and the corresponding probability density function (pdf) $f(x)=\dfrac{dF(x)}{dx}$. In this situation the joint pdf $f_{1,...,n}(x_1,...,x_n)$ of gos $X(1,n,m_1,k)$, ..., $X(n,n,m_n,k)$ can be written at once as

$$f_{1,...,n}(x_1,...,x_n) \quad = k \prod_{j=1}^{n-1} \gamma_j \prod_{i=1}^{n-1} (1-F(x_i))^{m_i} f(x_i)(1-F(x_n))^{k-1} f(x_n),$$

for $F^{-1}(0)<x_1<\cdots<x_n<F^{-1}(1)$,

and

$$f_{1,...,n}(x_1,...,x_n) = 0, \text{ otherwise,} \qquad (6.1.1)$$

where where γ_j = k+n-j +$M_j \geq 1$ for aa j = 1,..., n-1, n$\geq$ 2, and $M_j = m_1+m_2+\ldots++n-1$.
Note that the joint pdf for the uniform gos $U(1,n,m_1,k)$, ..., $U(n,n,m_n,k)$ has the form

$$f_{1,...,n}(x_1,...,x_n)=k\prod_{j=1}^{n-1}\gamma_j \prod(1-x_i)^{m_i}(1-x_n)^{K-1}, 0<x_1<...<x_n<1,$$

and

$f_{1,...,n}(x_1,...,x_n) = 0$, otherwise.

Comparing (6.1.1) with the corresponding joint pdf's for order statistics and record values we easily find that order statistics are the partial case of gos under k=1 and m_j=0, j=1,2,...,n-1, as well as record values coincide with gos if k=1 and m_j=-1, j=1,2,...,n-1. One more important partial case of gos is the case, when m_j=-1, j=1,2,...,n-1 and k=1,2,... is an integer. In this situation we deal with k-th record values. Since the most important for us are the partial cases of gos that are connected with the case of identical values of parameters $m_1,m_2,\ldots$, in the sequel we will assume $m_1= m_2 = \ldots = m_{n-1} = m$. Then γ_j = k+(n-j)(m+1) and condition $\gamma_j \geq 1$, j=1,2,...,n-1, holds if m $\geq$(1-k)/(n-1) –1.

Integrating out $x_1,\ldots, x_{r-1}, x_{r+1},\ldots ,x_n$ from (6.1.1), we get the pdf $f_{r,n,m,k}$ of X(r,n,m,k) :

$$f_{r,n,m,k}(x) = \frac{c_{r-1}}{(r-1)!}(1-F(x))^{\gamma_r-1} f(x) g_m^{r-1}(F(x)),$$

for $F^{-1}(0)<x<F^{-1}(1)$,

and

$f_{r,n,m,k}(x) = 0$, otherwise, (6.1.2)

where

$$c_{r-1} = \prod_{j=1}^{r} \gamma_j$$

and $g_m(x)$ is defined for $x \in (0,1)$ as follows:

$$g_m(x) = \frac{1}{m+1}(1-(1-x)^{m+1}) \text{ , if } m \neq -1, \text{ and}$$

$g_m(x) = -\ln(1-x)$, if $m = -1$.

Since

$$\lim_{m \to -1} \frac{1}{m+1}(1-(1-x)^{m+1}) = -\ln(1-x),$$

we will write

$$g_m(x) = \frac{1}{m+1}(1-(1-x)^{m+1}),$$

for all m with

$$g_{-1}(x) = \lim_{m \to -1} g_m(x).$$

The joint pdf of X(r,n,m,k) and X(s,n,m,k), r<s, is expressed from (6.1.1) as

$f_{r,s,n,m,k}(x,y)$

$=$

$$\frac{c_{s-1}}{(r-1)!(s-r-1)!} (\overline{F}(x))^m (g_m(F(x)))^{r-1} [g_m(F(y)) - g_m(F(x))]^{s-r-1} (\overline{F}(y))^{\gamma_s - 1} f(x) f(y),$$

for $F^{-1}(0) < x < y < F^{-1}(1)$,

and

$f_{r,s,n,m,k}(x,y) = 0$, otherwise, (6.1.3)

where $\overline{F}(x) = 1 - F(x)$.

Now consider some properties of generalized order statistics.

Theorem 6.1.1.

The sequence of gos X(1,n,m,k),…,X(n,n,m,k), which correspond to a continuous distribution function F, forms a Markov chain.

Proof.

For the sake of simplicity we prove theorem 6.1.1 only for the absolutely continuous distributions. It can easily be shown from (6.1.1) that the joint pdf of X(1,n,m,k), X(2,n,m,k),…, X(r,n,m,k), X(s,n,m,k) for $r < s$ is

$f_{1,2,\ldots,r,s,m,n,k}(x_1, x_2, \ldots, x_r, x_s)$

$$= \frac{c_{s-1}}{(s-r-1)!}\left\{\prod_{j=1}^{r}(\overline{F}(x_j))^m f(x_j)\right\}\{g_m(F(x_s)) - g_m(F(x_r))\}^{s-r-1}(\overline{F}(x_s))^{\gamma_s - 1} f(x_s),$$

for $F^{-1}(0) < x_1 < \cdots < x_r < x_s << F^{-1}(1)$,

and

$f_{1,2,\ldots,r,s,m,n,k}(x_1, x_2, \ldots, x_r, x_s) = 0$, otherwise. (6.1.4)

The joint pdf of X(1,n,m,k), X(2,n,m,k),…, X(r,n,m,k) is given as

$$f_{1,2,\ldots,r,m,n,k} = c_{r-1}\left\{\prod_{j=1}^{r-1}(\overline{F}(x_j))^m f(x_j)\right\}(\overline{F}(x_r))^{\gamma_r - 1} f(x_r),$$

for $F^{-1}(0) < x_1 < \cdots < x_r < F^{-1}(1)$,

and

$f_{1,2,\ldots,r,m,n,k} = 0$, otherwise. (6.1.5)

From (6.1.4) and (6.1.5) we obtain that the conditional pdf of X(s,n,m,k) given X(1,n,m,k) $= x_1$, X(2,n,m,k) $= x_2$,,..., X(r,n,m,k) $= x_r$ is expressed as follows:

$f_{s|1,2,..,r}(x_s|x_1,x_2,,\ldots,x_r)$

$= f_{1,2,\ldots,r,s,m,n,k}(x_1,x_2,\ldots,x_r,x_s)/\ f_{1,2,\ldots,r,m,n,k}(x_1,x_2,\ldots,x_r)$

$$= \frac{c_{s-1}\{g_m(F(x_s)) - g_m(F(x_r))\}^{s-r-1}}{(s-r-1)!\,c_{r-1}\ (\bar{F}(x_r))^{\gamma_r - m - 1}}(F(x_s))^{\gamma_s - 1}\ f(x_s),$$

for $F^{-1}(0) < x_r < x_s < F^{-1}(1)$,

and

$f_{s|1,2,..,r}(x_s|x_1,x_2,,\ldots,x_r) = 0$, otherwise. (6.1.6)

It follows from (6.1.3) and (6.1.2) that the RHS of (6.1.6) coincides with the pdf $f_{s|r}(x_s|x_r)$ of X(s,n,m,k) given X(r,n,m,k) $= x_r$. Thus the sequence {X(r,n,m,k), r =1,2,...,n} forms a Markov chain.
Mention that for $F^{-1}(0) < x_r < y << F^{-1}(1)$, the conditional pdf $f_{r+1|r}(x_{r+1}|\ x_r)$ of X(r+1,n,m,k) given X(r,n,m,k) $= x_r$ is expressed as

$$f_{r+1|r}(x_{r+1}|\ x_r) = \gamma_{r+1}\left(\frac{\bar{F}(x_{r+1})}{\bar{F}(x_r)}\right)^{\gamma_{r+1}-1}\frac{f(x_{r+1})}{\bar{F}(x_r)}$$

and the conditional probability P(X(r+1,n,m,k) $\geq$ y| X(r,n,m,k) = x) has the form

$$P(X(r+1,n,m,k) \geq y|\ X(r,n,m,k) = x) = \left(\frac{\bar{F}(y)}{\bar{F}(x)}\right)^{\gamma_{r+1}}.$$

Moments of generalized order statistics

Equalities (6.1.2) and (6.1.3) enable us to obtain single and joint moments of gos.
The mean of X(r,n,m,k) is defined as

$$\mu^{1}_{r,n,m,k} = EX(r,n,m,k) = \int_{-\infty}^{\infty} \frac{c_{r-1}}{(r-1)!} x(1-F(x))^{k+(n-r)(m+1)-1} g_m^{r-1}(F(x))\, dF(x). \tag{6.1.7}$$

The s-th moment of X(r,n,m,k) is

$$\mu^{s}_{r,n,m,k} = E(X(r,n,m,k))^s = \int_{-\infty}^{\infty} \frac{c_{r-1}}{(r-1)!} x^s (1-F(x))^{k+(n-r)(m+1)-1} g_m^{r-1}(F(x))\, dF(x). \tag{6.1.8}$$

The variance $\sigma^2_{r,r,m,n}$ of X(r,n,m,k) is defined as

$$\sigma^2_{r,r,m,n} = \mu^2_{r,n,m,k} - (\mu^1_{r,,n,m,k})^2 \tag{6.1.9}$$

The joint (m_1, m_2)th moment of X (r,n,m,k) and X(s,n,m,k) is given as follows:

$$\mu^{m_1 m_2}_{r,s,n,m,k} = \mathrm{E}(\mathrm{X}(\mathrm{r,n,m,k}))^{\mathrm{m}_1} (\mathrm{X}(\mathrm{s,n,m,k}))^{m_2} = \int_{-\infty}^{\infty} \int_{-\infty}^{y} x^{m_1} y^{m_2} \mathrm{f}_{\mathrm{r,s,n,m,k}}(x,y)\, dxdy \quad , \tag{6.1.10}$$

with joint pdf $\mathrm{f}_{\mathrm{r,s,n,m,k}}$ given in (6.1.3).

The covariance $\sigma^2_{\mathrm{r,s,n,m,k}}$ of X(r, n, m, k) and X (s, n, m, k) is expressed as

$$\sigma^2_{\mathrm{r,s,n,m,k}} = \mu^{22}_{r,s,n,m,k} - \mu^1_{r,n,m,k}\, \mu^1_{s,n,m,k}, \tag{6.1.11}$$

and the correlation coefficient of X(r,n,m,k) and X(s,n,m,k) is given by

$$\rho_{r,s,n,m,k} = \frac{\sigma^2_{r,s,m,n,k}}{\sqrt{\sigma^2_{r,n,m,k}\, \sigma^2_{s,n,m,k}}}. \tag{6.1.12}$$

Now we will compare moments of gos X(r,n,m,k) based on a cdf F and moments of a random variable X, having the cdf F. The following theorem (Kamps (1995)) is a generalization of Sen's theorem (Sen (1959)), which has been proved for the classical order statistics.

Theorem 6.1.2.

Let the probability density function of X be continuos and $E(|X|^{\delta}) < \infty$ for some $\delta > 0$. Then

$E(|X(r,n,m,k)|^{\alpha}) < \infty$,

if $r \geq \alpha/\delta$ and one of the next restrictions holds:

a) $k+(n-r)m \geq \alpha/\delta$, $m>-1$;
b) $k+(n-r)m > \alpha/\delta$, $m=-1$;
c) $k+(n-1)m \geq \alpha/\delta$, $m<-1$.

Representations for the exponential generalized order statistics

Ordered random variables generally are dependent but there are some representations which help to express them via independent r.v.'s. Let ξ_1 , ξ_2 ,... be independent exponential E(0,1) r.v.'s. We know from the previous chapters of the book that there are classical representations for the exponential order statistics $Z_{r,n}$ and the exponential record values Z(r).

Representation 6.1.1

$$\{ Z_{r,n},\ 1 \leq r \leq n\} \overset{d}{=} \{ \sum_{j=1}^{r} \frac{\xi_j}{n-j+1},\ 1 \leq r \leq n\}. \tag{6.1.13}$$

Representation 6.1.2.

$$\{Z(r),\ 1 \leq r \leq n\} \overset{d}{=} \{ \sum_{j=1}^{r} \xi_j,\ 1 \leq r \leq n\}. \tag{6.1.14}$$

Let Z(r,n,m,k) be the gos X(r,n,m,k) based on the exponential with pdf $F(x) = \max\{0. 1-e^{-x}\}$. The following result (see Kamps (1995) holds for exponential gos Z(r,n,m,k).

Representation 6.1.3.

$$\{Z(r,n,m,k),\ 1 \leq r \leq n\} \overset{d}{=} \{\sum_{j=1}^{r} \frac{\xi_j}{\gamma_j},\ 1 \leq r \leq n\}, \tag{6.1.15}$$

where $\gamma_j=k+(n-j)(m+1)$, j=1,2,...,n.

Proof.

From (6.1.1) we obtain that the joint pdf for the exponential gos has the form

$$f_{Z(1,n,m,k),\dots,Z(n,n,m,k)}(x_1,\dots,x_n) = \left(\prod_{j=1}^{n} \gamma_j\right)\exp\left\{-\sum_{i=1}^{n-1} (m+1)x_i - kx_n\right\} =$$

$$\left(\prod_{j=1}^{n} \gamma_j\right)\exp\left\{-\sum_{i=1}^{n} \gamma_i(x_i - x_{i-1})\right\} \qquad (6.1.16)$$

for $0 = x_0 < x_1 < \dots < x_n$.

We also know that ξ_1, ξ_2,... are independent exponential E(0,1) r.v.'s and hence the joint pdf $g(v_1,\dots,v_n)$ of random variables $v_j = \xi_j/\delta_j$, $j = 1,2,\dots,n$ is expressed as

$$g(v_1,\dots,v_n) = \left(\prod_{j=1}^{n} \gamma_j\right)\exp\left\{-\sum_{i=1}^{n} \gamma_i v_i\right\},\ v_1>0,\dots,v_n>0. \qquad (6.1.17).$$

To get now the joint pdf $h(x_1,x_2,\dots,x_n)$ of sums $\eta_j = v_1+\dots+v_j$, $j=1,2,\dots,n$, one has to note that $v_1=\eta_1$ and $v_j=\eta_j-\eta_{j-1}$, $j=2,3,\dots$. The corresponding transformation of co-ordinates $v_1 = x_1$, $v_2 = x_2-x_1,\dots,v_n= x_n-x_{n-1}$ with unit Jacobian enables us to obtain $h(x_1,x_2,\dots,x_n)$ from (6.1.17):

$$h(x_1,x_2,\dots,x_n) = \left(\prod_{j=1}^{n} \gamma_j\right)\exp\left\{-\sum_{i=1}^{n} \gamma_i(x_i - x_{i-1})\right\},\ 0 = x_0<x_1<\dots<x_n. \qquad (6.1.18)$$

From (6.1.16) and (6.1.18) we get that distributions of vectors

$$\{Z(r,n,m,k),\ 1\le r\le n\} \text{ and } \left\{\sum_{j=1}^{r} \frac{\xi_j}{\gamma_j},\ 1\le r\le n\right\}$$

coincide.

Corollary 6.1.1.

From the statement of the theorem 6.1.2 we can obtain immediately that random variables

$\gamma_1 Z(1,n,m,k)$, $\gamma_2(Z(2,n,m,k)-Z(1,n,m,k)),\ldots$, $\gamma_r(Z(r,n,m,k)-Z(r-1,n,m,k)),\ldots,$

$\gamma_n(Z(n,n,m,k)-Z(n-1,n,m,k))$

are independent and have the standard exponential E(0,1) distribution.

It comes from definitions of generalized order statistics that

$$\{X(1,n,m,k), \ldots, X(n,n,m,k); F\} \overset{d}{=} \{G(U(1,n,m,k)), \ldots, G(U(n,n,m,k))\}, \qquad (6.1.19)$$

where $X(r,n,m,k)$, $r = 1,\ldots,n$, denote gos based on a cdf F, $U(r,n,m,k)$, $r = 1,\ldots,n$, are the uniform gos, corresponding to the cdf $F(x)=x$, $0<x<1$, and $G(x)=F^{-1}(x)$ is the inverse of the cdf F. Note that if $F(x)=\max\{0,1-e^{-x}\}$ then (6.1.19) is equivalent to the following relation, which ties the exponential $Z(r,n,m,k)$ and the uniform $U(r,n,m,k)$ gos:

$$\{1-\exp(-Z(r,n,m,k)),\ r = 1,\ldots,n\} \overset{d}{=} \{U(r,n,m,k),\ r = 1,\ldots,n\}. \qquad (6.1.20)$$

Now denote $H(x) = F^{-1}(1-e^{-x})$. We get immediately from (6.1.15), (6.1.19) and (6.1.20) the following representation, which is valid for arbitrary gos $X(r,n,m,k)$.

Representation 6.1.4.

$$\{X(r,n,m,k), 1\le r\le n; F\} \overset{d}{=} \{H(\sum_{j=1}^{r} \frac{\xi_j}{\gamma_j}),\ 1\le r\le n\}, \qquad (6.1.21)$$

where $\gamma_j = k+(n-j)(m+1)$, $j=1,\ldots,n$.
Recall here also the similar representation for order statistics $X_{r,n}$ and record values $X(r)$, corresponding to a cdf F.

Representation 6.1.5.

$$\{X_{r,n},\ 1\le r\le n; F\} \overset{d}{=} \{H(\sum_{j=1}^{r} \frac{\xi_j}{n-j+1}),\ 1\le r\le n\}. \qquad (6.1.22)$$

Representation 6.1.6.

$$\{X(r), 1\le r\le n; F\} \overset{d}{=} \{H(\sum_{j=1}^{r} \xi_j),\ 1\le r\le n\}. \qquad (6.1.23)$$

Order statistics with non-integral sample size

Rohatgi and Saleh (1988) have brought in the so-called order statistics with non-integral sample size. These r.v.'s have no practical interpretations, but they can be interpreted (see, Kamps (1995)) as order statistics or records in some non classical models. Order statistics with non-integral sample size α present a partial case of gos, when $m_1=\ldots=m_{n-1}=0$ and $k = \alpha-n+1$ has any positive value. In this case $\gamma_r=\alpha-r+1$ and the joint pdf of order statistics $X_{r,\alpha} = X(r,n,0, \alpha-n+1)$, $r = 1,\ldots,n$, with non-integral sample size α , based on a pdf f(x), is given as

$$f_{1,\ldots,n}(x_1,\ldots,x_n)= (\prod_{i=1}^{n} (\alpha - i + 1)f(x_i))(1-F(x_n))^{\alpha-n}. \tag{6.1.24}$$

The pdf of $X_{r,\alpha}$ has the form

$$f_{r,\alpha}(x) = r\binom{\alpha}{r}F^{r-1}(x)(1-F(x))^{\alpha-r}f(x), \tag{6.1.25}$$

where

$$\binom{\alpha}{r}= \alpha(\alpha-1)\ldots(\alpha-r+1)/r!$$

is positive for all $r =1,2,\ldots,n$, since $\alpha > n-1$.
The main aim for us to introduce order statistics with non- integral values of the sample size α is to avoid the discontinuity of this parameter. Note that representations 6.1.1 and 6.1.5 stay valid if an integer n is substituted for any value $\alpha > 0$.

6.2. Relationships between Generalized Order Statistics, Order Statistics and Record Values

As one can see, generalized order statistics include some cases of ordered random variables, the most important of which are the classical order statistics (m=0, k=1) and record values (m =-1, k=1). In this section we will restrict ourselves by consideration of gos X(r,n,m,k), r=1,2,…,n, with parameters k=1 and $-1<m<0$. These generalized order statistics can be regarded as a "bridge" between records (m = -1) and order statistics (m = 0) and their probability structure enables us to explain similarity and discordance of different properties of records X(n) and order statistics $X_{r,n}$..

At first we will give some examples of such similarity and discordance.

Example 6.2.1..

Let F(x)=max {0,1-exp(-(x-a)/b)}. It is known (see Sukhatme (1937), Tata (1969)) that in this case r.v.'s

$$X_{1,n}, X_{2,n}-X_{1,n}, \ldots \qquad (6.2.1)$$

as well as differences

$$X(1), X(2)-X(1),\ldots \qquad (6.2.2)$$

are independent. Moreover, the independence of r.v.'s (6.2.1) and (6.2.2) under some mild restrictions on cdf F characterizes the exponential distribution.

Example 6.2.2.

Let F(x)=max {0,1-exp(-x/b)}. Then the following relation holds (see Rossberg (1972))

$$(n-r)(X_{r+1,n}-X_{r,n}) \stackrel{d}{=} nX_{1,n} \qquad (6.2.3)$$

and this property characterizes the distribution F(x) = max {0,1-exp(-x/b)}.

Analogously, the equality

$$X(r+1)-X(r) \stackrel{d}{=} X(1) \qquad (6.2.4)$$

characterizes the same distribution.

It happens that such analogy between characterizations fails in some other situations.

Example 6.2.3.

Lin (1988) proved that if $E(|X|^p)$ for some p>2, then

$$(EX(r))^2 \leq (r/(r-1))E(X(r-1))^2,\ 2\leq r \leq n, \qquad (6.2.5)$$

and the equality in (6.2.5) holds for the exponential distribution F(x) = max {0,1-exp(-x/b)} only. Meanwhile Lin has obtained that if $EX_1^2 < \infty$ then an analogous inequality is attained for order statistics as well:

$$E(X_{r,n})^2 \leq (rn/(r-1)(n+1))E(X_{r-1,n-1})^2,\ 2 \leq r\leq n, \qquad (6.2.6)$$

but now the equality in (6.2.6) is attained for degenerate and uniform U(0,b), b>0, distribution.

Example 6.2.4.

Compare the following results (theorems 6.2.1 and 6.2.2) for order statistics and records.

Theorem 6.2.1 (*Szekely, Mori* (1985)).

Let $0<Var(X_{r,n}) < \infty$ for some $r = i,j$, where $1 \leq i < j \leq n$. Then the correlation coefficient $\rho(X_{i,n,}X_{j,n})$ satisfies the following inequality:

$$\rho\,(X(i)\,X(j)) \leq \left(\frac{i(n+1-j)}{j(n-i+1))}\right)^{1/2} \tag{6.2.7}$$

and the equality in (6.2.7) holds if and only if X_1 has the uniform distribution.

Theorem 6.2.2. (*Nevzorov (1992)*).

Let $0<Var(Xr,n)< \infty$ for some $r = i,j$, where $1\leq i<j$. Then

$$\rho(X(i))\,X(j) \leq (i/j)^{1/2} \tag{6.2.8}$$

and the equality in (6.2.8) holds if and only if

$F(x) = \max\{0,1\text{-}\exp(-x/b)\}$.

Note that in all examples given above, properties of record values characterize the exponential distribution, meanwhile in two situations the analogous properties of order statistics correspond to exponential distributions, as it can be expected, and in other two examples we unexpectedly have got characterizations of uniform distributions. Two questions appear in this context. What is the distribution family, which connects exponential and uniform types of distributions? Can we predict what distribution is characterized by the same property of records if we know a characterization of some distribution based on a property of order statistics?

To answer the first question one has to take the so-called generalized uniform distribution (Proctor (1987))

$$F(x) = 1-(1-(x-b)/a)^c,\ b<x<b+a,\ -\infty<b<\infty,\ a>0,\ c>0. \tag{6.2.9}$$

If c=1 in (6.2.9), then
$F(x)=(x-b)/a$, $b<x<b+a$,

and one gets the uniform type of distributions.
Taking *ac* instead of *a* and letting *c* to infinity one has the exponential distribution function
$F(x)=1-\exp\{-(x-a)/b\}$, $x>a$.

One can see that distribution family (6.2.9) presents a bridge between exponential and uniform distributions. Kamps (1995) considered example 6.2.3 and constructed a family of characterizations for the generalized uniform distribution, based on inequalities of the types (6.2.5) and (6.2.6), which were obtained for gos. This family of characterizing theorems includes both the results of Lin given in example 6.2.3. Below we will get the corresponding characterization of distribution family (6.2.9) by the maximal value of correlation coefficient between two generalized order statistics. This new result will tie the theorem of Szekely and Mori and Nevzorov's theorem cited in example 6.2.4.

Both the results connected with generalizations of the situations given in examples 6.2.3 and 6.2.4 show that a characterization of the uniform distribution by a property of order statistics can imply the corresponding characterization of the exponential distribution based on records property. We will try to describe situations when the same distribution can be characterized, as in examples 6.2.1 and 6.2.2, by identical properties of records and order statistics. We also will give an explanation why "exponential" record characterizations are generated by different ("exponential" and "uniform") characterizations based on properties of order statistics.

It happens that there is one more distribution family, which includes exponential and uniform distributions. Both of these types of distributions belong to the domain of attraction of the exponential distribution, which is one of extreme (minimal) value distribution. A distribution function F is said to belong to the domain of minimal attraction of a nondegenerate cdf H if there exist sequences { an } and {bn>0} such that

$$1-(1-F(an+bnx))^n \to H(x),\ n \to \infty. \qquad (6.2.10)$$

If (6.2.10) holds , we will write $F\in D(H)$.
As it is known from the theory of extremes (see chapter 2), there are only three types of asymptotic distributions L for minimal values. They are

$$H_1(x)=H_{1,\gamma}(x)=1-\exp(-(-x)^{-\gamma}),\ \text{if } x<0, \text{ and } H1(x)=1,\ \text{if } x>0,\ \gamma>0; \qquad (6.2.11)$$

$$H_2(x) = H_{2,\gamma}(x)=1-\exp(-x^{\gamma}),\ \text{if } x>0,\ \gamma < 0, \qquad (6.2.12)$$

$$\text{And } H_3(x)=1-\exp(-\exp x),\ -\infty < x < \infty . \qquad (6.2.13)$$

For both the types of distribution

$F(x)=(x-b)/a,\ b<x<b+a,$

and

$F(x)=1-\exp\{-(x-a)/b\},\ x>a,$

one can write that $F\in D(H_{2,1})$, where $H_{2,1}(x)=1-\exp(-x)$. It will be shown below that for any pair (F, H) such that $F\in D(H)$, one can expect that a characterization of the cdf F by a property of order statistics will imply the characterization of the cdf H based on the analogous property of record values.
Similarity of properties of order statistics and record values for exponential and some other distribution families

To clear the correspondence between exponential records and order statistics we will give some corollaries of the representations for records, order statistics and generalized order statistics listed in section 6.1. The first of them evidently follows from (6.1.21) and (6.1.23).

Corollary 6.2.1.

For any continuous d.f. F ,

$$\{X(r),\ 1\le r\le n\ ;\ F\} \stackrel{d}{=} \lim\ \{\ X(r,n,m,1),\ 1\le r\le n\ ;\ F\},\ m=-1. \tag{6.2.14}$$

Corollary 6.2.2.

For any continuous d.f. F , k1 and $-1< m\le 0$,

$$\{X(r,n,m,k),\ 1\le r\le n\ ;\ F\} \stackrel{d}{=} \{Xr,n+k/(m+1)-1,\ 1\le r\le n\ ;\ F_m\}, \tag{6.2.15}$$

where order statistics with non-integral sample size $n+k/(m+1)-1$ correspond to the cdf

$$F_m(x)=1-(1-F(x))^{m+1}. \tag{6.2.16}$$

Equality (6.2.15) follows from the fact that

$$H\left(\sum_{j=1}^{r}\frac{\xi_j}{k+(n-j)(m+1)}\right)=H\left(\frac{1}{m+1}\sum_{j=1}^{r}\frac{\xi_j}{k/(m+1)+(n-j)}\right)=H_m\left(\sum_{j=1}^{r}\frac{\xi_j}{\alpha-j+1}\right),$$

where

$H_m(x)=H(x/(m+1)) = F^{-1}(1-\exp(-x/(m+1)))$, and $\alpha =.\ n+k/(m+1)-1$.

Now one can check that

$$H_m(x)=Fm^{-1}(1-\exp(-x)),$$

where F_m^{-1} is the inverse function for the cdf. $F_m(x)=1-(1-F(x))^{m+1}$.

Equalities (6.2.14) and (6.2.15) immediately imply the following result, which enables us to express distributions of record values via limit distributions of order statistics with non-integral sample sizes.

Theorem 6.2.3.

Let F be a continuous cdf. Then,

$$\{X(r),),\ 1 \le r \le n\ ;\ F\} \overset{d}{=} \lim \{X_{r,n+1/(m+1)-1}),\ 1 \le r \le n\ ;\ F_m\}, \qquad (6.2.17)$$

as $m \to -1$, where

$$F_m(x)=1-(1-F(x))^{m+1}.$$

Choosing a sequence $mN=1/(N+1)-1$, $N=1,2,\dots$, one can express distributions of record values via distributions of the classical order statistics.

Theorem 6.2.4.

As $N \to \infty$,

$$\{X(r),\ 1 \le r \le n\ ;\ F\} \overset{d}{=} \lim \{X_{r.n+N}\ 1 \le r \le n\ ;\ F_N\}, \quad (6.2.18)$$

where

$$F_N(x)=1-(1-F(x))^{1/(N+1)}$$

Comment 6.2.1.

Representation (6.2.18) and examples 6.2.1 and 6.2.2 explain why for some distributions a certain property of order statistics implies the corresponding property of records. Let $F(x)=1-\exp\{-(x-a)/b\}$, then for any $N=1,2,\dots$, $F_N(x)=1-\exp\{-(x-a)/b_N)$, where $b_N = b(N+1)$, also is the exponential cdf. From the well-known property of spacings of the exponential order statistics one obtains that for any n = 2,3,... and N = 1,2,... the random variables $\{X_{1,n+N}, X_{2,n+N}- X_{1,n+N}, ...,X_{n,n+N} -X_{n-1,n+N}, F_N\}$are independent for any exponential cdf F. The independence of r.v.'s X(1), X(2)-X(1),..., X(n)-X(n-1) follows now from (6.2.18), rewritten in the form

$$\{X(1), X(2)-X(1),\ldots,X(n)-X(n-1); F\} \overset{d}{=}$$

$$\lim\{ X_{1,n+N}, X_{2,n+N}- X_{1,n+N}, ...,X_{n,n+N} -X_{n-1,n+N}, F_N \}$$

Unfortunately we cannot get immediately from (6.2.18) that the independence of X(1), X(2)-X(1),..., X(n)-X(n-1) characterizes the exponential type of distributions. In fact, even if we know that for any N the components of the vectors $\{ X_{1,n+N}, X_{2,n+N}- X_{1,n+N}, ...,X_{n,n+N} -X_{n-1,n+N}, F_N)$ are dependent in the case when the cdf F is not exponential, it does not mean that this dependence will be saved for the limit distributions as N . Hence we can only propose and then check by other methods that inter-record values X(r)-X(r-1) are dependent for distributions which differ from the exponential distributions.

The same approach can be used in example 6.2.2, when F(x)=1-exp(-x/b) and FN(x)=1-exp{-x/b(N+1)}. If (6.2.3) is valid, then

$$\{X_{1, n+N}; F_N\} \overset{d}{=} \{(n+N-r)/(n+N)(X_{r+1,n+N}-X_{r,n+N});F_N\}, \qquad (6.2.19)$$

for any r =1,2,...,n and N =1,2,.... From (6.2.18) and (6.2.19) one obtains that

$$X(1) \overset{d}{=} X(r+1)-X(r), r=1,2,...,$$

if F(x)=1-exp(-x/b). Moreover, equalities (6.2.18) and (6.2.19) give arguments to propose (and then to check) that relation (6.2.4) characterizes the exponential distributions.

Indeed, the latter facts are well known but these examples help to illustrate the given method, which enables us to obtain results for record values from the analogous results for order statistics.

Besides exponential, there are some other distribution families, for which the inheritance of some properties of order statistics by record values becomes evident if to apply Theorem 6.2.4. Amongst such distributions we can mention the following:

F(x)=1-exp(-exp(x-a)) , for which $F_N(x)=F(x-\ln(N+1))$;

$F(x)=1-\exp(-(x/b)^{\alpha})$, x>0, b>0, α >0, with $F_N(x)=F(x/(N+1)^{1/\alpha})$ (which includes the exponential distribution, if α=1);

$F(x)=1-\exp(-(-x/b)^{-\alpha})$, $x<0, b>0$, $\alpha>0$, for which $F_N(x)=F(x(N+1)^{1/\alpha})$;
$F(x)=F^{(\alpha)}(x)=1-((x-a)/b)-$, $x>a$, $b>0$, $\alpha>0$, with $F_N(x)=F^{(\alpha/(N+1))}(x)$.

Properties of records and order statistics connected with asymptotic distributions of minimal values.
There is a correspondence between properties of record values and order statistics, which can be explained if to apply the theory of extremes. Above we have shown that both the distributions, exponential and uniform, which are characterized correspondingly by the same properties of order statistics and records (see examples 6.2.3 and 6.2.4), are members of the same family of distributions. They both belong to the domain of minimal attraction of the exponential distribution (see definition (6.2.10)). In fact, if $F(x)=1-\exp(-x)$, $x>0$, then

$$1-(1-F(x/n))^n = H_{2,1}(x),\ n=1,2,\ldots,$$

and $F\in D(H_{2,1})$, where $H_{2,1}(x)=\max\{0,1-\exp(-x)\}$. Analogously, if $F(x)=x$, $0<x<1$, then

$$1-(1-F(x/n))^n=1-(1-x/n)^n\to 1-\exp(-x) = H_{2,1}(x),\ n\to\infty$$

and $F\in D(H_{2,1})$. It happens that these similar asymptotic properties of the exponential and uniform distributions is a reason, why the analogous properties of record values and order statistics in examples 6.2.3 and 6.2.4 gives characterizations of different distributions.

The following result connects order statistics for a cdf F and record values for a cdf H, when F belongs to the domain of minimal attraction of the non-degenerate cdf H.

Theorem 6.2.5.

Let F be a continuous cdf and $F\in D(H)$, with centering and normalizing constants $\{a_n\}$ and $\{b_n\}$, where H is one of the cdf's $H_{1,\gamma}(x)$, $H_{2,\gamma}(x)$ and $H_3(x)$ defined in (6.2.11)-(6.2.13). Then for any fixed $j=1,2,\ldots$,

$$\{X(j),\ 1\le j\le r;\ H\} \overset{d}{=} \lim \{(X_{j,n}-a_n)/b_n,\ 1\le j\le r;\ F\}, \tag{6.2.20}$$

as $n\to\infty$.

Comment 6.2.2.

In fact, the statement of theorem 6.2.5 follows from the similar result of Nagaraja (1982) for maximal order statistics and lower records, but we want to suggest another simple way to prove (6.2.20) based on representations for records and order statistics.

Proof of Theorem 6.2.5.

The joint distribution function of record values X(1), X(2),..., X(r), corresponding to the cdf H, can be expressed via r.v.'s 1 , 2 ,...from representation 6.1.6, in the following form:

$T(x_1,x_2,...,x_r)=P\{X(1) \le x_1, X(2) \le x_2,..., X(r) \le x_r\}=$

$$P\{\xi_1 \le R(x_1), \xi_1+\xi_2 \le R(x_2),..., \xi_1+\xi_2+...+\xi_r \le R(x_r)\}, \qquad (6.2.21)$$

where $R(x)=-\ln(1-H(x))$.

Contrariwise , the representation 4 implies that

$T_n(x_1,x_2,...,x_r) =P\{ X_{1,n} -a_n \le x_1 b_n, X_{2,n} -a_n \le x_2 b_n,..., X_{r,n} -a_n \le x_r b_n\}$

$$= P\{\xi_1 \le R_n(x_1), \xi_1+\xi_2\frac{n}{n-1} \le R_n(x_2),..., \xi_1+\xi_2\frac{n}{n-1}+...+\xi_r\frac{n}{n-r+1} \le R_n(x_r)\}, \qquad (6.2.22)$$

where $R_n(x)= -n\ln(1-F(a_n+b_n x))$.

By comparing (6.2.21) and (6.2.22) one gets that

$| T(x_1,x_2,...,x_r)- T_n(x_1,x_2,...,x_r)| \le S_1+S_2,$

where

$$S_1= \sup_{x_1,...,x_r} | P\{\xi_1 \le R_n(x_1), \xi_1+\xi_2\frac{n}{n-1} \le R_n(x_2),..., \xi_1+\xi_2\frac{n}{n-1}+...+\xi_r\frac{n}{n-r+1} \le R_n(x_r)\} -$$

$$P\{\xi_1 \le R_n(x_1), \xi_1+\xi_2 \le R_n(x_2),..., \xi_1+\xi_2+...+\xi_r \le R_n(x_r)\}|,$$

and

$$S_2= \sup_{x_1,...,x_r} | P\{\xi_1 \le R_n(x_1), \xi_1+\xi_2 \le R_n(x_2),..., \xi_1+\xi_2+...+\xi_r \le R_n(x_r)\}-$$

$$P\{\xi_1 \le R(x_1), \xi_1+\xi_2 \le R(x_2),..., \xi_1+\xi_2+...+\xi_r \le R(x_r)\}|.$$

Since $\xi_1, \xi_2,..., \xi_r$ are independent exponentially E(0,1) distributed r.v.'s it is easy to be

convinced that for any fixed r the vector

$$(\xi_1, \xi_1+\xi_2\frac{n}{n-1}, \ldots, \xi_1+\xi_2\frac{n}{n-1}+\ldots+\xi_r\frac{n}{n-r+1})$$

converges in distribution as $n\to\infty$ to the vector $(\xi_1, \xi_1+\xi_2, \ldots, \xi_1+\xi_2+\ldots+\xi_r)$. Hence $S_1\to 0$, if $n\to\infty$. Now let us recall that $1-(1-F(a_n+b_nx))^n\to L(x)$, $n\to\infty$. It implies that

$$R_n(x)=-\ln((1-F(a_n+b_n x))^n)\to -\ln(1-L(x))=R(x),\ n\to\infty,$$

and then $S_2\to 0$. It completes the proof of the theorem.

Comment 6.2.3.

Theorem 4 explains the reason why we can have "exponential" record characterizations generated by different ("exponential" and "uniform") characterizations based on properties of order statistics . Both the distributions, exponential and uniform, belong to the domain of minimal attraction of the exponential distribution. Hence, for wide class of functionals Φ, (6.2.20) implies the convergence

$$\Phi(X_n)\to\Phi(X), \tag{6.2.23}$$

as $n\to\infty$, where

$$X_n=\{(X_{j,n}-a_n)/b_n, 1\le j\le r; F\},\quad X=\{X(j), 1\le j\le r; H\},$$

F is the exponential or uniform d.f. and H is the exponential d.f. Relation (6.2.23) immediately holds for continuous bounded functionals Φ, but in other situations we need to check this relation because it can fail if one has no additional restrictions on the initial distributions. In any case, now it seems rather natural that some property is valid for exponential records if it is true for exponential or uniform normalized order statistics. Below we give one more example of such agreement between the analogous properties of order statistics and record values for another pair (F,H) of cdf.'s.

Example 6.2.5.

Let $F(x)=\exp(x)$, $x<0$. Then $F\in D(L_3)$. It is known that components of the vector

$$(X_{2,n}-X_{1,n}, 2(X_{3,n}-X_{2,n}), \ldots, r(X_{r+1,n}-X_{r,n}); F)$$

are independent and have the same exponential E(0,1) distribution for any $r=1,2,\ldots$, and $n>r$. In this situation the r.v. $X_{1,n}$, unlike the case $F(x)=1-\exp(-x)$, $x>0$, is not independent with the given spacings of order statistics. The statement of theorem 4 immediately implies that for any $r=2,3,\ldots$ r.v.'s

$X(2)-X(1), 2(X(3)-X(2)),..., r(X(r+1)-X(r))$

are also independent and identically E(0,1) distributed for d.f. $H_3(x)=1-\exp(-\exp(x))$ (Houchens (1984)).

6.3. CHARACTERIZATIONS OF DISTRIBUTIONS BY PROPERTIES OF GENERALIZED ORDER STATISTICS.

Corollary 6.2.2 gives the tools to investigate generalized order statistics. We can rewrite (6.2.15) in the following way:

$$\{X(r,n,m,k), 1\le r\le n; F_{1/(m+1)}\} \overset{d}{=} \{X_{r,n+k/(m+1)-1}, 1\le r\le n; F\}, -1<m\le 0, \qquad (6.3.1)$$

where

$$F_{1/(m+1)}(x)=1-(1-F(x))^{1/(m+1)}. \qquad (6.3.2)$$

Very often we can results for classical order statistics rewrite (practically without any changing in proof) for order statistics with non-integral sample sizes. Then relation (6.3.1) enables us to extend these results for the case of general order statistics.

In the statistical literature there are a lot of results for the classical order statistics and only a few results are available for order statistics with non-integral sample size, but in the most of situations it is enough only to analyze the proof of a result for order statistics $X_{r,n}$ to make sure that this result can be reformulated for order statistics $X_{r,\alpha}$ with any parameter α. For instance, looking through the proof of Theorem 6.2.1 one can check that the parameter n does not need to be an integer. Hence Theorem 6.2.1 can be given in the following more general form.

Theorem 6.3.1.

Let $0<Var(X_{r,\alpha})<\infty$ for some $r=i,j$, where $1\le i<j\le\alpha$. Then the correlation coefficient $\rho(X_{i,\alpha},X_{j,\alpha})$ satisfies the following inequality:

$$\rho(X_{i,\alpha}, X_{j,\alpha}) \le (i(\alpha+1-j)/j(\alpha-i+1))^{1/2}, \qquad (6.3.3)$$

and the equality in (6.3.3) holds iff X_1 has the uniform distribution.

For the uniform cdf $F(x)=(x-a)/b$, $a<x<a+b$ and the cdf's $F_{1/(m+1)}$, $-1<m\le 0$, defined in (6.3.2) are of the form

$F_{1/(m+1)}(x)=1-(1-(x-a)/b)^{1/(m+1)}$, $a<x<a+b$. (6.3.4)

From (6.3.1) and (6.3.3) , with $\alpha= n+k/(m+1)-1$, one immediately gets a new generalizations of Theorem 6.2.1.

Theorem 6.3.2.

Let $0<Var(X(r,n,m,k))<\infty$ for some r=i,j, where $1\le i<j \le n$. Then for any $-1<m\le 0$ and $k\ge 1$, the correlation coefficient $\rho(X(i,n,m,k),X(j,n,m,k))$ satisfies the following inequality:

$$\rho(X(i,n,m,k),X(j,n,m,k)) \le (i(n+ k/(m+1) -j)/j(n+ k/(m+1) -i))^{1/2}, \quad (6.3.5)$$

and the equality in (6.3.5) holds iff gos X(r,n,m,k) correspond to cdf

$$F(x,m)= F_{1/(m+1)}(x)=1-(1-(x-a)/b)^{1/(m+1)} .$$

Comment 6.3.1.

If m in Theorem 6.3.2 tends to -1, then the RHS of (6.3.5) transforms into $(i/j)^{1/2}$, the cdf's F(x,m) , b being written as b/(m+1), converge to the exponential d.f. $F_{-1}(x)=1-\exp(-(x-a)/b)$, and we get the statement of Theorem 6.2.2 as the limit form of Theorem 6.3.2 (under k=1). Taking m=0 in Theorem 6.3.2 one comes to Theorem 6.2.1.

In Example 6.2.3 we mentioned the inequality

$$(EX_{r,n})^2 \le (rn/(r-1)(n+1))EX^2_{r-1,n-1} , \; 2 \le r \le n,$$

which was given by Lin (1988) under condition that $EX^2<\infty$. It has been shown that this inequality becomes equality for the uniform distribution. The simple proof of this fact based on applying the Cauchy-Schwarz inequality is also applicable for order statistics with non-integral sample size and it is easy to obtain the following result.

Theorem 6.3.3.

If $0<EX^2<\infty$, then

$$(EX_{r,\alpha})^2 \le (r\alpha/(r-1)(\alpha+1))EX^2_{r-1,\alpha-1} , \; 2 \le r \le \alpha, \quad (6.3.6)$$

and the equality in (6.3.6) is attained for the uniform U(0,b), b>0, distributions.

Now it follows from (6.3.1) that analogous result is valid for generalized order statistics with $-1<m\leq 0$ (more general results for moments of gos are given in Kamps (1995)).

Theorem 6.3.4.

If for some m, $-1<m\leq 0$, the relation $0<E(X(j,n,m,k))^2<\infty$ is valid for j=r-1 and j=r, then

$$(EX(r,n,m,k))^2 \leq (r((m+1)(n-1)+k)/(r-1)(n(m+1)+k))E(X(r-1,n,m,k))^2,\ 2 \leq r \leq n, \qquad (6.3.7)$$

and the equality in (6.3.6) is attained for generalized order statistics, which correspond to cdf $F_m(x)=1-(1-x/b)^{1/(m+1)}$, b>0.

Comment 6.3.2.

If m in Theorem 6.3.2 tends to -1, then the coefficient on the RHS of (6.3.7) transforms into r/(r-1), the cdf's F(x,m) , b being written as b/(m+1), converge to the exponential cdf. $F_{-1}(x)=1-\exp(-x/b)$, and we get relation (6.2.5) as the limiting form of (6.3.7) (under k=1). Taking m=0 and k=1 in (6.3.7) one gets inequality (6.2.6).

In Example 6.2.2 we discussed Rossberg's (1972) relation (6.2.3)

$$(1-r/n)(X_{r+1,n}-X_{r,n}) \overset{d}{=} X_{1,n}\,,$$

which characterizes the distribution F(x)=max {0,1-exp(-x/b)}.
There are some modifications of (6.2.3) (see, Ahsanullah (1976, 1977) and some analogous theorems for record values in the spirit of (6.2.4). For example, Ahsanullah (1977) has proved that if X_1 has a density function f and a monotone failure rate h(x)=f(x)/(1-F(x)) then the next two statements are equivalent:

(a) X_1 has d.f. F(x)= max {0,1-exp(-x/b)};

(b) for any $1\leq m<n$, $X(n)-X(m) \overset{d}{=} X(n-m)$. (6.3.8)

Below we will give some characterizations of exponential distributions based on properties of spacings of generalized order statistics. In the sequel we will denote

$$D(1,n,m,k) = \gamma_1 X(1,n,m,k)$$

and

$$D(r,n,m,k)= \gamma_r(X(r,n,m,k) - X(r-1,n,m,k)),\ r=2,..,n,$$

where γ_r=k+(n-r)(m+1), normalized spacings of gos.

To simplify formulations of our results, together with generalized order statistics based on a cdf F we will consider a random variable X having the same cdf F.

The following result develops the statements of example 6.2.1.

Theorem 6.3.5 .

Let X be a non-negative random variable having an absolutely continuous (with respect to Lebesgue measure) strictly increasing cdf F(x) with F(0) = 0 and F(x)< 1 for all x>0. Then the following properties are equivalent

(a) X has an exponential distribution, E(0);

(b) for $1 < r \leq n$, the statistics {X(r,n,m,k) - X(r-1,n,m,k)} and X(r-1, n,m,k) are independent.

Proof.

Integrating out $x_1,\ldots, x_{r-2}, x_{r+1}, \ldots, x_n$, and using the transformation

U = X(r-1,n,m,k) and W = (X(r,n,m,k) - X(r-1,n,m,k)),

we get on simplification the joint pdf $f_{UW}(u,w)$ of U and W as

$$f_{UW}(u,w)=\frac{c_{r-2}}{(r-2)!}(1-F(u))^m g_m^{r-2}(F(u))(1-F(u+w))^{\gamma_r-1} f(u) f(u+w), \qquad (6.3.9)$$

for $0<u, w<\infty$, and $f_{UW}(u,w) = 0$, otherwise, where

$$c_{r-1} = \prod_{j=1}^{r} \gamma_j, \; r=1,2,\ldots,n, \; \gamma_j=k+(n-j)(m+1), \; j=1,2,\ldots,n;$$

$$g_m(x)=\frac{1}{m+1}(1-(1-x)^{m+1}), \text{ if } m\neq-1,$$

and

$$g_m(x)= -\ln(1-x), \text{ if } m = -1.$$

Let X have the exponential E(0) distribution. Then on simplification, we get from (6.3.9) that

$$f_{UW}(u,w)=\frac{c_{r-2}}{\sigma^2(r-2)!}g_m^{r-2}(1-e^{-u/\sigma})e^{-\gamma_{r-1}u/\sigma}e^{-\gamma_r w/\sigma},\ u>0,w>0. \tag{6.3.10}$$

It follows from (6.3.10) that in this case W and U are independently distributed.

One obtains from (6.1.2) that the pdf $f_U(u)$ of the random variable U =X(r-1, n, m, k) has the form

$$f_U(u)=\frac{c_{r-2}}{(r-2)!}(1-F(u))^{-1+\gamma_{r-1}}\, g_m^{r-2}\,(F(u))\, f(u)\,. \tag{6.3.11}$$

Using (6.2.9) and (6.3.11) and the relation $\gamma_{r-1}=\gamma_r+m+1$, we get the conditional distribution of $f_{W|U}(W|\ U=u)$ as

$$f_{W|U}(w|\ U=u) = \left((\bar{F}(u+w))/(\bar{F}(u))\right)^{\gamma_r-1} f(u+w)(\bar{F}(u))^{-1} \tag{6.3.12}$$

for w>0 ,u>0, where $\bar{F}(u)$=1-F(u).

Integrating the expression in (6.3.12) with respect to w, we obtain that the conditional probability P{W>w|U=u} has the form

$$\bar{F}_{W|w}(w_1\ |U=u)=\left(\bar{F}(u+w)/(\bar{F}(u))\right)^{\gamma_r}/\gamma_r,\ u>0,w>0 \tag{6.3.13}$$

Suppose that W and U are independent. Then we get from (6.3.13) that

$$\left(\bar{F}(u+w)/(\bar{F}(u))\right)^{\gamma_r} = G(w),$$

for all u>0 and w>0, where G(w) is a function of w only. Now taking limit u $\rightarrow$ 0, we get that

$$G(w) = \left(\bar{F}(w)\right)^{\gamma_r}.$$

Hence for all w >0 and u>0 , we obtain

$$\bar{F}(u+w)= \bar{F}(u)\,\bar{F}(w)\,. \tag{6.3.14}$$

The solution of equation (6.3.14) is (see Aczél (1966)) $\bar{F}(u)= e^{-u/\sigma}$, where σ is arbitrary. Since F(x) = 1- $\bar{F}(x)$ is a distribution function, σ must be positive.

If k = 1 and m = 0, then from Theorem 6.3.5, we obtain the result of Rossberg (1972) for order statistics. If k = 1 and m = -1, then Theorem 6.3.5 presents the result of Ahsanullah

(1978) giving the corresponding characterizations of the exponential distribution by the independence of record statistics X(r-1) and X(r)-X(r-1).

Remark 6.3.1.

From theorem 6.3.5 it is easy to see that for the exponential $E(0,\sigma)$ distribution, the regressional function $\varphi(x)=E(X(r+1,m,n,k)|X(r,n,n,k) = x)$ is a linear function of x. It can be shown that this is a characteristic property of the exponential distribution.

The following is based on identical distributions of the standardized spacings of generalized order statistics. To prove the next theorem we need the monotonicity of a failure rate (hazard rate). We define the failure rate of a random variable X as $h(x)=f(x)/(1-F(x))$, where F(x) is the cdf and f(x) is the pdf of X.

Theorem 6.3.6.

Let X be a non-negative r.v. having an absolutely continuous (with respect to Lebesgue measure) strictly increasing distribution function F(x) for all $x > 0$, $F(0) = 0$ and $F(x) < 1$ for all $x>0$. Then the following properties are equivalent.

(a) X has an exponential distribution with pdf $f(x) = \sigma^{-1}\exp(-x/\sigma)$, $0 <x<\infty$, $\sigma > 0$;
(b) the cdf F of X has increasing or decreasing failure rate and there exist integers r and n , $1\le r<n$, such that the normalized spacings D(r,n,m,k) and D(r+1,n,m,k), are identically distributed for some $m\ge -1$.

Proof.

It is easy to show that (a) $\Rightarrow$ (b). We will prove that (b) $\Rightarrow$ (a).
It follows from (6.3.9) that the pdf of D(r)= D(r,n,m,k) for $r\ge 2$ can be written as

$$f_{D(r)}(x)=\frac{c_{r-2}}{(r-2)!}\int_0^\infty (\bar{F}(y))^m f(y) g_m^{r-2}(F(y))(\bar{F}(y+\frac{x}{\gamma_r}))^{\gamma_r-1} f(y+\frac{x}{\gamma_r})\,dy. \qquad (6.3.15)$$

Then the cdf of D(r,n,m,k) is

$$F_{D(r)}(x)=1-\frac{c_{r-2}}{(r-2)!}\int_x^\infty\int_0^\infty (\bar{F}(y))^m f(y) g_m^{r-2}(F(y))(\bar{F}(y+\frac{u}{\gamma_r}))^{\gamma_r-1} f(y+\frac{u}{\gamma_r})dydu=$$

$$1-\frac{c_{r-2}}{(r-2)!}\int_0^\infty (\bar{F}(y))^m f(y) g_m^{r-2}(F(y))\int_x^\infty (\bar{F}(y+\frac{u}{\gamma_r}))^{\gamma_r-1} f(y+\frac{u}{\gamma_r})du\,dy=$$

$$1-\frac{c_{r-2}}{(r-2)!}\int_0^\infty (\bar{F}(y))^m f(y) g_m^{r-2}(F(y))(\bar{F}(y+\frac{x}{\gamma_r}))^{\gamma_r}\,dy. \tag{6.3.16}$$

and analogously the cdf of D(r+1,n,m,k) can be written respectively as

$$F_{D(r+1)}(x)=1-\frac{c_{r-1}}{(r-1)!}\int_0^\infty (\bar{F}(y))^m f(y) g_m^{r-1}(F(y))(\bar{F}(y+\frac{x}{\gamma_{r+1}}))^{\gamma_{r+1}}\,dy. \tag{6.3.17}$$

Taking into account the definition of the function g_m one can see that the RHS of (6.3.16) can be transformed as follows:

$$1-\frac{c_{r-2}}{(r-2)!}\int_0^\infty (\bar{F}(y))^m f(y) g_m^{r-2}(F(y))(\bar{F}(y+\frac{x}{\gamma_r}))^{\gamma_r}\,dy=$$

$$1-\frac{c_{r-2}}{(r-1)!}\int_0^\infty (\bar{F}(y+\frac{x}{\gamma_r}))^{\gamma_r}\,d(g_m^{r-1}(F(y)))=$$

$$1-\frac{\gamma_r c_{r-2}}{(r-1)!}\int_0^\infty g_m^{r-1}(F(y))(\bar{F}(y+\frac{x}{\gamma_r}))^{\gamma_r-1} f(y+\frac{x}{\gamma_r})dy=$$

$$1-\frac{c_{r-1}}{(r-1)!}\int_0^\infty g_m^{r-1}(F(y))(\bar{F}(y+\frac{x}{\gamma_r}))^{\gamma_r-1} f(y+\frac{x}{\gamma_r})dy. \tag{6.3.18}$$

Let statement (b) of the theorem be valid. Moreover suppose that F has increasing failure rate h(x). Then $\ln\bar{F}$ is concave and for m≥ -1, we have

$$\ln(\bar{F}(y+\frac{x}{\gamma_r}))=\ln(\bar{F}(\frac{(m+1)y}{\gamma_r}+\frac{\gamma_{r+1}}{\gamma_r}(y+\frac{x}{\gamma_{r+1}}))$$

$$\geq\frac{m+1}{\gamma_r}\ln(\bar{F}(y))+\frac{\gamma_{r+1}}{\gamma_r}\ln\bar{F}(y+\frac{x}{\gamma_{r+1}}).$$

Thus,

$$(\bar{F}(y+\frac{x}{\gamma_r}))^{\gamma_r}\geq(\bar{F}(y))^{m+1}(\bar{F}(y+\frac{x}{\gamma_{r+1}}))^{\gamma_{r+1}}. \tag{6.3.19}$$

Since X(r,n,m,k) and X(r+1,n,m,k) are identically distributed and $\gamma_r=\gamma_{r+1}+m+1$, we get from (6.3.16) - (6.3.19), that

$$0=\frac{c_{r-1}}{(r-1)!}\int_0^\infty g_m^{r-1}(F(y))[(\overline{F}(y))^m f(y)(\overline{F}(y+\frac{x}{\gamma_{r+1}}))^{\gamma_{r+1}} -$$

$$(\overline{F}(y+\frac{x}{\gamma_r}))^{\gamma_r-1} f(y+\frac{x}{\gamma_r})dy \leq$$

$$\leq \frac{c_{r-1}}{(r-1)!}\int_0^\infty g_m^{r-1}(F(y))[(\overline{F}(y))^{m+1}(\overline{F}(y+\frac{x}{\gamma_{r+1}}))^{\gamma_{r+1}}[h(y)-h(y+\frac{x}{\gamma_{r+1}})]\,dy \qquad (6.3.20)$$

Since

$h(y) \geq h(y+x/\gamma_{r+1})$

for any $x>0$, the RHS of (6.3.20) takes zero value provided that

$$h(y)-h(y+\frac{x}{\gamma_{r+1}})=0$$

for almost all $y>0$ and all $x>0$. The constancy of the failure rate characterizes the exponential distribution. Since additionally $F(0)=0$ and F is a strictly increasing for $x>0$ distribution function, we obtain that $F(x) = 1-e^{-x/\sigma}$, $x \geq 0$, where σ is any arbitrary positive number. The case, when h(x) is a decreasing failure rate can be considered analogously.

Remark 6.3.2.

The condition of monotone failure rate in the Theorem 6.3.6 can be replaced by NBU (New Better than Used) and NWU (New Worse than Used) properties. A distribution function F is said to be NBU (NWU) if $\overline{F}(x+y) \leq (\geq) \overline{F}(x)\overline{F}(y)$ for all x,y and x+y within the support of F.

Remark 6.3.3.

The condition of identical distribution of the normalized spacings of generalized order statistics in Theorem 6.3.6 can be replaced by the equality of their expectations.

The proof of the following two theorems are similar to that of Theorem 6.3.6.

Theorem 6.3.7.

Let X be a non negative random variable having an absolutely continuous (with respect to Lebesgue measure) strictly increasing distribution function F(x) for all x > 0, F(0) =0 and F(x) < 1 for all x > 0. Then the following two properties are equivalent.

(a) X has an exponential distribution with the pdf $f(x) = \sigma^{-1}\exp(-x/\sigma)$,x>0, σ > 0;
(b) X has a monotone failure rate and for one k, one r , one n and m≥ -1, , the statistics D(r,n,m,k) and $\gamma_r X(1,n-r+1,m,k)$ are identically distributed.

Theorem 6.3.8.

Let X be a non-negative random variable having an absolutely continuous (with respect to Lebesgue measure) strictly increasing distribution function F(x) for all x > 0 and F(x) < 1 for all x>0. Then the following properties are equivalent.

(a) X has an exponential distribution with the pdf $f(x) = \sigma^{-1}\exp(-x/\sigma)$, 0<x<∞, σ >0;
(b) X has a monotone failure rate and for one k, one r, one n and m≥ -1, (k ≥ 1, m is a real number, n and r are integers , 1<r≤n) the normalized spacing {k +(n-r)(m+1)} {X(r,n,m,k) - X(r-1,n,m,k)} and random variable X are identically distributed.

REFERENCES

1. Aczēl, J. (1966). Lectures on Functional Equations and Their Applications.Academic Pres, New York, NY..

2. Ahsanullah, M (1976). On a characterization of the exponential distribution by order statistics. J. Appl. Prob,13 , 818-822.

3. Ahsanullah, M. (1977). A characteristic property of the exponential distribution. Ann.Statist., 5 (3), 580-582.

4. Ahsanullah (1978a). A Characterization of the Exponential Distribution by Spacing. J. Appl Prob. 15,650-653.

5. Ahsanullah (1978b). Record Values and the Exponential Distribution. Ann. Inst. Stat. Math.30,429-433.

6. Ahsanullah, M. (1995a). Record Statistics. Nova Science Publishers, Inc. Commack, NY, USA

7. Ahsanullah, M. (1995b). The generalized Order Statistics and a Characteristic Property of the Exponential Distribution. Pak. J. Statist. 11(3), 215-218.

8. Ahsanullah, M. (1996). The generalized order statistics from two parameteruniform distribution. Commun.Statist, - Theory Meth., 25(10), 2311-2318.

9. Ahsanullah, M. (2000a).On Generalized Order Statistics from two parameter Exponential Distribution. Perspectives in Statoistical Sciences. 1-10. Eds. A;K. Basu, J.K. Ghosh, P.K. Sen and B.K.Sinha. Oxford University Press.

10. Ahsanullah, M. (2000b). Generalized Order Statistics from Exponential Distribution. J. Stat. Plan. Inf. 85, 85-91.

11. Arnold, B.C. (1983). The Pareto Distributions. International Cooperative Publishing House,Fairland, MD, USA.

12. Arnold,B.C., N. Balakrishnan and H.N. Nagaraja (1992). A first course in order statistics. Wiley, New York, NY.

13. Balkema, A.A. and L.De Haan (1974). Residual Life at great age. Annals of Probability. 2, 792- 804.

14. Bryson, M.C. (1974). Heavy tailed distributions: Properties and tests. Technometrics 16, 61-68

15. David, H.A .(981). Order Statistics, 2nd edition, John Wiley, New York, NY. USA

16. Galambos, J. (1987)). The asymptotic theory of extreme order statistics. Kreiger, FL

17. Graybill, F. (1969). Introduction to Matrices with applications in Statistics. Wadsworth Publishing Company, Inc. Belmont, CA. USA.

18. Gradshteyn, I.S. and Ryzhik, I.M. (1980). Tables of Integers, Series and Products. Academic Press Inc. New York.

19. Kulldoroff, G. and Vannman, K. (1973). Estimation of the location and scale parameter of a Pareto distribution by linear functions of order statistics. Journal of the American Statistical Association, 68, 218-227.Corrigenda,70, 494.

20. Houchens , R.L. (1984). Record value theory and inferences. Ph.D. Thesis, University of California, Riverside.

21. Kabe, D.G. (1972). On moments of order statistics from the Pareto distribution. Skand. Aktuar. 55, 179-181.

22. Kamps, U. (1995). A concept of generalized ordered statistics, B.B. Teubner, Stuttgart, Germany.

23. Lin, G.D. (1988). Characterizations of uniform distributions and of exponential Distributions Sankhya 50 A, 64-69

24. Llyod, E.H. (1952). Least squares estimation of location and scale parameters using order statistics, Biometrika, 39, 88-95.

25. Lomax, K.S. (1954). Business failures. Another example of the analysis of failure data. Journal of American Statistical Association. 49, 847-852.

26. Malik, H.J. (1966). Exact moments of order statistics from the Pareto distribution. Skand. Aktuar. 49, 144-157.

27. Nagaraja, H.N. (1982). Record Values and Extreme Value Distribution. J. Appl. Prob. 19, 233-259.

28. Nevzorov , V.B. (1987). Records. Theory Probab. and Appl., v.32, 201-228.

29. Nevzorov, V.B. (1992). A characterization of exponential distributions by correlation between records. Mathematical methods of statistics,1, 49-54.

30. Nevzorov, V.B (2000). Some distributions of induced records. Biometrical Journal, 42,1069-1081..

31. Proctor, J.W. (1987). Estimation of two generalized curves covering the Pearson system. Proceedings of ASA Section on Statistical Computing. 287-292.

32. Quandt, R.E. (1966). Old and new methods of estimation and the Pareto distribution. Metrika 10, 55-82.

33. Rohatgi, V.K.and A,K.Saleh (1988).A class of distributions connected to order statistics with non integer sample size. Commun.Statist.-Theory Meth., 17, 2005-2012.

34. Rossberg, H.J. (1972). Characterization of the exponential and the Pareto distributions by means of some properties of the distributions which the differences and quotients of order statistics are subject to. Mathematish Operationsforschung und Statistik, Series Statistics, 3, 207-215.

35. Sen, P.K. (1959). On the Moments of Sample Quantiles. Cal. Statist. Assoc. Bull. 9, 1-19.

36. Sukhatme, P.V. (1937). Tests of significance for samples of the χ^2 population with two degrees of freedom. Ann. Eugenics 8, 52-56.

37. *Szekely*, G.J. and T.F. Mori (1985)). An extremal property of rectangular distributions. Stat.& Probab. Lett., 3, 107-109.

38. Tata, M.N. (1969). On outstanding values in a sequence of random variables. Z.Wahrscheinlichkeitsth. verw. Geb. 12, 9-20.

INDEX

O

P

R

S

T

U

W